AF581649

Sustainable Development Processes for Renewable Energy Technology

Sustainable Development Processes for Renewable Energy Technology

Editors

Sergey Zhironkin
Radim Rybár

MDPI • Basel • Beijing • Wuhan • Barcelona • Belgrade • Manchester • Tokyo • Cluj • Tianjin

Editors
Sergey Zhironkin
National Research Tomsk
Polytechnic University
Russia

Radim Rybár
Technical University of Kosice
Slovakia

Editorial Office
MDPI
St. Alban-Anlage 66
4052 Basel, Switzerland

This is a reprint of articles from the Special Issue published online in the open access journal *Processes* (ISSN 2227-9717) (available at: https://www.mdpi.com/journal/processes/special_issues/Energy_Renewable_Technology).

For citation purposes, cite each article independently as indicated on the article page online and as indicated below:

LastName, A.A.; LastName, B.B.; LastName, C.C. Article Title. *Journal Name* **Year**, *Volume Number*, Page Range.

ISBN 978-3-0365-6203-2 (Hbk)
ISBN 978-3-0365-6204-9 (PDF)

Contents

About the Editors

Sergey Zhironkin

Sergey Zhironkin, Prof., Dr. Sc., Professor of the open-pit mining department of T.F. Gorbachev Kuzbass State Technical University (Kemerovo, Russian Federation).

Research interests: open-pit mining; sustainable development; innovation mining technologies; mining clusters' neo-industrial development.

Author of more than 200 research works, tutorials, textbooks, and course books.

Radim Rybár

Prof. Radim Rybár studied at the Faculty of Mining, Ecology, Process Control and Geotechnologies of the Technical University in Košice, specializing in Deposit Mining – land resource utilization and protection. He received his PhD degree in 2002, and he is a professor at the Institute of Earth Resources. He teaches the following subjects: traditional energy sources, alternative energy sources, alternative energy technologies, and energy sources and transformations. In his scientific and research activities, he deals with the use of renewable energy sources.

Preface to "Sustainable Development Processes for Renewable Energy Technology"

Currently, the production of energy from traditional fossil sources is beginning to give way to renewable energy technologies. New technological, economic, and environmental requirements raise the question of how to make renewable energy sources a driver of sustainable development for the coming decades. Solar and wind energy, the recycling of synthetic materials, and new methods for their utilization can not only significantly reduce the burden on the environment, but also give a new impetus to the development of modern industry and help find a new balance between the needs of people and the capabilities of nature.

One of the tools for achieving these goals is a multidisciplinary approach to solving key problems and an analogous approach to the related scientific research process.

Sergey Zhironkin and Radim Rybár
Editors

MDPI

Editorial

Sustainable Development Processes for Renewable Energy Technology

Sergey Zhironkin [1,2,3] and Radim Rybár [4,*]

1 Department of Trade and Marketing, Siberian Federal University, 79 Svobodny Av., 660041 Krasnoyarsk, Russia; zhironkin@inbox.ru
2 Department of Open Pit Mining, T.F. Gorbachev Kuzbass State Technical University, 28 Vesennya St., 650000 Kemerovo, Russia
3 School of Core Engineering Education, National Research Tomsk Polytechnic University, 30 Lenina St., 634050 Tomsk, Russia
4 Faculty of Mining, Ecology, Process Technologies and Geotechnology, Institute of Earth Sources, Technical University of Košice, Letná 9, 042 00 Kosice, Slovakia
* Correspondence: radim.rybar@tuke.sk

1. Introduction

Currently, the production of energy from traditional fossil sources is beginning to give way to renewable energy technologies. New technological, economic, and environmental requirements raise the question of how to make renewable energy sources a driver of sustainable development for the coming decades. Solar and wind energy, the recycling of synthetic materials, and new methods for their utilization can not only significantly reduce the burden on the environment, but also give a new impetus to the development of modern industry and help find a new balance between the needs of people and the capabilities of nature.

To enhance the international dissemination of these ideas, it is necessary to unite the efforts of researchers in the fields of traditional and modern energy, waste-free production, and ecology in multidisciplinary discussions of urgent problems of transition to the use of renewable energy.

In order to gain new ideas and insights in these areas worldwide, this Special Issue, entitled "Sustainable Development Processes for Renewable Energy Technology", has been prepared. This Special Issue mainly covers original research integrating the principles of sustainable development, the production and recycling of new materials, and the use of renewable energy.

In this Special Issue, 11 original articles were accepted and published, including 1 review and 10 scientific articles. These published research papers appropriately cover the intended breadth of understanding of the complex interdisciplinary issues that the field of Sustainable Development Processes for Renewable Energy Technology presents.

2. Papers Presented in the Special Issue

The review presented by Beer et al. [1] deals with the assessment of the energy mix of five Central European countries in the context of achieving their carbon-neutral or carbon-negative future. The evolution of the assessed energy mixes as well as GHG emissions is presented in a long-term perspective, which allows the assessment and comparison of trends and approaches towards carbon-free energy sectors with each other.

The first scientific article presented by Feckova Skrabuulakova et al. [2] examines the level of readiness of Visegrad Group countries for the widespread use of electric vehicles, introducing a new quantifier—the electromobility infrastructure coefficient of the countries, K—comprising a number of indices that store specific information on the state of readiness for electromobility. The results show, on the one hand, a good position of Slovakia within the Visegrad Group, on the other hand a significant deficit behind the EU leaders.

Citation: Zhironkin, S.; Rybár, R. Sustainable Development Processes for Renewable Energy Technology. *Processes* **2022**, *10*, 1363. https://doi.org/10.3390/pr10071363

Received: 24 June 2022
Accepted: 7 July 2022
Published: 13 July 2022

Papurello et al. [3], they modelled a microgeneration power system consisting of a solar concentrator system coupled to a Stirling engine, with CFD tools, by way of using two methods for comparison: the empirical first-order Beale equation and the Schmidt isotherm method. The results showed the level of electrical power generated for given concentrator parameters and solar power input level.

Teplicka et al. [4], present aspects of performance management in mining companies and their importance for gaining competitiveness in the market of mining companies in the direction of sustainable development and economic growth through performance indicators of mining processes after the implementation of strategic innovation evaluated through Pareto analysis. The results show the improvement of mining processes through the application of innovative mining space and the reduction of environmental impacts of resource exploitation.

The findings of Zhou et al. [5] propose a simple approach to introduce oxygen vacancies into a commercially used anode material for lithium-ion LTO batteries. The presented results show that LTO containing free oxygen exhibits much better performance than the sample before H_2 treatment, especially at high current rates. The results also indicate an improved and fast electrochemical kinetic process.

Wittenberger et al. [6] proposes a process for creating electronic monitoring and graphical mapping of the current technical condition of gas wells in the East Slovakian Lowland using the graphical software tool ArcGIS, using a stepwise algorithm extending the application potential of the method also for oil, geothermal, and hydrogeological wells, while the defined technical patterns were incorporated into the design.

Hong et al. [7] discuss the preparation of a 3D high-level nitrogen-doped metal-free electrocatalyst for high-performance fuel cells and metal–air batteries. Results have been achieved that suggest an increase in electrocatalytic performance compared to the commercial Pt/C form and most metal-free carbon materials in alkaline media, which was achieved due to the high active nitrogen group content, large surface area, and 3D hierarchical porous network structure of the material.

Rybár et al. [8] presents a case study describing the process of creating and validating the benefits of two Innovative Learning Tools aimed at more effective knowledge acquisition in the interdisciplinary field of earth resource extraction with links to the status of renewable energy. The opinions and attitudes of both students and educators towards the tools were surveyed, and some research questions related to this form of knowledge acquisition were validated. The presented results a potential of the educational form as well as the attractive content that goes beyond conventional educational subjects, with its connections.

Seňova et al. [9] propose the use of geothermal resource in the conditions of central Europe, for the purpose of heat supply to the population, while calculating the energy balance, they consider three scenarios, depending on the temperature level of the output medium. The results show the economic and environmental justification for the use of low- to medium-temperature sources in the given conditions and in the current energy-environmental and economic framework.

Szurgacz et al. [10] present the design of a driven section of support (hydraulic reinforcement) used in a system for deep reservoir mining. The research itself was focused on the analysis of the bearing capacity of selected elements of the proposed driven support section with the application of software using the finite element method. The results led to the design of a model with a strength value that meets safety standards.

Rybárová et al. [11] investigated the possibilities of the fragmentation of cultural, especially agricultural landscapes, which are characterized by a high proportion of large land units used for growing cereals and crops subsequently used as energy sources. A fragmentation method based on the reconstruction of dividing lines, mainly formed by dirt roads, based on historical mapping was proposed. The results show and quantify to what extent it is possible to achieve denser landscape fragmentation in this way, to create dividing green belts, to increase the resilience of the environment to water and wind erosion, and so on.

3. Conclusions

We anticipate that the development of the field of renewable resource technologies and the overall use of raw material and energy resources, and, in a broader sense, the Earth's resources in general, in accordance with the principles of sustainability will become increasingly significant and fundamental to the overall picture of the future of human society, also in view of the fact that the current global climate, environmental, energy and raw material problems of mankind are becoming more and more pressing and fundamental. One important prerequisite for the success of such endeavors is the breaking down of the boundaries of the individual disciplines or fields involved and the need for an interdisciplinary, holistic approach to the overall process. It is hoped that this Special Issue creates a suitable platform for an appropriately broad forum.

Finally, we would like to thank our authors, reviewers and editors who have contributed significantly to the success of our Special Issue. We also hope that the individual published papers will be of benefit to readers in their future scholarly work.

Author Contributions: All authors contributed equally to this manuscript. All authors have read and agreed to the published version of the manuscript.

Funding: This paper was created in connection with the project KEGA—048TUKE-4/2021: Universal educational-competitive platform and project RM@Schools-ESEE. RawMaterials@Schools-ESEE: 19069.

Institutional Review Board Statement: Not applicable.

Informed Consent Statement: Not applicable.

Data Availability Statement: Not applicable.

Conflicts of Interest: The authors declare no conflict of interest.

References

1. Beer, M.; Rybár, R. Development Process of Energy Mix towards Neutral Carbon Future of the Slovak Republic: A Review. *Processes* **2021**, *9*, 1263. [CrossRef]
2. Feckova Skrabulakova, E.; Ivanova, M.; Rosova, A.; Gresova, E.; Sofranko, M.; Ferencz, V. On Electromobility Development and the Calculation of the Infrastructural Country Electromobility Coefficient. *Processes* **2021**, *9*, 222. [CrossRef]
3. Papurello, D.; Bertino, D.; Santarelli, M. CFD Performance Analysis of a Dish-Stirling System for Microgeneration. *Processes* **2021**, *9*, 1142. [CrossRef]
4. Teplická, K.; Khouri, S.; Beer, M.; Rybárová, J. Evaluation of the Performance of Mining Processes after the Strategic Innovation for Sustainable Development. *Processes* **2021**, *9*, 1374. [CrossRef]
5. Zhou, Y.; Xiao, S.; Li, Z.; Li, X.; Liu, J.; Wu, R.; Chen, J. Introducing Oxygen Vacancies in $Li_4Ti_5O_{12}$ via Hydrogen Reduction for High-Power Lithium-Ion Batteries. *Processes* **2021**, *9*, 1655. [CrossRef]
6. Wittenberger, G.; Čambal, J.; Skvarekova, E.; Senova, A.; Kanuchova, I. Understanding Slovakian Gas Well Performance and Capability through ArcGIS System Mapping. *Processes* **2021**, *9*, 1850. [CrossRef]
7. Hong, W.; Wang, X.; Zheng, H.; Li, R.; Wu, R.; Chen, J.S. Molten-Salt-Assisted Synthesis of Nitrogen-Doped Carbon Nanosheets Derived from Biomass Waste of Gingko Shells as Efficient Catalyst for Oxygen Reduction Reaction. *Processes* **2021**, *9*, 2124. [CrossRef]
8. Rybár, R.; Gabániová, L'.; Rybárová, J.; Beer, M.; Bednárová, L. Designing a Tool for an Innovative, Interdisciplinary Learning Process Based on a Comprehensive Understanding of Sourcing: A Case Study. *Processes* **2022**, *10*, 9. [CrossRef]
9. Senova, A.; Skvarekova, E.; Wittenberger, G.; Rybarova, J. The Use of Geothermal Energy for Heating Buildings as an Option for Sustainable Urban Development in Slovakia. *Processes* **2022**, *10*, 289. [CrossRef]
10. Szurgacz, D.; Trzop, K.; Gil, J.; Zhironkin, S.; Pokorný, J.; Gondek, H. Numerical Study for Determining the Strength Limits of a Powered Longwall Support. *Processes* **2022**, *10*, 527. [CrossRef]
11. Rybárová, J.; Gabániová, L'.; Bednárová, L.; Rybár, R.; Beer, M. Strengthening the Mitigation of Climate Change Impacts in Slovakia through the Disaggregation of Cultural Landscapes. *Processes* **2022**, *10*, 658. [CrossRef]

MDPI

Review

Development Process of Energy Mix towards Neutral Carbon Future of the Slovak Republic: A Review

Martin Beer * and Radim Rybár

Faculty of Mining, Ecology, Process Technologies and Geotechnology, Institute of Earth Sources, Technical University of Košice, Letná 9, 042 00 Košice, Slovakia; radim.rybar@tuke.sk
* Correspondence: martin.beer@tuke.sk; Tel.: +421-055-602-2398

Abstract: Global climate change is putting humanity under pressure, which in many areas poses an unprecedented threat to society as we know it. In an effort to mitigate its effects, it is necessary to reduce the overall production of greenhouse gases and thus, dependence on fossil fuels in all areas of human activities. The presented paper deals with an evaluation of energy mix of the Slovak Republic and four selected neighboring countries in the context of achieving their carbon neutral or carbon negative future. The development of the evaluated energy mixes as well as greenhouse gas emissions is presented from a long-term perspective, which makes it possible to evaluate and compare mutual trends and approaches to emission-free energy sectors.

Keywords: renewable energy; carbon neutral; energy mix; Slovak Republic

Citation: Beer, M.; Rybár, R. Development Process of Energy Mix towards Neutral Carbon Future of the Slovak Republic: A Review. *Processes* **2021**, *9*, 1263. https://doi.org/10.3390/pr9081263

Academic Editors: Enrique Rosales-Asensio and Ambra Giovannelli

Received: 21 June 2021
Accepted: 20 July 2021
Published: 21 July 2021

1. Introduction

The defining feature of recent decades is turbulence. Whether it is geopolitical, economic, or environmental instability, it is possible, to a certain extent, to connect this high variability, respective uncertainty with global climate change [1–4]. Global climate change is a phenomenon that came to the forefront of interest in the 1960s and is becoming increasingly important [5,6]. Impacts of climate change can be observed on all continents, in all oceans and in the highest layers of the atmosphere [7–10]. Climate change affects all levels of the ecological pyramid and affects the food chain to an extent that was previously known only from prehistoric cataclysmic events [11–14]. All manifestations of climate change expose the entire biological sphere to adaptive pressure, which causes not only the extinction of various species but mainly a change in the species composition of a certain geographical area [15–18]. For human society, the greatest threat is the loss of living space due to rising sea levels [19–22], the lack of drinking water for personal consumption [23–25] as well as agriculture and industry [26,27], extreme weather events [28,29] and long-term changes in temperature profile [30,31], spread of new diseases [32–34], famine [35,36], migration [37,38], deteriorating air quality [39,40], etc.

The importance of the topic of climate change is also evident from summary of scientific papers published in database ScienceDirect.com (accessed 25 May 2021), where 663,186 publications which contained the keywords climate change, have been published since 1998. To this number we can add 288,071 publications with the keywords global warming, which was used mainly in the past. At present, the preferred term is global climate change, which better captures the essence of the problem. Figure 1 shows the development of the number of publications on a given topic in the years 1998–2020. However, if we included publications dealing with the indirect consequences of climate change, or publications dealing with approaches and methods of mitigating the impacts of climate change, the resulting number would be an order of magnitude higher.

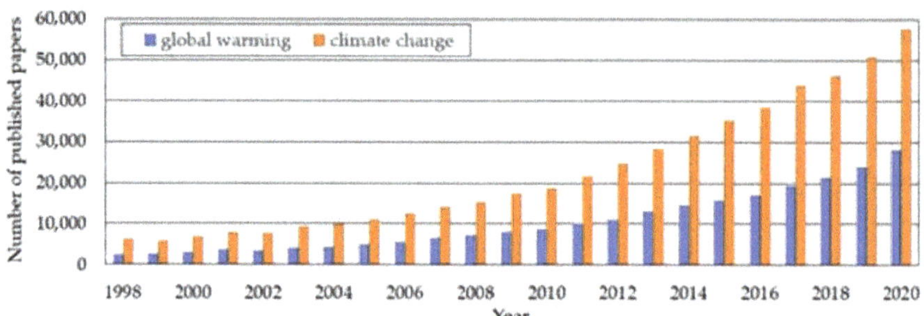

Figure 1. Number of published papers with keywords "climate change" and "global warming" (1998–2020).

At present, the vast majority of the world's scientific organizations, as well as the leading scientists in the field of climate, no longer doubt the reasons that contributed to the gradual development of climate change. All the reasons can be attributed to the anthropogenic activities associated with increasing the level of CO_2 in the atmosphere [41–45]. This increase can be dated to the beginning of the industrial revolution [46] at the turn of the 18th and 19th centuries, when mankind began to use fossil fuels in previously unknown scales [47]. The evolution of CO_2 emissions (from fossil fuel combustion and cement production), primary energy consumption and global temperature anomaly between 1800 and 2019 is shown in Figure 2. From all three graphs, it is possible to identify interrelated trends. The first one is the logical link between the increase in CO_2 emissions and the consumption of primary energy sources. The second is the increase in the value of the temperature anomaly, where it is interesting to observe the growth rate in the last half century, which is almost identical to the growth of primary energy consumption and CO_2 emissions.

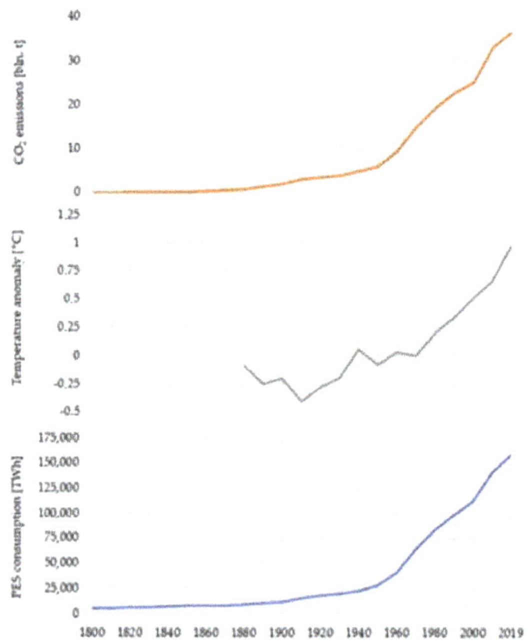

Figure 2. Development of primary energy consumption, CO_2 emissions and global temperature anomaly (1800–2019) [48–50].

One possible way to mitigate the effects of climate change is to keep global temperature rise to 1.5 °C [51] compared to the pre-industrial revolution time. This can be done by adjusting the consumption and structure of primary energy sources, especially in electricity generation or transportation [52–57]. The presented paper deals with the long-term development of the energy mix of Slovakia as well as the surrounding countries from the point of view of a carbon neutral or carbon negative future. The comparison evaluates individual approaches to reducing CO_2 emissions and structural changes in energy mixes and related changes in the structure of primary energy consumption, which can be an example of good practice in the field.

2. Global Energy Review

Current global yearly primary energy consumption is just below 160,000 TWh [49], while most of energy requirements is still covered by consumption of fossil fuels (approximately 86%). Despite the positive examples of reducing dependence on fossil fuels using renewable and low-emission sources that can be seen in some geographical locations (or countries) [58,59] or specific sectors [60], it is clear that global energy consumption is mainly addressed by fossil fuels. Figure 3 shows the development of primary energy consumption from 1800 to 2019. Conversion of outputs is based on the BP methodology, considering a 0.359 efficiency of fossil fuel conversion [61].

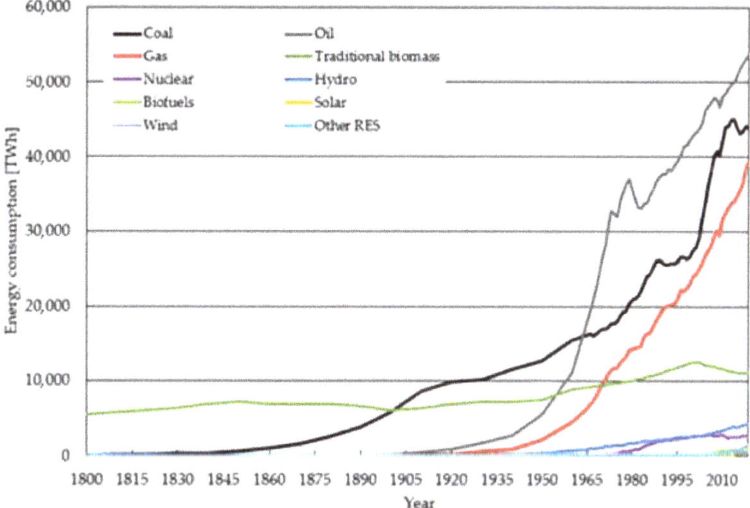

Figure 3. Consumption of primary energy by source (1800–2019) [48,49].

In the graph, all sources, except for traditional biomass and coal, show the growing character of the development of their consumption. The steepest growth can be seen in the consumption of natural gas, which is caused by several factors. The first is the environmental aspect, where the combustion of natural gas generates less CO_2 emissions per unit of energy produced than other fossil fuels [62]. This fact is the driving force in the construction of relatively costly power plants with gas combustion turbines, which replace the coal thermal power plants in a large part of the so-called developed, or Western countries. Gas combustion turbine power plants have a fast start-up [63] (compared to steam turbine power plants) and can provide back-up sources for renewable energy sources (RES), whose prediction of electricity production is in some cases difficult, which can lead to grid instability and endangering the supply of electricity to the final consumer [64,65]. Another factor accelerating global natural gas consumption is the increasing share of LPG and LNG use, especially in freight transport [66,67], where gradual exit from the use of diesel engines producing increased emissions is needed. The last significant factor underpinning the increase in natural gas consumption is the increase in hydrogen use [68]. Hydrogen production is currently carried out mainly in two ways, in the first we can talk about the green hydrogen, which is produced by the electrolysis of water, using electricity produced from renewable sources [69]. The second way is less ecological production of hydrogen by steam reforming of methane. At present, 95% of hydrogen is produced in the second, less environmentally friendly way [70].

The essential non-fossil primary energy source has long been the use of traditional biomass. The term traditional biomass mainly means the use of forest bulk wood. This energy source is typical for the developing countries and forms the basic energy pillar of many people who use wood biomass for food preparation, water disinfection or ensuring thermal comfort in their homes. The course of the development of the consumption of this energy source has an interesting character, which is based on the dramatic increase in the population of the so-called third countries since the Second World War. This population increase was caused by a slight increase in living standards (availability of at least minimal health care, food aid from developed countries, political reforms, improved economic efficiency) and was necessarily covered by an increase in traditional biomass consumption, which in many areas caused significant environmental problems associated with deforestation. The peak around 2000 and the gradual decline is paradoxically also associated with rising living standards, where in many countries (especially Southeast Asia) living standards have risen to such an extent that the use of only traditional biomass is no longer necessary, and the population has access to other more advanced fuels. The short stagnation evident from the recent period is caused not only by a decrease in the rate of population growth but also by a slowdown in the increase in the living standards of the population, which is not a desirable phenomenon.

The overall share of RES in primary energy consumption is at a very small level, but a closer look shows exponential growth and thus it is possible to expect an increase in their role in the future. The exception is the use of hydropower, where the proportion is higher than in case of solar and wind energy, but the potential of large rivers to build large hydropower electricity plants is in many parts of the world very small [71] and no exponential increase in utilization of hydropower can be expected.

Based on the above facts, it is possible to assume that the future increase in primary energy consumption will be covered mainly by renewable sources, natural gas and nuclear energy. The situation is quite different if we look only at the electricity production depicted in Figure 4. In 2020 almost 26,000 TWh of electricity was generated [49]. In the production of electricity, the situation with the share of fossil fuels does not look so dominant. The use of hydro and nuclear energy has a significant share in the production of electricity, and the solar and wind energy is also gaining in importance, the growth of which is also exponential in this case.

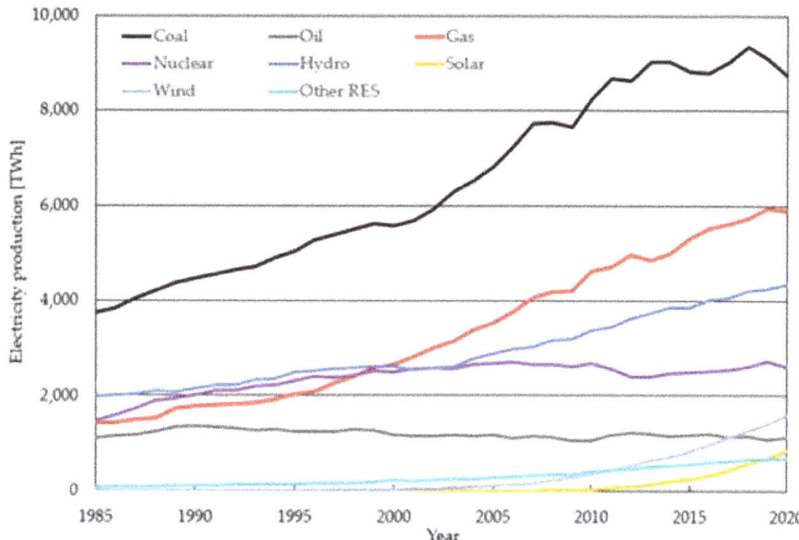

Figure 4. Global electricity production by source (1985–2020) [49].

Based on presented statistics and evaluation of the development of key factors in CO_2 emissions, primary energy consumption and structure of used energy sources, it is unrealistic to expect a significant reduction in energy consumption on a global scale. At present, there are still several countries that are achieving a population boom. There are also a number of countries (especially in Africa) still waiting for their development opportunity. In these cases, it is impossible to expect that population growth and the desired increase in living standards will not be supported by an increase in the consumption of available and cheap primary energy sources.

One of the ways to mitigate or slow down the effects of climate change is to reduce energy consumption at the local level (this is happening, for example, in Western European countries), or to adjust the energy mix so that most of the consumed energy has a low-carbon origin. In some sectors, such as industry or transport (especially aviation and shipping) it is a difficult and costly problem, but in the production of electricity with use of present technologies and methods it is a solvable problem. Especially in the current efforts "to green" a large number of processes and technologies by their gradual electrification, the electricity generation sector will become increasingly important. The possibilities of adjusting the energy mix with regard to reducing CO_2 emissions are illustrated in the example of the Slovak Republic, which is compared within neighboring countries grouped in the so-called Visegrad Four (Slovakia, Czechia, Poland, Hungary) and extended by the close neighboring country of Austria, which often forms a comprehensive group with the Visegrad Four countries on an international scale.

3. General View of Slovak Republic Energy Mix and Emissions

The Slovak Republic was established in 1993 by the split of Czechoslovakia, which created two independent countries, the Slovak Republic, and the Czech Republic. Since 2000 Slovakia has been a member of the Organization for Economic Cooperation and Development (OECD), since 2004 a member of the European Union, and since 2007 a member of the International Energy Agency (IEA). The Slovak Republic is a country in a temperate climate zone with a predominantly mountainous relief, which fundamentally determines the hydrological conditions. Approximately 5.5 million inhabitants live in Slovakia, almost half of whom live in urban settlements. The Slovak Republic is an industrial-agricultural state with a market economy. The basic building blocks of industry are the production of electrical components, the automotive and metallurgical industries. Gross domestic product per capita is USD 34 815 [72].

The industrial use of energy resources occurred with the development of mining and metal processing in the 12th, respectively, 13th century. Another significant milestone was the post-war transformation of the country based on the planned economy, when it was necessary to ensure sufficient electricity and fuels for the growing heavy industry. Electrification in the territory of the Slovak Republic (then still in the territory of Austria-Hungary) began in 1884 in Bratislava and ended in 1960 (already as part of the Czechoslovak Socialist Republic), when the last village was connected to the electricity grid system.

3.1. Historical Development and Current Situation in Energy Sector

The current energy policy is ruled by a government document approved in 2014, which sets out the goals and priorities of the energy sector by 2035 and by 2050. The strategic goal is to achieve competitive low-carbon energy ensuring a secure, reliable, and efficient supply of all forms of energy at affordable prices, taking into account consumer protection and sustainable development. The updated Integrated National Energy and Climate Plan for 2021–2030 takes into account more realistic targets as well as the incorporation of comments and climate commitments towards the European Union. This document will be approved in II.Q of 2021.

The main goal across the European Union is to reduce greenhouse gas emissions by 40% (compared to 1990) by 2030, another goal is to achieve a share of RES in total primary energy consumption of 32%, but each member state must reach the share of RES in transport at least 14% [73]. The Slovak Republic has set indicators for reducing emissions by 20% and the share of RES in final consumption at 20%. In the electricity generation sector, the target for RES is then set at 27.3% by a subsequent conversion, taking into account technical possibilities and real potential [73]. At present, the share of RES in primary energy consumption is 19.7%, and in transport 8.9%.

Primary energy consumption in 2020 was 183 TWh [49] and electricity 28.61 TWh [74], Figure 5 shows the historical development of these fundamental energy characteristics beginning in 1965 and 1985, respectively. It should be noted that the data before 1993 refer to the territory of the Slovak Republic, then included in Czechoslovakia.

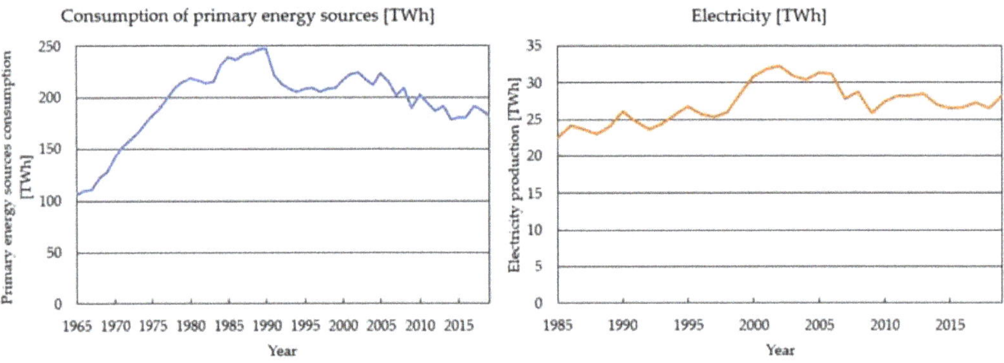

Figure 5. Primary energy consumption and electricity production of the Slovak republic [49].

The graph of primary energy consumption has shown a sharp increase since the beginning of the period under review, which culminated around 1990, when a revolutionary wave took place in the country, changing the political establishment from communism to democratic establishment. The change also included the transition from a planned economy to a market-driven economy. This change was negative for many sectors of the economy, but especially for heavy and military industry [75]. Another factor was the rise of the environmental movement and the adoption of laws to increase environmental protection. In the following years of rebirth and independence, we can talk about stagnation in terms of energy consumption. A significant increase, which is also visible in the graph of electricity production, occurred in the years around the 2000, when the country entered a period of economic growth thanks to the right-wing pro-reform government [76]. In this period (1998/1999), two reactor blocks of the Mochovce nuclear power plant were also launched, which contributed to energy stability at the time. Another more turbulent period characterized by a drop in consumption or electricity production is associated with the economic crisis at the turn of 2008/2009. The current development is more or less stable with only small fluctuations in both directions. The structure of energy consumed as well as its long-term development are presented in Figure 6.

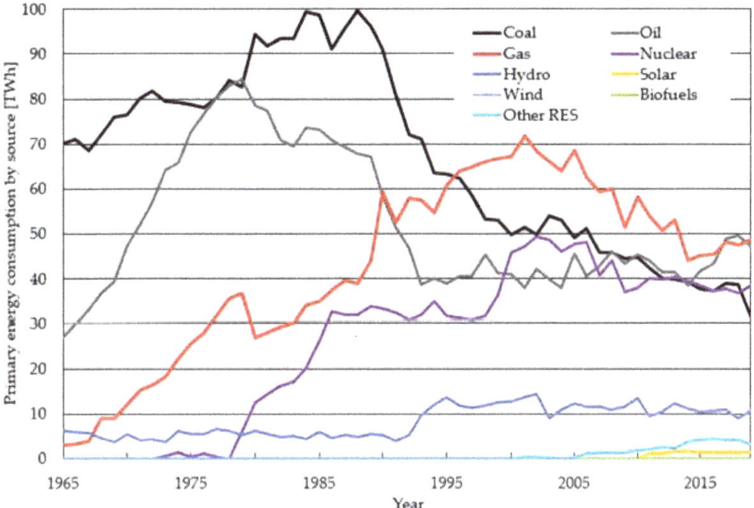

Figure 6. Historical development of primary energy consumption of the Slovak Republic (1965–2019) [49].

From the point of view of long-term development and efforts towards decarbonization, the massive drop in coal consumption is a positive factor. For other fossil fuels, i.e., natural gas and oil, their consumption has stabilized with only small fluctuations after slight declining periods. An important role is represented by the use of nuclear energy, in which we can identify four significant milestones, which are clearly visible in the graphical representation of development. The first is the launching of the first two reactors of the Jaslovské Bohunice nuclear power plant in 1978 and 1980, the second is the commissioning of third and fourth reactors in 1984 and 1985. Other nuclear reactors were launched in the territory of independent Slovakia in 1998 and 1999 at the Mochovce nuclear power plant. So far, the last milestone is the shutdown of the first two reactors of the Jaslovské Bohunice nuclear power plant, which ended its operation in 2006 and 2008, respectively. From the point of view of hydropower, the most significant event is the building and launching of the Gabčíkovo hydroelectric power plant in 1992.

Other renewables contribute to the total primary energy consumption to a small extent, with a significant percentage in this group representing the use of solar energy and biomass in the form of biofuels but also traditional wood biomass.

The installed capacity of the power system is around 7.7 GW, while the average load of the power system is at the level of 3.5 GW and a maximum of 4.5 GW [74]. From the point of view of the stability of the electricity distribution system, the most important power plants are listed in Table 1. The structure of installed capacity and electricity generation is shown in Figure 7.

Table 1. Key power plants in the Slovak Republic.

Name of the Power Plant	Installed Power	Type
Jaslovské Bohunice V2	2 × 505 MWe	Nuclear
Mochovce 1 and 2	2 × 470 MWe	Nuclear
Gabčíkovo	720 MW	Hydro
Liptovská Mara	198 MW	Pumped-storage hydro
Čierny Váh	735 MW	Pumped-storage hydro
Nováky	266 MWe	Coal thermal
Vojany	2 × 110 MWe	Coal thermal

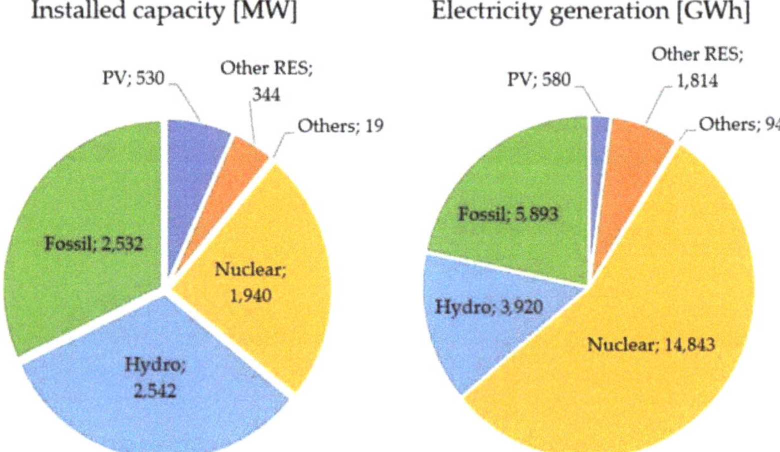

Figure 7. Structure of Slovak installed capacity and electricity generation in 2019 [73].

The Slovak Republic does not have significant reserves of fossil fuels and more than 90% are imports, especially from the countries of the former Soviet Union. The transport of coal is carried out by a broad-gauge railway connected to the railway network of the countries of the former Soviet Union. Oil and gas are transported via Druzhba oil pipeline and the Bratstvo (Brotherhood) gas pipeline which passes through the Slovak Republic [77]. The territory of the Slovak Republic is also connected by pipelines with the Adria oil pipeline and gas pipelines of the surrounding countries (especially the connection to Hungary and Austria). As part of increasing energy security, natural gas storage facilities were built in natural rock structures with a total capacity of 35.6 TWh (calculated using the energy density of natural gas).

3.2. Greenhouse Gas Emissions

The latest actualized national greenhouse gas inventory in 2016 shows total greenhouse gas emissions of 41,037 Gg CO_2 eq. [73]. This represents a decrease of 44.5% compared to the 1990 reference year, which is the basic benchmark for European Union strategies. In total, between 1991 and 2015, emissions did not exceed the level of 1990 for a single year, with exception of F-gases (fluorinated gases: HFC, PFC, SF6). A graph expressing trends in greenhouse gas emissions (in relative terms) is shown in Figure 8.

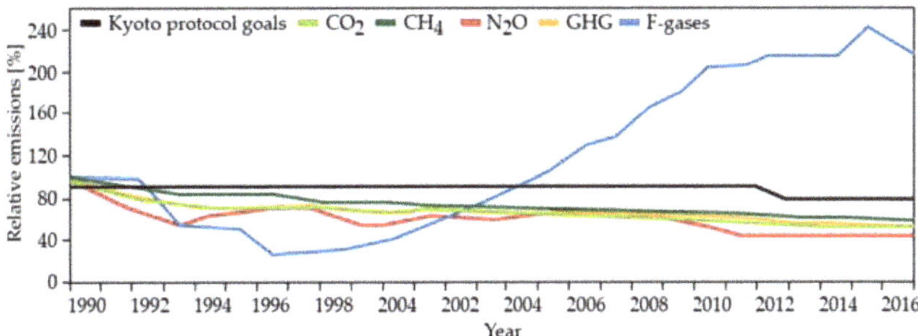

Figure 8. Comparison of relative greenhouse gas emissions trends with Kyoto protocol goals [73].

The reduction in emissions by 2005 was mainly due to industrial and technological restructuring linked to the transition from coal and oil to natural gas. The driving force behind this reduction was adoption of new environmental laws. After 2005, the amount of greenhouse gas emissions has been declining mainly due to the restructuring of the economy and the transition to less energy-intensive industrial sectors. The sector in which there is an annual increase in the amount of greenhouse gases emissions is transportation. Anthropogenic sources of carbon dioxide emissions in 2016 were at the level of 33,996 Gg CO_2 (which is a decrease of 45.2% compared to 1990). Other components of the greenhouse gases such as methane decreased by 39.1% and N_2O by 56% compared to 1990 [73].

Total anthropogenic emissions of the F-gases showed a significant increase during the considered period. The trends of individual emissions according to the 1990 level are shown in Figure 9.

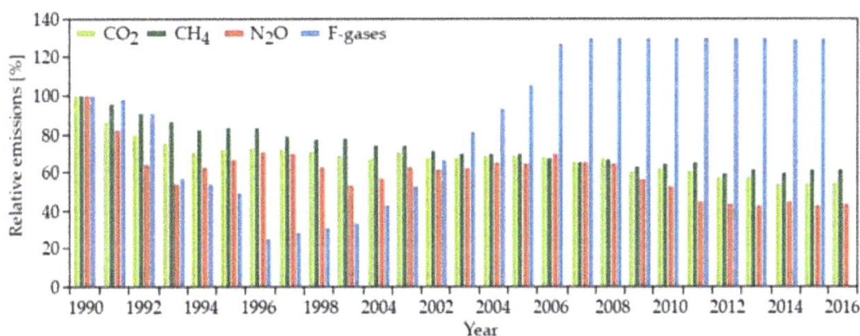

Figure 9. Trends of individual emissions according to the 1990 level [73].

The most important sector responsible for total greenhouse gas emissions is energy, including transport, with 67%, while transport is responsible for 16% of total emissions. Another significant source of emissions is the industrial processes and product use sector, which accounts for 23% of total greenhouse gas emissions. In this sector, emissions are produced mainly from the technological process of mineral processing, chemical production and the production of steel and iron. Agriculture accounts for 6% and waste treatment for 4% [73].

In terms of carbon dioxide emissions, a key sector is road transport using diesel engines, which is responsible for 24% of total CO_2 emissions, followed by fuel combustion (17%) and iron production (16%). More than an absolute expression of the number of emissions, it is more interesting to look at the changes in the trend compared to 1990, where road transport shows an increase of 30%, fuel combustion a decrease of 35% and iron production by 50% [73].

3.3. Projections of Future Development of Key Energy and Ecology Indicators

The future development of the energy mix of the Slovak Republic will depend mainly on the effort to comply with the established ecological goals of the EU and the effort to ensure cheap, ecological, and stable electricity for industry and the population. A significant shaping factor will also be the reduction of emissions in the transport sector, where high legislative support for electromobility can be expected, which may endanger the stability of the electricity system. The future threat to the stability of the electricity system is based on both an increase in consumption points with high power-outputs levels but also an increasing share of renewable sources in the energy mix, which are characterized by fluctuations in production. The long-term forecast of the development of the use of renewable energy sources assumes that the share of RES in total energy consumption in 2040 will be at the level of 22%, and the share in electricity production at the level of 25%. The forecast for the development of this key area is shown in Figure 10.

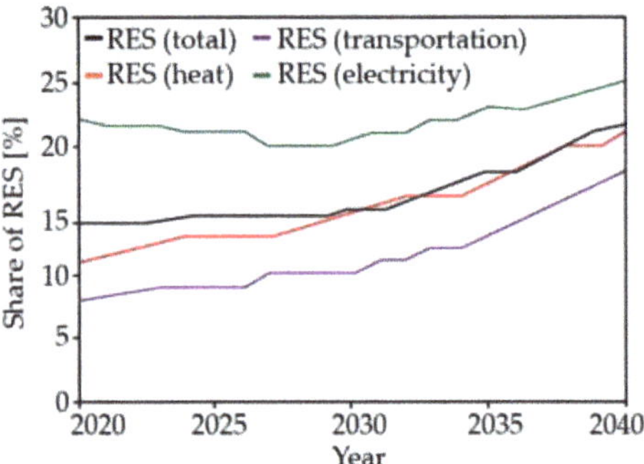

Figure 10. Future trends of RES share in total energy mix [73].

The most fundamental change in the energy mix, which will define the energy sector of the Slovak Republic for several decades, is the expected launch of the third and fourth reactors of the Mochovce nuclear power plant with an installed capacity of 2 × 470 MW. The start of commercial operation of the third reactor is expected in IV. Q 2021, the fourth reactor should be launched in 2023 according to the current pace of work. After the completion of these energy sources, Slovakia will become an export country, from the point of view of electricity production. In 2030, the installed capacity is expected to increase to 8720 MW, of which RES will contribute to 4630 MW. The energy concept assumes an increase in the load on the electricity system up to 5250 MW [73].

The future concept does not envisage the use of the Nováky thermal power plant, the Vojany thermal power plant will be used only if its conversion to the use of more environmentally friendly fuels (wood chips, waste) is an economically advantageous decision and the price of electricity produced in this way will be adequate for its use in the distribution network. The high price of gas and the low price of electricity do not indicate the development of the use of the steam-gas cycle. The hydropotential of the Slovak Republic is used to 71% and theoretically it would be possible to build new hydropower electricity plants with a total installed capacity of 241 MW and a production of 1900 GWh [73], but high investment intensity, negative public attitude reflected in complicated EIA process mean a low probability of significant development of this energy sector.

The period around 2030, when there will be a loss of financial support for photovoltaic power plants built in 2010–2012, seems unstable. In total, up to 530 MW of installed capacity is at risk, which may be missing both in the Slovakia's EU targets for RES and in the electricity production system, of course only in the case if owners decide to end their operation after the loss of the government financial support.

In the long term, the concept also envisages the construction of another nuclear power source, but it does not envisage it until the end of operation of last two reactors of the Jaslovské Bohunice nuclear power plant after 2045. The new nuclear power plant should have an installed capacity of 1200 MW and should be located near the decommissioned nuclear power plant in Jaslovské Bohunice. However, such a large installed capacity would currently represent a high load on electricity systems, which is why the construction of a 560 MW pumped-storage hydroelectricity power plant is being considered in the future, which should be able to work with a weekly cycle by accumulating the weekend surplus from nuclear power plants and use it during demand peak in working days. This pumped storage power plant would also play an important role in balancing the production of future wind and photovoltaic power plants.

The development of the amount of greenhouse gas emissions within the Integrated National Energy and Climate Plan of the Slovak Republic is projected according to two scenarios—WEM (scenario with the application of existing measures) and WAM (scenario with the application of other measures). Projections of greenhouse gas emissions prepared according to the WEM scenario, which is equivalent to the EU reference scenario 2016 (EU 2016 RS) for the years 2015–2040 (with a view to 2050) are shown in Figure 11. Year 2016 was designated as the reference year for modeling greenhouse gas emissions for all scenarios for which verified data sets from the national greenhouse gas emissions inventory were available. The scenario is based on the logic of the EU 2016 RS scenario and includes policies and measures adopted and implemented at EU and national level by the end of 2016, including the measures needed to achieve the 2020 renewable energy and energy efficiency targets.

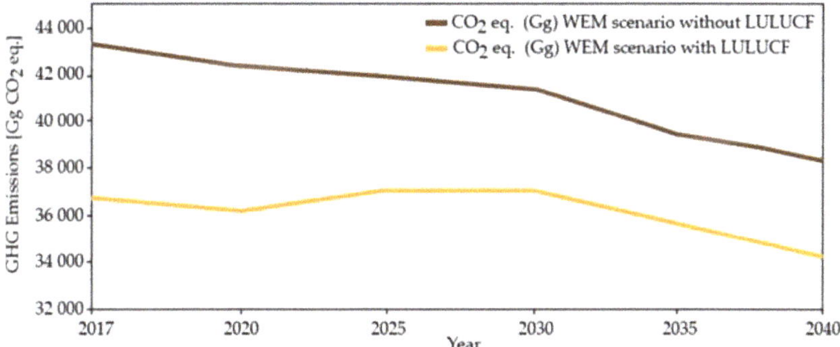

Figure 11. Projections of greenhouse gas emissions prepared according to the WEM scenario [73].

The WAM scenario includes ways to achieve various combinations of ambitious energy efficiency, renewable energy and 2030 emission reduction targets. The WAM scenario analyzes the possibility of achieving the EU's 2050 emission reduction targets (i.e., carbon neutrality). The scenario includes Slovakia's participation in the EU emissions trading system after 2020 and intermediate targets for renewable energy sources and energy efficiency, construction of new capacities for nuclear energy production, while maintaining its key role in the production mix. The projection of emissions according to the WAM scenario is shown in Figure 12.

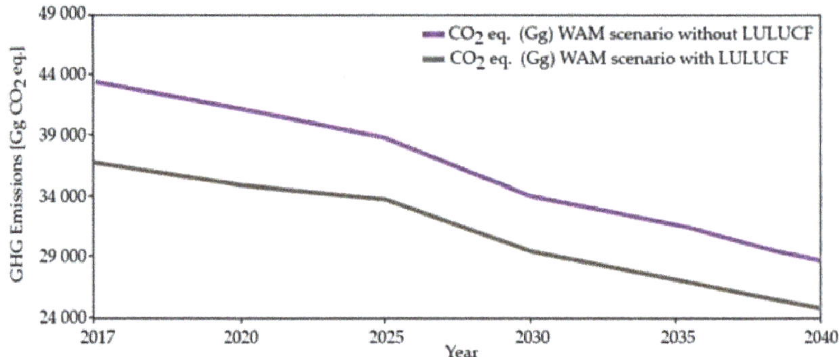

Figure 12. The projection of emissions according to the WAM scenario [73].

Both scenarios work with the division of total emissions with the implementation or omission of LULUCF measures, i.e., land use, land-use change, and forestry. These are measures that can, by their nature, reduce greenhouse gas emissions, by capturing CO_2 as biomass grows.

4. Comparison of the Energy Mixes of the "Visegrad Four" Countries (Extended by Austria)

The energy mix of the Slovak Republic is relatively unconventional in other European countries with a high share of nuclear energy in electricity production (approximately 55%). Only countries such as France (67%), Ukraine (52%), Hungary (47%) and Bulgaria (41%) have the closest values of the share of nuclear energy [49]. For better clarity, the energy mix of Slovakia will be compared with other countries of the Visegrad Four (Czech Republic, Hungary, Poland), which will be extended by Austria. In addition to energy mixes, available indicators quantifying greenhouse gas emissions will also be compared. The most current essential economic and energy indicators characterizing the compared countries are summarized in Table 2.

Table 2. Essential economic and energy indicators of compared countries [49,72,78].

Indicator	Slovakia	Czechia	Poland	Hungary	Austria
Number of populations (mil.)	5.45	10.72	37.80	9.63	9.04
GDP per capita (USD)	34,815	42,956	35,957	35,088	57,891
Electricity consumption per capita (kWh)	5241	7534	4159	3500	7716
Employment in industry (%)	36.34	37.87	31.33	30.18	25.62
Employment in services (%)	60.75	59.27	58.11	64.84	70.07
Employment in agriculture (%)	2.91	2.87	10.56	4.98	4.31

Figure 13 and Table 3 show the structure as well as the absolute per capita consumption of primary energy of each country. The size of energy consumption reflects the position of countries in terms of their economic performance as well as the structure of the economy. In this comparison, Austria has the highest primary energy consumption of 45,862 kWh per capita, followed by the Czech Republic (43,940 kWh per capita), Slovakia (33,171 kWh per capita) and Poland with 30,880 kWh per capita, the last country in this comparison is Hungary with 28,278 kWh per capita.

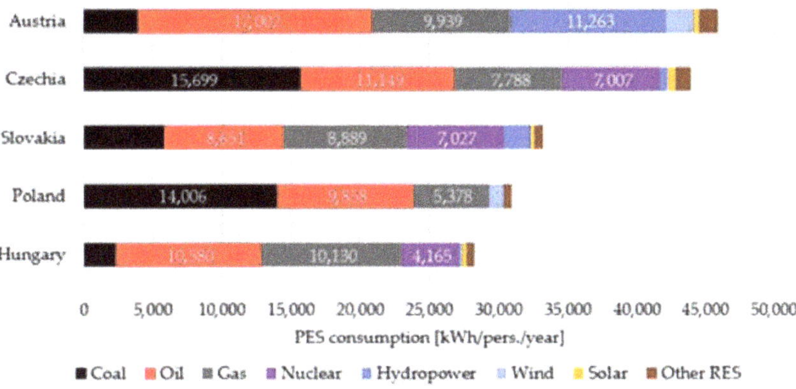

Figure 13. Structure of per capita primary energy consumption of compared countries in 2019 [49].

Table 3. Per capita primary energy consumption by source of the compared countries in 2019 [49].

Country	Primary Energy Consumption by Source (kWh/pers./year)							
	Coal	Oil	Gas	Nuclear	Hydro	Wind	Solar	Other RES
Austria	3904	17,002	9939	0	11,263	2056	374	1324
Czechia	15,699	11,149	7788	7007	465	162	529	1141
Slovakia	5813	8651	8889	7027	1927	1	267	596
Poland	14,006	9858	5378	0	128	983	47	480
Hungary	2258	10,580	10,130	4165	56	186	354	549

All the considered countries are members of the European Union and are subject to the same obligations arising from the effort to reduce greenhouse gas emissions. From this point of view, an interesting view of the share of low-emission sources in the total primary energy consumption is shown in Figure 14. The category of low-emission sources includes the use of nuclear energy and renewable energy sources (except for the use of traditional biomass). A positive finding is the trend of increasing use of low-emission sources among all countries. By synthesizing the information from Figures 13 and 14, we get a realistic idea of the low-emission approach, as well as the mutual position of countries. The worst position in this comparison is Poland, whose low-emission energy sources represent only about 6% of total primary energy consumption. Poland has long been a country without a nuclear energy source and uses domestic coal reserves in many thermal power plants. Renewable sources are mainly represented by the use of wind energy and other RES (geothermal energy in the southern part of the country). With more than 18% following Hungary, it has a lower use of renewables than Poland, but generates electricity at the Paks nuclear power plant with an installed capacity of 4 × 500 MW. The third place belongs to the Czech Republic, which uses 21% of low-emission sources, in the case of the Czech Republic the situation is again positively affected by the existence of a pair of nuclear power plants Dukovany (4 × 510 MW) and Temelín (2 × 1082 MW) which are supplemented by the use of renewable sources energy in the form of hydropower, solar energy and biomass. Slovakia follows with a significant leap with almost 30% of low-emission sources, which is caused by the massive share of nuclear energy in electricity production, but also by the use of hydropower. In the first place is Austria, which ensures 33% of primary energy consumption through low emission sources. In the case of Austria, which does not have a nuclear power plant, as was the case in previous countries (except for Poland), this share is more interesting. In terms of its geographical location and morphology, Austria has significant potential for the use of hydropower in the Alpine region, as well as for the use of wind energy in the eastern flat part of the country. Significant state support also enables the efficient use of solar energy and biomass.

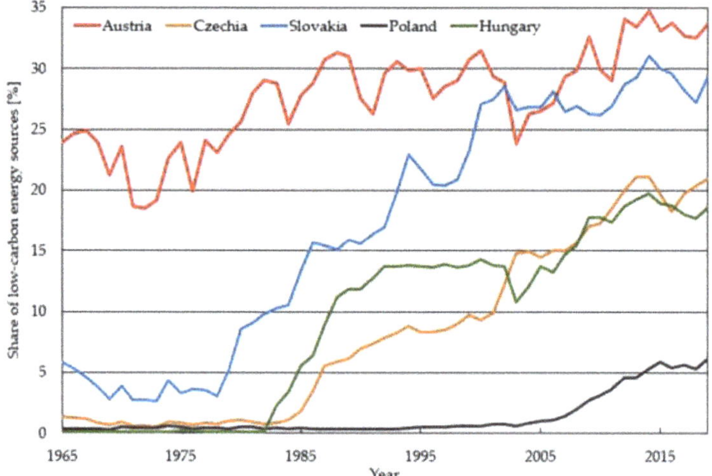

Figure 14. Share of low-emission sources in primary energy consumption (1965–2019) [49].

In the production of electricity, where the development is shown in Figure 15, the situation is different. Slovakia and Austria are in first and second place with almost 80% use of low-emission sources for electricity production, with Hungary following 62%, the Czech Republic 49% and Poland 16%. In this comparison, two factors come to the fore—the use of nuclear energy (Slovakia, Czechia, Hungary) and the existence of significant potential for RES in the country (Austria).

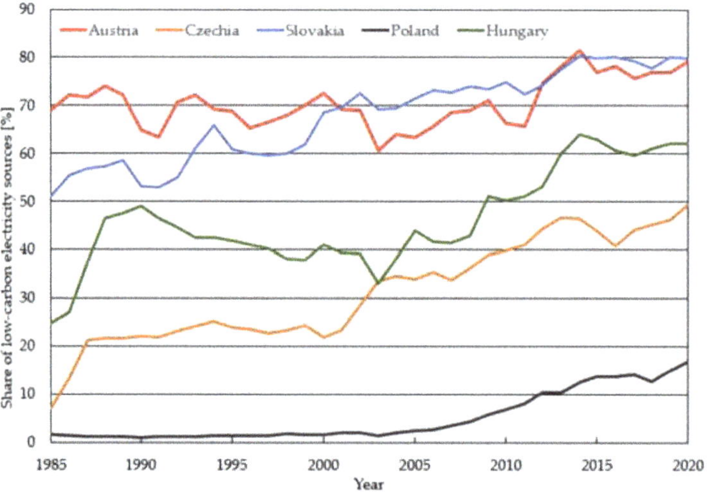

Figure 15. Share of low-emission sources in electricity production (1965–2019) [49].

Unsurprisingly, the biggest polluter in terms of CO_2 emissions is Poland, which emitted 322 million tons of carbon dioxide into the atmosphere in 2019. It is followed by the Czech Republic with more than 100 million tons, Austria 68 million tons, Hungary

49 million tons and Slovakia at last place with 33 million tons. Looking at the last period of development shown in Figure 16 (left), it is possible to speak of a steady course with relatively small fluctuations caused, for example, by the decline in economic activity in 2008, when the global economic crisis swept the world. However, the absolute values do not necessarily indicate the real state of the issue, if we look at the number of emissions per capita, the Czech Republic will significantly come to the fore. The development of this indicator is shown in Figure 16 (right).

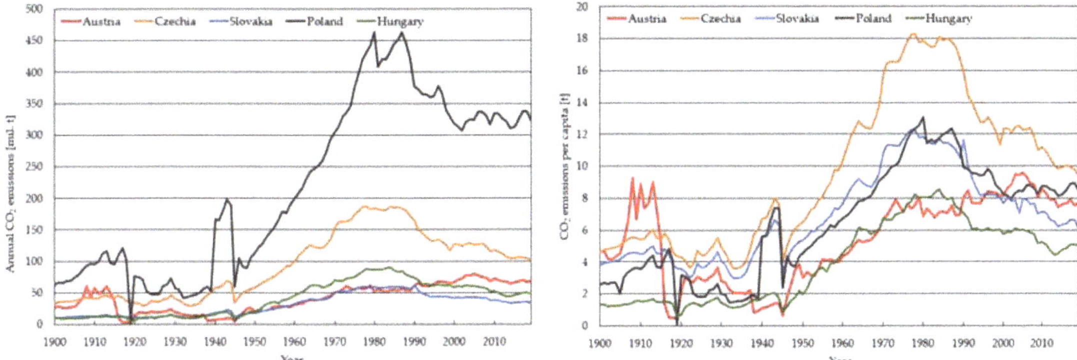

Figure 16. Historical development of total (**left**) and per capita (**right**) carbon dioxide emissions [79].

The above summary and the development of key indicators have highlighted several interesting findings in the approach to ensuring a balance between the supply of stable and cheap electricity and, on the other hand, reducing CO_2 emissions. From the point of view of efforts to move away from the direct use of fossil fuels, especially in transportation sector and their replacement by electromobility or hydrogen, the provision of stable low-emission sources of electricity is a very important factor. On the example of the compared countries, we see two directions in this approach.

The first is the building of large hydro and wind energy sources (marginally also solar farms) in Austria. In this case, there is a synergistic effect of the work of pumped storage hydropower plants (more than 5 GW of installed capacity) built in the Alpine region, with large wind farms in the eastern part of the country. Pumped storage power stations allow to accumulate the electricity produced by wind power at a time of large supply but low demand for electricity. Later, at electricity load peak times, pumped storage plants supply this energy to the distribution network. It should be noted that Austria has specific possibilities in terms of geomorphology and hydrology. Nevertheless, total CO_2 emissions as well as per capita emissions are higher than, for example, in Slovakia or Hungary, which is caused by high standard of living of the population of Austria. With a high standard of living, there is always an increase in energy consumption, whether in the housing sector, but also in transport and leisure. Reducing emissions in these segments will be a difficult task for Austria's government, as it will have to persuade the population to reduce frequency of their energy intense habits or to pay more for the same but more environmentally friendly activities.

The second approach is a significant share of nuclear energy, as we see in the case of Slovakia, where nuclear energy contributes more than 50% to electricity production and the low-emission part of the energy mix is supplemented by hydropower and only marginally by other renewable energy sources with unstable nature of the operation. On the example of these countries, we see a completely different approach to the structure of the energy mix as well as to its further direction. It is likely that Austria will address the increase in energy consumption as well as the agenda for reducing greenhouse gas emissions mainly from renewable energy sources, as nuclear energy is not popular in the country. Resistance to nuclear energy peaked in 1978, when residents in a referendum rejected the launch of the completed nuclear power plant Zwentendorf [80], which is a unique situation in the global perspective. On the contrary, from the above presented information it is clear that the Slovak Republic expects to launch and build additional nuclear resources. In this view, it is an interesting cross-border relationship between a strong opponent and a supporter, which brings complications in the form of behind-the-scenes as well as open political struggle on the international stage. In efforts to achieve a low-emission or carbon-neutral future, both countries rely on proven practices or on the overall potential of the territory, which translates into technically usable renewable energy potential, which is estimated at between 219 and 359 TWh/year [81] for Austria and 56 TWh/year for Slovakia [82].

If we accept the fact that the low-emission future of most processes (especially transport and industry) is associated with their massive electrification, both countries will have to continue to use stable sources of electricity. From the comparison of the potential of RES, it is clear that a massive retreat from nuclear energy in Slovakia is not technically possible and would pose a threat to the stability of the transmission and distribution system (taking into account the small potential for the expansion of stable hydropower electricity sources). Based on available information, other Visegrad Four countries are also planning a low-emission future of the energy mix based on the model of Slovakia. Thus, building a strong base in nuclear energy (new planned 1000 MW nuclear power plant in Poland, extending the life of nuclear reactors in the Czech Republic and building a new nuclear power plant until 2037, new nuclear reactors in Hungary in 2027, etc.), complemented by strong renewable source and a smaller proportion of other renewables of regional importance, which complement the overall energy mix and which are easily replaced by hot reserves (e.g., gas power plants) of small installed capacities or regulated by pumped storage. In the case of Slovakia, the use of hydropower is a strong renewable source; in the case of Poland, it could be wind farms in the north of the country in the Baltic region, and Hungary could use the potential in wind or solar energy. More problematic is the Czech Republic, which, due to its geographical location and morphology, does not have significant potential in one major and stable renewable energy source. An approach embodying the interconnection of nuclear energy and renewables is thus likely to dominate the evaluated area of the east part of the European Union, which will allow operation of a stable interconnected electricity system.

However, a similar approach is not only dominated by the countries of the former "Eastern bloc", the situation has long been similar in France but also in Finland, which is completing the construction of new nuclear reactors. The second approach in the form of the widest possible use of renewables is represented in Norway, Denmark, or Sweden, which use the huge potential of renewables (hydropower, wind energy, biomass energy). Germany has a slightly different position, which does not have a dominant, strong and stable renewable resource. It seeks to combine many sources, which contribute to the complexity of electricity system management as well as enormous financial costs for energy sources themselves. Complication also brings the building of new lines of high-voltage transmission system that must transfer electricity from the Baltic Sea wind farms to the industrial south.

The success of individual approaches in terms of reducing greenhouse gas emissions, i.e., the promotion of nuclear energy or the promotion of exclusively renewable energy sources, will be possible to evaluate only in the future, when it will be possible to compare the predicted scenarios with reality. However, we are already seeing a few examples of countries that have nuclear-based energy (France), nuclear energy and renewables (Sweden) and whose emissions are among the lowest in Europe. On the contrary, countries like Germany with a huge installed capacity in renewable energy sources, despite enormous efforts and financial costs, still produce a large amount of greenhouse gas emissions. On this basis, it can be predicted that the European energy sector (as well as global energy sector) will not be able to achieve the required reductions in greenhouse gas emissions without the use of nuclear energy, as not every country has the potential and opportunities to use renewable energy sources. It is these countries that will have to rely on nuclear energy when switching from fossil fuels. Closely related to this is the speed of transition to low-emission sources. The construction of photovoltaic power plants or wind farms is relatively timesaving in comparison with the construction of nuclear power plants or hydroelectric power plants. This fact is visible especially in the construction of nuclear power plants in Europe, when almost every currently constructed project is not only extended but also more expensive (for example, Flamanville, Olkiluoto, Mochovce). Delays and frequent construction errors that need to be corrected may to some extent be linked to the inexperience of companies with similar types of contracts, as there has been a slowdown in the construction of nuclear power plants in the past and know-how acquired over the years has been lost. However, these problems can be solved by the increased intensity of construction of these energy sources as well as the introduction of modular construction of reactors of uniform European design. However, even in trouble-free construction, it is often necessary to consider a construction time of 7–10 years. The transition to zero-emission energy therefore brings problems of long construction times of energy resources as well as investment costs. These investment costs will have an impact on energy users in the form of increasing final prices due to various surcharges and charges for energy consumed or reserved capacity. However, it is the increase in prices that can be another impetus in the effort to reduce the consumption of primary energy sources, which is a highly desirable phenomenon. In addition to the undesirable financial consequences, the transition to zero-emission energy is accompanied by fears of possible threats to the stability of the electricity system due to a geographical imbalance in electricity production and consumption.

Both presented approaches have their advantages and disadvantages. The question remains what the costs will be for rebuilding the energy mixes of the countries, which will translate into a final price for the consumer and how stable the European common distribution network will operate. These are all questions that we will receive answers to in the next decade.

5. Conclusions

Global climate change brings many negative effects that fundamentally change the living environment and way of life of human society. Negative impacts in the form of loss of living space due to rising sea levels, lack of drinking water for personal consumption as well as agriculture and industry, extreme weather events and long-term changes in temperature profile, spread of new diseases, famine, migration or deteriorating air quality expose to adaptive pressure large amounts of biological communities, including humans. Today, most scientific organizations and leading scientists in the field of climate already attribute the global climate change to anthropogenic human activity associated with the use of fossil fuels and rising levels of carbon dioxide in the atmosphere. According to various studies, it is important to keep the global temperature rise to 1.5 °C compared to the temperature before the Industrial Revolution. This is also the aim of the efforts of the countries of the European Union, which have set themselves an ambitious goal in reducing emissions by 2050 and in the wider use of renewable energy sources. In addition to energy savings, a change in the structure of the energy mix is a fundamental prerequisite for success on the path to a carbon-neutral or negative future. The presented article dealt with the possibilities of changing the structure of used energy sources in the long term. The comparison of selected countries—Slovakia, Czechia, Poland, Hungary, and Austria—brought two different approaches to the issue. The first is the massive use of the potential of renewable energy sources in the form of hydropower and wind energy in Austria. In this case, the country uses to the maximum extent geomorphological and hydrogeological conditions, while not yet fulfilling the technical potential for their use. The use of hydropower in the production of electricity is a stable element that positively affects the security of the transmission system, and in addition, pumped storage hydropower plants can properly regulate the volatile production of electricity from large wind farms in the east of the country. On the contrary, other countries do not have such a significant potential in renewables, and in the pursuit of emission-free energy they must rely on the use and further development of nuclear energy, which is supplemented to a lesser extent by renewables. From this point of view, an interesting example is the case of the Slovak Republic, which uses a significant share of nuclear and hydropower to produce electricity. An emission-free approach to electricity generation will be crucial for the future, as most efforts to greening processes is their electrification, whether it is passenger and freight transport, various types of industrial processes or the production of hydrogen as the green fuel of the future.

Author Contributions: Conceptualization, methodology, investigation, resources, data curation, writing—original draft preparation, writing—review and editing, visualization, M.B., R.R. All authors have read and agreed to the published version of the manuscript.

Funding: This paper was created in connection with the project VEGA 1/0290/21 Study of the behavior of heterogeneous structures based on PCM and metal foams as heat accumulators with application potential in technologies for obtaining and processing of the earth resources. This research was funded by Scientific Grant Agency of the Ministry of Education, Science, Research and Sport of the Slovak Republic (VEGA). This paper was created in connection with the project 048TUKE-4/2021 Universal educational-competitive platform.

Conflicts of Interest: The authors declare no conflict of interest.

References

1. Kumar, V. Development of Precise Indices for Assessing the Potential Impacts of Climate Change. *Atmosphere* **2020**, *11*, 1231. [CrossRef]
2. Nightingale, A.J. Power and politics in climate change adaptation efforts: Struggles over authority and recognition in the context of political instability. *Geoforum* **2017**, *84*, 11–20. [CrossRef]
3. Brigode, P.; Oudin, L.; Perrin, C. Hydrological model parameter instability: A source of additional uncertainty in estimating the hydrological impacts of climate change? *J. Hydrol.* **2013**, *476*, 410–425. [CrossRef]
4. Se Min Suh, S.; Chapman, D.A.; Lickel, B. The role of psychological research in understanding and responding to links between climate change and conflict. *Curr. Opin. Psychol.* **2021**, *42*, 43–48.
5. Heimann, M. Charles David Keeling 1928–2005. *Nature* **2005**, *437*, 331. [CrossRef]
6. Pales, J.C.; Keeling, C.D. The concentration of atmospheric carbon dioxide in Hawaii. *J. Geophys. Res.* **1965**, *70*, 6053–6076. [CrossRef]
7. Godebo, T.R.; Jeuland, M.A.; Paul, C.J.; Belachew, D.L.; McCornick, P.G. Water Quality Threats, Perceptions of Climate Change and Behavioral Responses among Farmers in the Ethiopian Rift Valley. *Climate* **2021**, *9*, 92. [CrossRef]
8. You, Q.; Kang, S.; Pepin, N.; Flügel, W.A.; Sanchez-Lorenzo, A.; Yan, Y.; Zhang, Y. Climate warming and associated changes in atmospheric circulation in the eastern and central Tibetan Plateau from a homogenized dataset. *Glob. Planet. Chang.* **2010**, *72*, 11–24. [CrossRef]
9. Adkins, J.F.; Ingersoll, A.P.; Pasquero, C. Rapid climate change and conditional instability of the glacial deep ocean from the thermobaric effect and geothermal heating. *Quat. Sci. Rev.* **2005**, *24*, 581–594. [CrossRef]
10. Hunt, C. Instability of response to climate change by tropical lowland forest: Evidence from the Great Cave of Niah, Sarawak. *Quat. Int.* **2012**, *279–280*, 212. [CrossRef]
11. Butler, C.J. A Review of the Effects of Climate Change on Chelonians. *Diversity* **2019**, *11*, 138. [CrossRef]
12. Shafer, C.L. From non-static vignettes to unprecedented change: The U.S. National Park System, climate impacts and animal dispersal. *Environ. Sci. Policy* **2014**, *40*, 26–35. [CrossRef]
13. Lovejoy, T.E.; Hannah, L.J. *Climate Change and Biological Diversity*; Yale University Press: New Haven, CT, USA, 2005.
14. Ani, C.J.; Robson, B. Responses of marine ecosystems to climate change impacts and their treatment in biogeochemical ecosystem models. *Mar. Pollut. Bull.* **2021**, *166*, 112223. [CrossRef]
15. Gauthier, S.; May, B.; Vasseur, L. Ecosystem-Based Adaptation to Protect Avian Species in Coastal Communities in the Greater Niagara Region, Canada. *Climate* **2021**, *9*, 91. [CrossRef]
16. Lazic, D.; Hipp, A.L.; Carlson, J.E.; Gailing, O. Use of Genomic Resources to Assess Adaptive Divergence and Introgression in Oaks. *Forests* **2021**, *12*, 690. [CrossRef]
17. Łubek, A.; Kukwa, M.; Jaroszewicz, B.; Czortek, P. Shifts in Lichen Species and Functional Diversity in a Primeval Forest Ecosystem as a Response to Environmental Changes. *Forests* **2021**, *12*, 686. [CrossRef]
18. Melamed, Y.; Kislev, M.; Weiss, E.; Simchoni, O. Extinction of water plants in the Hula Valley: Evidence for climate change. *J. Hum. Evol.* **2011**, *60*, 320–327. [CrossRef]
19. Robert, R.; Schleyer-Lindenmann, A. How ready are we to cope with climate change? Extent of adaptation to sea level rise and coastal risks in local planning documents of southern France. *Land Use Policy* **2021**, *104*, 105354. [CrossRef]

20. Tehrani, M.J.; Helfer, F.; Jenkins, G. Impacts of climate change and sea level rise on catchment management: A multi-model ensemble analysis of the Nerang River catchment, Australia. *Sci. Total Environ.* **2021**, *777*, 146223. [CrossRef]
21. Ranjbar, M.H.; Etemad-Shahidi, A.; Kamranzad, B. Modeling the combined impact of climate change and sea-level rise on general circulation and residence time in a semi-enclosed sea. *Sci. Total Environ.* **2020**, *740*, 140073. [CrossRef]
22. Mallin, M.A.F. From sea-level rise to seabed grabbing: The political economy of climate change in Kiribati. *Mar. Policy* **2018**, *97*, 244–252. [CrossRef]
23. Leveque, B.; Burnet, J.-B.; Dorner, S.; Bichai, F. Impact of climate change on the vulnerability of drinking water intakes in a northern region. *Sustain. Cities Soc.* **2021**, *66*, 102656. [CrossRef]
24. Hashempour, Y.; Nasseri, M.; Mohseni-Bandpei, A.; Motesaddi, S.; Eslamizadeh, M. Assessing vulnerability to climate change for total organic carbon in a system of drinking water supply. *Sustain. Cities Soc.* **2020**, *53*, 101904. [CrossRef]
25. Delpla, I.; Jung, A.-V.; Baures, E.; Clement, M.; Thomas, O. Impacts of climate change on surface water quality in relation to drinking water production. *Environ. Int.* **2009**, *35*, 1225–1233. [CrossRef] [PubMed]
26. Grusson, Y.; Wesström, I.; Joel, A. Impact of climate change on Swedish agriculture: Growing season rain deficit and irrigation need. *Agric. Water Manag.* **2021**, *251*, 106858. [CrossRef]
27. Golfam, P.; Ashofteh, P.S.; Loáiciga, H.A. Modeling adaptation policies to increase the synergies of the water-climate-agriculture nexus under climate change. *Environ. Dev.* **2021**, *37*, 100612. [CrossRef]
28. Pastor-Paz, J.; Noy, I.; Sin, I.; Sood, A.; Fleming-Munoz, D.; Owen, S. Projecting the effect of climate change on residential property damages caused by extreme weather events. *J. Environ. Manag.* **2020**, *276*, 111012. [CrossRef] [PubMed]
29. Calheiros, T.; Pereira, M.G.; Nunes, J.P. Assessing impacts of future climate change on extreme fire weather and pyro-regions in Iberian Peninsula. *Sci. Total Environ.* **2021**, *754*, 142233. [CrossRef]
30. Yaduvanshi, A.; Nkemelang, T.; Bendapudi, R.; New, M. Temperature and rainfall extremes change under current and future global warming levels across Indian climate zones. *Weather Clim. Extrem.* **2021**, *31*, 100291. [CrossRef]
31. Vaughan, A. Global warming will drive extreme wet weather in summer. *New Sci.* **2019**, *243*, 9. [CrossRef]
32. Sorgho, R.; Jungmann, M.; Souares, A.; Danquah, I.; Sauerborn, R. Climate Change, Health Risks, and Vulnerabilities in Burkina Faso: A Qualitative Study on the Perceptions of National Policymakers. *Int. J. Environ. Res. Public Health* **2021**, *18*, 4972. [CrossRef]
33. Santos-Guzman, J.; Gonzalez-Salazar, F.; Martínez-Ozuna, G.; Jimenez, V.; Luviano, A.; Palazuelos, D.; Fernandez-Flores, R.I.; Manzano-Camarillo, M.; Picazzo-Palencia, E.; Gasca-Sanchez, F.; et al. Epidemiologic Impacts in Acute Infectious Disease Associated with Catastrophic Climate Events Related to Global Warming in the Northeast of Mexico. *Int. J. Environ. Res. Public Health* **2021**, *18*, 4433. [CrossRef]
34. Kurane, I. The Effect of Global Warming on Infectious Diseases. *Osong Public Health Res. Perspect.* **2010**, *1*, 4–9. [CrossRef] [PubMed]
35. Khanal, U.; Wilson, C.; Rahman, S.; Lee, B.L.; Hoang, V.N. Smallholder farmers' adaptation to climate change and its potential contribution to UN's sustainable development goals of zero hunger and no poverty. *J. Clean. Prod.* **2021**, *281*, 124999. [CrossRef]
36. Woodward, A.; Porter, J.R. Food, hunger, health, and climate change. *Lancet* **2016**, *387*, 1886–1887. [CrossRef]
37. Hermans, K.; McLeman, R. Climate change, drought, land degradation and migration: Exploring the linkages. *Curr. Opin. Environ. Sustain.* **2021**, *50*, 236–244. [CrossRef]
38. Stoler, J.; Brewis, A.; Kangmennang, J.; Keough, S.B.; Pearson, A.L.; Rosinger, A.Y.; Stauber, C.; Stevenson, E.G.J. Connecting the dots between climate change, household water insecurity, and migration. *Curr. Opin. Environ. Sustain.* **2021**, *51*, 36–41. [CrossRef]
39. Puškár, M.; Jahnátek, A.; Kádárová, J.; Soltesova, N.; Kovanic, L.; Krivosudska, J. Environmental study focused on the suitability of vehicle certifications using the new European driving cycle (NEDC) with regard to the affair "dieselgate" and the risks of NOx emissions in urban destinations. *Air Qual. Atmos. Health* **2019**, *12*, 251–257. [CrossRef]
40. Sinay, J.; Puškár, M.; Kopas, M. Reduction of the NOx emissions in vehicle diesel engine in order to fulfill future rules concerning emissions released into air. *Sci. Total Environ.* **2018**, *624*, 1421–1428. [CrossRef] [PubMed]
41. Cook, J.; Oreskes, N.; Doran, P.T.; Anderegg, W.R.L.; Verheggen, B.; Maibach, E.W.; Carlton, J.S.; Lewandowsky, S.; Skuce, A.G.; Green, S.A.; et al. Consensus on consensus: A synthesis of consensus estimates on human-caused global warming. *Environ. Res. Lett.* **2016**, *11*, 048002. [CrossRef]
42. Cook, J.; Nuccitelli, D.; Green, S.A.; Richardson, M.; Winkler, B.; Painting, R.; Way, R.; Jacobs, P.; Skuce, A. Quantifying the consensus on anthropogenic global warming in the scientific literature. *Environ. Res. Lett.* **2013**, *8*, 024024. [CrossRef]
43. Anderegg, W.R.L. Expert Credibility in Climate Change. *Proc. Natl. Acad. Sci. USA* **2010**, *107*, 12107–12109. [CrossRef]
44. Doran, P.T.; Zimmerman, M.K. Examining the Scientific Consensus on Climate Change. *EOS Trans. Am. Geophys. Union* **2009**, *90*, 22–23. [CrossRef]
45. List of Worldwide Scientific Organizations Accepted that Climate Change Has Been Caused by Human Action. Available online: http://www.opr.ca.gov/facts/list-of-scientific-organizations.html (accessed on 7 May 2021).
46. Abram, N.J.; McGregor, H.V.; Tierney, J.E.; Evans, M.N.; McKay, N.P.; Kaufman, D.S. The PAGES 2k Consortium. Early onset of industrial-era warming across the oceans and continents. *Nature* **2016**, *536*, 411–418. [CrossRef]
47. Wrigley, E.A. Energy and the English Industrial Revolution. *Philos. Trans. R. Soc. A* **2013**, *371*, 20110568. [CrossRef]
48. Smil, V. *Energy Transitions: Global and National Perspectives*, 2nd ed.; Praeger: Westport, CT, USA, 2016.
49. BP Statistical Review of World Energy. Available online: https://www.bp.com/en/global/corporate/energy-economics/statistical-review-of-world-energy.html (accessed on 5 May 2021).

50. Global Temperature Anomaly. NASA's Goddard Institute for Space Studies. Available online: https://climate.nasa.gov/vital-signs/global-temperature/ (accessed on 1 May 2021).
51. Masson-Delmotte, V.; Zhai, P.; Pörtner, H.O.; Roberts, D.; Skea, J.; Shukla, P.R.; Pirani, A.; Moufouma-Okia, W.; Péan, C.; Pidcock, R.; et al. (Eds.) *Intergovernmental Panel on Climate Change, 2018: Global Warming of 1.5 °C. An IPCC Special Report on the Impacts of Global Warming of 1.5 °C Above Pre-Industrial Levels and Related Global Greenhouse Gas Emission Pathways, in the Context of Strengthening the Global Response to the Threat of Climate Change, Sustainable Development, and Efforts to Eradicate Poverty*; Intergovernmental Panel on Climate Change: Geneva, Switzerland, 2018.
52. Gawlik, L.; Mokrzycki, E. Changes in the Structure of Electricity Generation in Poland in View of the EU Climate Package. *Energies* **2019**, *12*, 3323. [CrossRef]
53. Son, D.; Kim, J.; Jeong, B. Optimal Operational Strategy for Power Producers in Korea Considering Renewable Portfolio Standards and Emissions Trading Schemes. *Energies* **2019**, *12*, 1667. [CrossRef]
54. Ranzani, A.; Bonato, M.; Patro, E.R.; Gaudard, L.; De Michele, C. Hydropower Future: Between Climate Change, Renewable Deployment, Carbon and Fuel Prices. *Water* **2018**, *10*, 1197. [CrossRef]
55. Jesse, B.-J.; Morgenthaler, S.; Gillessen, B.; Burges, S.; Kuckshinrichs, W. Potential for Optimization in European Power Plant Fleet Operation. *Energies* **2020**, *13*, 718. [CrossRef]
56. Ghandriz, T.; Jacobson, B.; Islam, M.; Hellgren, J.; Laine, L. Transportation-Mission-Based Optimization of Heterogeneous Heavy-Vehicle Fleet Including Electrified Propulsion. *Energies* **2021**, *14*, 3221. [CrossRef]
57. Drożdż, W.; Elżanowski, F.; Dowejko, J.; Brożyński, B. Hydrogen Technology on the Polish Electromobility Market. Legal, Economic, and Social Aspects. *Energies* **2021**, *14*, 2357. [CrossRef]
58. Korberg, A.D.; Skov, I.R.; Mathiesen, B.V. The role of biogas and biogas-derived fuels in a 100% renewable energy system in Denmark. *Energy* **2020**, *199*, 117426. [CrossRef]
59. Mortensen, A.W.; Mathiesen, B.V.; Hansen, A.B.; Pedersen, S.L.; Grandal, R.D.; Wenzel, H. The role of electrification and hydrogen in breaking the biomass bottleneck of the renewable energy system—A study on the Danish energy system. *Appl. Energy* **2020**, *275*, 115331. [CrossRef]
60. Figenbaum, E.; Assum, T.; Kolbenstvedt, M. Electromobility in Norway: Experiences and Opportunities. *Res. Transp. Econ.* **2015**, *50*, 29–38. [CrossRef]
61. BP Statistical Review of World Energy Methodology. Available online: https://www.bp.com/en/global/corporate/energy-economics/statistical-review-of-world-energy/using-the-review/methodology.html#accordion_primary-energy-methodology (accessed on 12 July 2021).
62. Demirel, Y. *Energy Production, Conversion, Storage, Conservation, and Coupling*, 2nd ed.; Springer: Cham, Switzerland, 2016; p. 445.
63. Cengel, Y.; Boles, M. *Thermodynamics: An Engineering Approach*, 8th ed.; McGraw-Hill Education: New York, NY, USA, 2014.
64. Impram, S.; Nese, S.V.; Oral, B. Challenges of renewable energy penetration on power system flexibility: A survey. *Energy Strategy Rev.* **2020**, *31*, 100539. [CrossRef]
65. Cole, W.; Frazier, A.W. Impacts of increasing penetration of renewable energy on the operation of the power sector. *Electr. J.* **2018**, *31*, 24–31. [CrossRef]
66. Peng, P.; Lu, F.; Cheng, S.; Yang, Y. Mapping the global liquefied natural gas trade network: A perspective of maritime transportation. *J. Clean. Prod.* **2021**, *283*, 124640. [CrossRef]
67. McFarlan, A. Techno-economic assessment of pathways for liquefied natural gas (LNG) to replace diesel in Canadian remote northern communities. *Sustain. Energy Technol. Assess.* **2020**, *42*, 100821.
68. The Future of Hydrogen. IEA Report. Available online: https://www.iea.org/reports/the-future-of-hydrogen (accessed on 25 April 2021).
69. Martino, M.; Ruocco, C.; Meloni, E.; Pullumbi, P.; Palma, V. Main Hydrogen Production Processes: An Overview. *Catalysts* **2021**, *11*, 547. [CrossRef]
70. Rapier, R. Life Cycle Emissions of Hydrogen. Available online: https://4thgeneration.energy/life-cycles-emissions-of-hydrogen/ (accessed on 25 April 2021).
71. Hoes, O.A.C.; Meijer, L.J.J.; van der Ent, R.J.; van de Giesen, N.C. Systematic high-resolution assessment of global hydropower potential. *PLoS ONE* **2017**, *12*, e0171844. [CrossRef]
72. International Monetary Fund. *World Economic Outlook—GDP Per Capita*; International Monetary Fund: Washington, DC, USA, 2021.
73. Slovak Ministry of Economy. *Integrated National Energy and Climate Plan (2021–2030)*; Slovak Ministry of Economy: Bratislava, Slovakia, 2019.
74. International Energy Agency—Slovak Republic Review. Available online: https://www.iea.org/countries/slovak-republic (accessed on 21 April 2021).
75. Morvay, A. *Economic Transformation: The Experience of Slovakia*; Repro-Print: Bratislava, Slovakia, 2005.
76. Kotulic, R.; Kravcakova Vozarova, I.; Nagy, J.; Huttmanova, E.; Vavrek, R. Performance of The Slovak Economy in Relation to Labor Productivity and Employment. *Procedia Econ. Financ.* **2015**, *23*, 970–975. [CrossRef]
77. IEA—Slovak Republic Energy System Overview. Available online: https://euagenda.eu/upload/publications/untitled-69956-ea.pdf (accessed on 7 April 2021).

78. International Labour Organization, ILOSTAT Database. Available online: http://data.worldbank.org/data-catalog/world-development-indicators (accessed on 7 April 2021).
79. Friedlingstein, P.; O'Sullivan, M.; Friedlingstein, P.; Jones, M.W.; Andrew, R.M.; Hauck, J.; Olsen, A.; Peters, W.; Pongratz, J.; Sitch, S. Global Carbon Budget 2020. *Earth Syst. Sci. Data.* **2020**, *12*, 3269–3340. [CrossRef]
80. Behrsin, I. Controversies of justice, scale, and siting: The uneven discourse of renewability in Austrian waste-to-energy development. *Energy Res. Soc. Sci.* **2020**, *59*, 101252. [CrossRef]
81. Geyer, R.; Knöttner, S.; Diendorfer, C.; Drexler-Schmid, G.; Alton, V. 100% Renewable Energy for Austria's Industry: Scenarios, Energy Carriers and Infrastructure Requirements. *Appl. Sci.* **2021**, *11*, 1819. [CrossRef]
82. Soltesova, K. Renewable Energy Sources in Slovakia (Slovak Energy Agency). Available online: www.sea.gov.sk/NLCCollection (accessed on 15 March 2021).

Article

On Electromobility Development and the Calculation of the Infrastructural Country Electromobility Coefficient

Erika Feckova Skrabulakova [1], Monika Ivanova [2], Andrea Rosova [1,*], Elena Gresova [1], Marian Sofranko [1] and Vojtech Ferencz [1]

[1] Faculty of Mining, Ecology, Process Control and Geotechnology, Technical University of Kosice, Letna 9, 04200 Kosice, Slovakia; erika.feckova.skrabulakova@tuke.sk (E.F.S.); elena.gresova@tuke.sk (E.G.); marian.sofranko@tuke.sk (M.S.); vojtech.ferencz@gmail.com (V.F.)

[2] Faculty of Humanities and Natural Sciences, University of Presov, 17. Novembra 15, 08116 Presov, Slovakia; monika.ivanova@unipo.sk

* Correspondence: andrea.rosova@tuke.sk; Tel.: +421-55-602-3144

Abstract: The question of electromobility is greatly discussed theme of the present especially in connection with the reduction of greenhouse gas emissions. In order to fulfill decarbonization targets, incentives of many countries lead to the support of electromobility. In this paper we ask to which extend are Visegrád Group countries prepared for the widespread utilization of electric cars and define a new coefficient K called the infrastructural country electromobility coefficient. Its computing is covered by appropriate analysis and calculations done previously. Several indices that keep particular information about the state of preparation for electromobility are defined and debated here, as well. Their product forms the coefficient K. Obtained results include outcomes and discussion regarding the level of infrastructural electromobility preparedness for the chosen states, among which we extra focus on the position of Slovakia compared to the European Union average and European electromobility leaders. Based on the data obtained, we found out that the stage of preparation of Slovakia for electromobility among Visegrad Group countries is rather good, although it is far behind the European Union leaders. We realized that there was a rapid growth of electromobility infrastructure in Slovak Republic in the last five years as its infrastructural country electromobility coefficient grew 334 times.

Keywords: charging stations; infrastructural country electromobility coefficient; electric vehicles; electromobility; infrastructure development

Citation: Feckova Skrabulakova, E.; Ivanova, M.; Rosova, A.; Gresova, E.; Sofranko, M.; Ferencz, V. On Electromobility Development and the Calculation of the Infrastructural Country Electromobility Coefficient. *Processes* **2021**, *9*, 222. https://doi.org/10.3390/pr9020222

Received: 21 December 2020
Accepted: 21 January 2021
Published: 25 January 2021

1. Introduction

The European decarbonization targets to reduce greenhouse gas emissions by around 20% with respect to 2008 levels until 2030 impose significant economical, social and technological challenges in which electromobility plays a key role [1]. Over 71% of the transport-related CO_2 emissions in Europe come from road transport while the majority of greenhouse gas emissions is associated with cars. Hence, the aim is to reduce, if not to forbid, the utilization of conventionally fueled vehicles in cities until 2050 [1]. Electric vehicles do not cause the local pollution. The widespread use of electric cars instead of conventionally fueled ones will reduce the greenhouse gas emissions as well as environmental impacts from transport. Therefore, the electromobility topic is super highly up to date. This is evidenced by a number of scientific studies that have been published in recent years.

While in 2014 the battery electric vehicles (BEVs) and plug-in hybrid vehicles (PHEVs) constituted only 0.5% of the total new vehicle registrations in the European Union [2], the world market of nowadays indicates the boom in sales of electric vehicles (EVs) and that the growth rate of worldwide sales is exponential [3]. Regions of China, Europe, and USA made up over 90% of global electric cars sales. The world statistics on battery electric cars plus plug-in hybrid electric cars sales can be find at Figure 1 (own adaptation based on [4]).

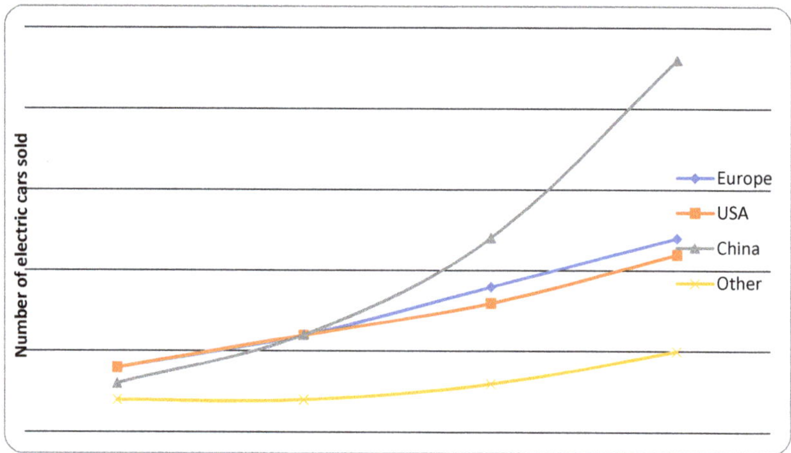

Figure 1. Global electric cars sales according to world's regions, (source: elaborated by authors based on [4]).

In this paper, we asked a question to which extend are countries of Visegrad Group (Czech Republic, Hungary, Poland, Slovak Republic) prepared for the widespread utilization of electric cars-in what follows, the car electromobility, or simply, electromobility. Before all we target on the electromobility preparation of Slovakia and compared it with V4 countries, European Union (EU) average, and selected electromobility leader countries of Europe. Aiming to reach our goal we have formulated the following 4 questions of research:

Q1: To which extend are Visegrád Group (V4) countries prepared for the widespread use of electric cars (car) electromobility (from the infrastructural point of view)?

Q2: Compared to European electromobility leaders and European Union average what is the situation in Visegrád Group countries considering the preparation for (car) electromobility?

Q3: Compared to V4 countries what is the situation in Slovakia as a member of Visegrád Group countries considering the preparation for (car) electromobility from the infrastructural point of view?

Q4: Compared to European electromobility leaders and European Union average what is the situation in Slovakia considering the preparation for (car) electromobility from the infrastructural point of view?

One of the main contributions of this paper is the data collection. Their processing, analyzing and visualization provided a comprehensive view of preparation for electromobility in the Visegrád Group countries from the infrastructural point of view. A comparison of their situation with European leaders in the field of electromobility and with the EU average made it possible to specify more in detail their position within Europe. Among V4 countries, the position of Slovakia has been investigated more in detail and compared to both situation in European electromobility leading countries and EU average. The drawing of the maps of fast charging stations' distribution in Slovakia showing the situation in both 2018 and beginning of 2020 help to understand the development and trend tendencies within this area in the last years. Another contribution is the introduction of the infrastructural country electromobility coefficient, the index K, and its calculation for selected countries, that allows the easy cross-countries comparison considering the state of their (car) electromobility preparation from the infrastructural point of view.

2. Literature Review

The low carbon energy technologies should be developed and implemented in a cost-effective way under the Strategic Energy Technology Plan of the European Commission [5].

In [6] by Straka et al. the possibilities and limitations of electromobiles utilizations have been studied. Authors focus on electric cars and charging stations as the key element in electromobility development. They discuss the finance, legislation, landowners, location, electric cars quantification, technological and other aspects that affect electromobility development. The important element of electromobility deployment and expanding has been studied in [7], as well. Shafiei et al. took into account a tax-induced electric vehicles transition here. Authors combine a technical-economic simulation model of an integrated energy transmission system with a macroeconomic model of general equilibrium. The impact of the new government's tax reform proposal is compared with the current vehicle taxation policy, as well as with other incentives for EVs, which include a tax exemption, and a ban on the sale of new diesel vehicles. They examine individual scenarios in a wide range of future changes in fuel prices and emission cost reductions. The results show that the overall macroeconomic benefits will be negligible, but, sustainable and beneficial in the long run. Although a tax-based technological solution to support electric vehicles deployment will allow for long-term reductions in greenhouse gas emissions, it will not be enough to meet the short-term climate goals. The study on electric vehicles transition, obstructions, advantages, and likewise the socio-technical aspects followed through the multi-mode interaction and the multilevel perspective configuration was performed by Lin and Sovacool in [8]. In their contribution, the conceptual framework of interaction with several regimes is visible. It illustrates the interaction between BEVs and other technologies, especially PHEVs. Authors summed up that BEVs may have an alignment and re-alignment path or a reconfiguration path depending on the interaction between different technologies.

An interesting view on the theme referred Ryghaug and Skjølsvold in [9]. Authors claim that the general awareness indicates that transition to the electromobility is the consequence of a focused policies package for demand stimulation. The country of interest was Norway. The policies were initially pursued to support the progress of a Norwegian electric vehicle industry. Their paper focused on how the policies work, how the electric cars are produced and how their effects adapt across geographical borders. The electric vehicle sales on the present and historical data basis in the Norway and Netherlands with the visions for the future have been analyzed in [10]. The specific dynamic model was used and upgraded here. Empirically, the model validation and evaluation were performed by executing three types of tests. Additionally, two particular model parts were detected for the further enhancement. Authors found out that in the connection with the transition to the new fossil fuels, new emission limits for manufacturers need to be introduced, as well. The findings suggest that only strong incentives have led to a high share of zero-emission vehicle sales in the Netherlands and Norway. Norwegian market was scrutinized also by Hovi et al. in [11], who screened the impressions from battery-electric truck consumers. This case study reported that perceptions were mostly positive. Nevertheless, the tailoring of use patterns signified quite solicited element. Already in 2013, the share of EVs on Norwegian new vehicle market was 5.8%. 80% out of these vehicles were owned by private possessors. Figenbaum et al. in [12] described the Norwegian electromobility story and gave some explanations to the Norwegian development. The inspiriting incentives, as well as interactions between private enterprises, non-government organizations, public authorities, and taxation system, that influenced the vehicle purchase, and supported the Norwegian electromobility diffusion, have been discussed here. Norway well managed the electromobility challenges and became European electromobility leader and the trendsetter for many countries. Authors point out the importance of social networks, which play an important role in the purchase of EVs. They found out that in average 36% of the EVs owners have friends who bought an electric vehicle after they had information about the experiences with electric vehicle from these friends and another 34% now consider buying an electric vehicle, what creates a potential for the market. The potential growth of EVs share in rural areas will be easier to be realized when the technological developments secure a longer range of electric vehicles, more charging stations are at the hand, and there is more of information on electric vehicles energy efficiency [12].

The analyses of the Norwegian electric vehicle story became an inspiration for other (not only European) countries, as Norwegian policies led to the world's fastest diffusion of EVs. An important role played the interaction of events in and between the niche, regime and landscape levels, how it shaped the Norwegian electro vehicles policies that led to the world's fastest diffusion of EVs, also at user adoption from a socio-technical perspective. The process started in urban regions but is now covering other locations, as well [13]. The incentives history of some countries, such as Norway, has been known for decades of years [14], while the stimuli of other states, such as Slovakia, are only very recent [15]. Clearly, new standards for fuel consumption and emissions, together with grants and changes in national vehicle taxation created more demand for electromobiles [16] of all kinds and greater cognizance of electromobility.

Except of a few European countries such as Denmark, France, Germany, Netherland, the amount of electric vehicles in Europe is limited—see the development of annual registrations of BEVs and PHEVs in Europe at Figure 2 (own adaptation based on [17]). Despite many environmental benefits, the penetration is rather poor what might be explained by certain factors that daunt eventual purchasers. Therefore, EU states raise stimuli in order to balance these factors and support electromobility. The most important policy instruments to promote the use of EVs are tax and infrastructure measures along with financial incentives for purchasing and supporting R&D projects. The available information allowed authors of [18] to conclude that higher penetration levels of EVs appear in countries where the registration tax, the ownership tax, or both taxes have developed into a partial green tax by including CO_2 emissions in the calculation of the final invoice. The countries with a more intensive use of EVs also fund charging stations to facilitate local electromobility [18]. The current challenges and perspectives of the development of electromobility both under the conditions of the EU and on a global scale were discussed in [19] too. The authors derived several recommendations concerning the development of electromobility in the EU. They point out to the importance of related services, such as the possibility of simple payments within the system, flat rate payment, management of vehicle charging over time, a detailed overview of the functionality and availability of charging stations or customer support. They consider electric vehicles being useful in meeting the goals to be achieved by introducing the zero-carbon urban logistics in the centers of major cities until 2030 [19].

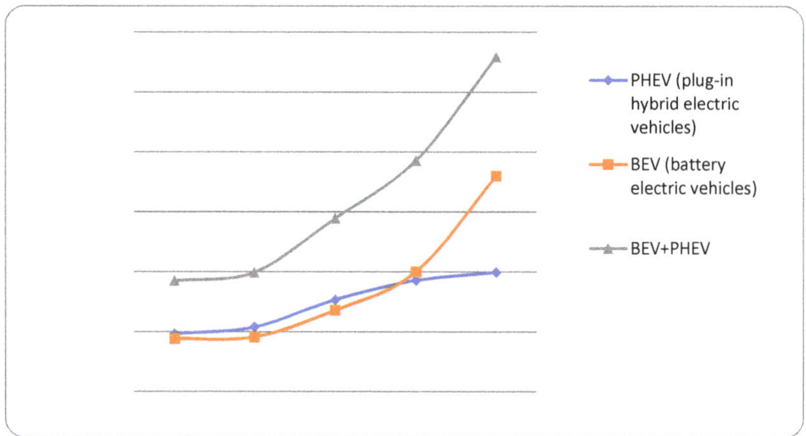

Figure 2. Annual registrations of battery electric vehicles and plug-in hybrid electric vehicles in Europe, (source: elaborated by authors based on [17]).

Johansen in his diploma thesis [20] sees electric mobility as the key technology that will replace fossil energy in the long run. He claims that "looking at the e-mobility research area, little research has been focused on measuring maturity within the field" and that during his literature research he found "only one piece of literature focusing on evaluating maturity on battery electric vehicles", namely [21]. In fact, beside other literature, he sources from 2 papers, [21,22], where the latter one presents an improvement of the approach presented in the first one. Based on the literature review Johansen in [20] understand the concept of electromobility as a complex system consisting of the following parts: smart grids, general infrastructure-intelligent charging and transport systems, policy aspects and regulation, economic aspects, sustainability, business models and market integration. Author in the thesis presents an electromobility maturity model-a framework for measuring electromobility readiness of countries. The chosen maturity indicators for e-mobility were sustainability, regulation, infrastructure, economic factors, consumer perception, and market integration. The model was constructed through an iterative process by going back and forth between through data to establish validity. It was evaluated by its application on Denmark and Germany. Based on the qualitative data it was found that the maturity of e-mobility is the same in each country, although it is strong in different indicators.

Papers [21,22] deal with the question how to identify whether the introduction of EVs is sustainable in megacities, as the potential for sustainable success depends on many local conditions such as energy mix, traffic and climate flow and even parameters like the human development index or corruption index. Authors focus only on BEVs and do not include any form of hybrid electric vehicles or two-wheelers in their calculations. They define the electromobility potential index—the parameter that should help the decision makers uncover the potential for sustainable and successful introduction of these vehicles in megacities, and answer the questions whether the city itself is ready for that and whether it provides supporting boundary conditions. Deriving three key topics–readiness, user acceptance, and sustainability, in [22] the five key performance indicators have been identified: consumption, environmental impact, costs, infrastructure and socio-demographic conditions. With the help of numerous formulae using quantitative data from many sources, authors derive the values of the key performance indicators and some other performance indicators in order to define the values of the electromobility potential index for 47 major cities. The modified composition of the electromobility potential index presented in [21] combines only four key performance indicators and their respective weightings. In order to have linearly independent data the consumption indicator has been removed, as its effects are already accounted in infrastructure indicator and costs indicator. The electromobility potential index has been in [21] recalculated for the set of all cities from [22] except of Riyadh, what effect in permutation of the sequence in which the cities stand according to descending electromobility potential index values. It was shown that in general, cities with well-developed infrastructure, high living standards and reliable governance show the best boundary conditions for successful and sustainable introduction of BEVs. The robustness of the electromobility potential index calculation has been confirmed in [21], as well. Higueras-Castillo et al. in [23] studied the marketing angle of view towards electromobility development. The crucial goal laid in the appreciation of the consumer attitudes towards electromobility. This recent study indicated the factors of perceived consumer effectiveness understood as the consumer's assessment of his/her ability to contribute to specific results of sustainable development via specific behaviors. The conclusions of this study should be facilitated in order to improve established marketing strategies. The results show that trust and external incentives are the main factors when buying a battery electric vehicle or plug-in hybrid electric vehicle. Perceived consumer effectiveness has a significant impact on the intent. The effect of green self-identity on attitude is noticeable in consumers with high levels of perceived consumer effectiveness [23]. According to 2018 statistics China is the world largest market with over 1,000,000 battery electric cars sold, followed by Europe (over 385,000 battery electric cars sold) and USA (over 361,000 battery electric cars sold).

At the other hand, the China's share is 4.5% and USA's share is 2.4%, while the world's highest market share, 46% (in 2018) for electric car sales, is associated to Norway [4].

Many studies pay attention to the manifold policy tools of individual countries that try to motivate moving to the BEVs and PHEVs. One of them is [24], where the emphasis was put on Greece. The work brought three policy assistance stimulants with regard on encouraging the utilization of the EVs until 2030. The results of the research show that the use of direct subsidies is more advantageous compared to the carbon taxation system. Similar stimulants have many European countries—see [25]. The consumer attitude towards electromobility has been studied in [26], as well. Authors focus here on electric vehicles, more concretely: plug-in hybrid electric vehicles, range extended electric vehicles, and battery electric vehicles. Although these electromobiles are perceived as secure and environmentally friendly, the high purchase prices form barriers for the widespread utilization of electric cars. Bühne et al. claim that these obstacles can be reduced by benefits in the shape of financial measures such as tax and energy cost reductions and dense network of charging stations [26]. On the model example of Germany authors discuss the use of different promotion measures by the national government that can enhance the market share of electromobiles and based on Norway experiences they postulate that tax exemptions can raise the stock of electromobiles quite quickly.

The development of the electromobility sector in the EU and its particular states with focus on Poland was estimated in [27]. The situation was compared with the one in Switzerland and Norway. Tucki et al. present a method for calculating primary energy which has been evaluated on the basis of methods proposed by the Fraunhofer Institute. Validation was performed based on a comparison of 4 proposed methods. It was proved that the method suggested by the authors is the most accurate. Increased demand for electric vehicles has been reflected in lower CO_2 emissions. Drożdż and Starzyński in [28] presented the legal regulations implemented in Poland in the area of electromobility. Similar issues were also addressed by Sendek and Matysiak. They monitor electromobility in Poland with the introduction of BEVs and the most important factors stimulating its development [29]. As the road transport is one of the major sources of air pollution in this country, the most important factors stimulating the current state of electromobility and its development were in focus of several research studies—see e.g., [30–33] and many others, where the country of the primary interest was Poland. Kupczyk et al. in [31] presented actual trends as well as forecasting changes regarding the markets of chosen alternative fuels and the market of EVs. The legal framework was considered in this paper, as well. The research has been formed on a score-based sector attractiveness method to compare the selected units-the biofuel and the electric car markets. Electric cars along with charging stations have been discussed by Kłos et al. in [32]. The effort to specify the impact on the demand for power and energy in the Polish electric power system caused by utilization of electric cars was the main point presented here. Kłos et al. consider it being necessary to take into account the development of electromobility both at the level of production and its transmission and distribution. The results of the research show that the impact of electromobility on the system depends mainly on the development of load curves associated with car charging. Authors prefer charging EVs during the night, what should be encouraged at the expense of discouraging fast charging during the day by an appropriate economic, legal, and technical environment. The connections between charging stations infrastructure and mentioned electric power system were analyzed by looking at the current conditions on the one hand, and by looking at the future and possible issues on the other hand. The potential negatives of the matter for the same country have been presented by Drożdż in [27]. Although no greenhouse gas emissions are generated directly from the operation of EVs, a significant amount is produced during the production in power plants. In some Central European countries, the efficiency of primary energy consumption for the end users is half of that in the most efficient countries. In the case of greenhouse gas production, there is up to 5.8 fold difference [33].

The environmental impact of electric power production considering the issue of electric vehicle operation has been discussed by Skrúcaný et al. in [34]. Authors show the ecological footprint of electric vehicle operations focusing on selected European countries: Austria, Czech Republic, Germany, Hungary, Poland, Slovakia, and Slovenia. The reduction of greenhouse gas emissions by increasing energy efficiency, the share of zero-emission electricity production, and encouraging electromobility in Hungary has been studied in [35]. The connections between CO_2 emissions in Austria, Czech Republic, Germany, Poland, and Slovak Republic and the public attitude to electromobility have been discussed by Jursová et al. in [36]. Authors say that the inhabitants of Czech and Polish cross border region are interested in electromobility. The results of the research confirm that they believe in its ecological contributions and they would be interested in buying an electric vehicle will not be the worries about its distance range, the existence of insufficient amount of charging points, the high purchasing price and the worries on electric cars reliability. The respondents indicate lack of information about this technology. Approximately 60% of respondents consider the investment to be insufficient for the massive application of electromobility for everyday life [36]. Igliński presented a research on electromobility development in Central Europe and Eastern Europe in [37], where a comparative study on electromobility development in Poland and another 10 countries including Slovakia, Hungary and Czech Republic can be found. The development of electromobility in the EU and in Slovakia has been studied by Daňo and Rehák in [38]. The paper has been published in 2018. Based on the data obtained in the previous time period they found out that one of the major problems hindering the development of electromobility in Slovakia is the lack of charging stations in the network. The only exception was the capital city-Bratislava. Another problem is the purchasing price of electric cars. It is at the level of 30,000 EUR, which is a high price for an average Slovak. Some recommendations that might increase the marketability of EVs in Slovakia can be found in [39] published by Rehák. The conditions for development of electromobility in Slovak Republic have been studied in [6,40], as well. On an example of the city Senec in [40] Hrudkay and Jaroš point out to a strategy for the development of electromobility with the aim of building some ecosystem with a gradual reduction of emissions from the local transport point of view. The options for the local micro electromobility in Liberec Region of Czech Republic and in the Middle Europe in general were discussed by Černohorský et al. in [41]. Authors highlight that the development of electromobility in post-communist countries is not sufficiently supported. They see opportunities in promoting personal mobility on a bicycle, which they compare with the local public transport and individual car transport, in terms of time and cost. The biggest problem in mechanical as well as electrical implementation of electric drive on a bicycle addresses Svetlík in [42].

Based on the literature review provided, one can sum up, that the sustainable traffic development is closely related to new technologies' development focusing on the greenhouse gas emission reduction and hence, searching for possibilities of the widespread utilization of alternative fueled vehicles. Within these vehicles, electric cars have an important place. The development of car electromobility depends on many different factors. Although factors like purchasing prices of electric cars, personal income, stimulants and incentives of government, benefits in the financial measures in the form of tax or energy cost reductions and similar ones play an important role here, before all the existence of the dense network of charging stations is crucial factor affecting the car electromobility. At the one hand, European countries with longer electromobility history have better electromobile infrastructure, but also a greater number of electric vehicles in use. Other ones, like e.g., V4 countries, are just at the beginning of their electromobility age, what is usually connected both with lacking electromobile infrastructure and small number of electric cars in use. Therefore, our questions on the extent to which these countries are prepared for the widespread utilization of car electromobility from the infrastructural point of view are very up to date.

3. Materials and Methods

In this paper we aim to answer the question to which extend are selected European countries prepared for electromobility in the EU context and boundness juncture to selected European countries.

In order to reach our goal, we asked four research questions. The answer to the all research questions assumed a broad literature review. Hence, we conducted a literature review on the topic, analyze the data obtained, summarize, process and visualize these data via standard techniques with the help of tables, maps, and charts. The review was conducted in a similar way as by Okorie et al. in [43] or Mohamad and Teh in [44] based on principles of systematic review. Denyer and Tranfield in [45] point out to the fact that systematic review has to be a self-contained research project and should not be regarded as a literature review in the traditional sense, as its aim is to find an answer to a clearly specified question. According to Moher et al. [46] systematic reviews should be systematically planned and build on a protocol that describes the rationale, hypothesis, and planned methods of the review. Following the ideas presented in [45,46] here we have searched articles in the Web of Science, Scopus, IEEE Xplore, and Google Scholar databases. In 2019 we set the end of 2014 as the starting point of our review as we aimed to work with the data that are at most 5 years old. Although via manual screening of cross-references of some articles we have found some more relevant publications exceeding our time range of revised articles. As the work has been finalized only in 2020, we have processed also some data that have been published until the end of March 2020. In order to get the most up to date data we searched also statistical data published at Eurostat's, European Automobile Manufacturers' Association's, World Bank Group's and many other web pages and online databases from trusted sources such as Slovak Investment and Trade Development Agency, Vienna Institute of Demography and International Institute for Applied Systems Analysis. The most up to date news about technical parameters of nowadays electric vehicles and firstly public presented plans about building new electromobility infrastructure we got from a clean mobility exhibition—the Slovak Electric Car Salon 2020 in Bratislava. This process led to the final list of data sources.

In order to obtain required information we have studied and cross-country compared several relevant parameters such as the total number of cars, BEVs, and PHEVs registered in country and its recomputation for 1000 inhabitants of the country, the total number of charging stations, its recomputation for 1 registered plug-in electromobile, its recomputation for the area of the country, and its recomputation for 1000 inhabitants of the country, the yearly percentage growth of associated infrastructure and so on. As the most important building blocks of this infrastructure are the fast charging stations, we have discussed their number per 100 km of highways in each country. Since we have focused on Slovakia, we have studied the overall distribution of the fast charging stations in Slovak Republic. The main problem was to determine some parameter that could be useful for cross-country comparison considering the electromobility preparation, as analyzing both the standard ways of measuring, counting, and displaying disparities between regions via statistical tools such as the Atkinson index, coefficient of variation, Gini coefficient, Hoover index, real convergence method, standard deviation, Theil index, and so on (see e.g., [47–49]) and less known ones via graph theory means (see e.g., [50]), we could not find the appropriate fitting one. Therefore, we have defined the infrastructural country electromobility coefficient, abbreviated as *K*–a parameter for infrastructural cross-countries (car) electromobility preparation comparison. For all the countries of selection we have counted and discussed the values of the introduced parameter K, defined as follows:

$$K = E_p \times N_k \times N_e \times N_p, \qquad (1)$$

E_p is the number of (BEVs + PHEVs) per 1000 inhabitants of the country,
N_k is the number of public charging stations per km^2 of the country,
N_e is the number of public charging points per plug-in electromobile,

N_p is the number of public charging points per 1000 inhabitants of the country.

Hence, clearly, K could be expressed in No∗km^{-2}, in SI units only km^{-2}, and defined alternatively:

$$K = N^3 \times 10^6 \times P^{-2} \times A^{-1}, \quad (2)$$

N is the number of public charging stations,
P is the number of inhabitants of the country,
A is the total area of the country.

Obviously, the higher values of K refer to the higher extent of preparation of the country for electromobility. Although K could be counted according to (2), we prefer the Equation (1) to do so, as each element of Equation (1): E_p, N_k, N_e, N_p, brings some particular information on countries electromobility preparation:

The higher is the normalized value of the number of BEVs and PHEVs already used by inhabitants of the country, the higher is the optimism about their worth utilization in daily life, the higher is the overall information about the benefits of EVs, the more positive is the overall environment towards electromobility, the more is the country prepared for the widespread utilization of BEVs and PHEVs. Therefore, K is directly proportional to E_p.

Although it is important to consider the exact spatial organization of the charging points within the country (as in the extreme case the charging stations can be concentrated only in urban areas and nowhere else), one can expect that similarly like in the case of mobile network providers, the large electric vehicle charging services providers would focus on good (in ideal case equidistant) coverage of the country by public charging points. Sure, the opportunities to charge a vehicle in big cities and the rest of the country might differ, but considering the text above, the average number of public charging stations per km^2 of the country, N_k, also gives partial information about the infrastructural preparedness of the country for EVs utilization. As the higher is the density of public charging stations' net, the more comfortable is the travelling using an electric car (even of a smaller range), K is directly proportional to N_k.

In a case of widespread utilization of EVs in the respective country, with the greater number of electric cars in use the demand on charging infrastructure grows. If there are more inhabitants of the country, there are more potential buyers of an electric car, each of which would like to use it in a comfort way, hence, without the fear where to charge it. Therefore, K is directly proportional to the number of public charging points per 1000 inhabitants of the country (N_p).

The lesser is the number of BEVs and PHEVs per one public charging point (the inverse value of the parameter N_e), the better is the public electromobile infrastructure and the better is country prepared for the widespread utilization of EVs. Therefore, K is indirectly proportional to the inverse value of N_e, or, conversely, K is directly proportional to N_e. In mathematics, being directly proportional means that as one amount increases, another amount increases at the same rate. Therefore, in our definition of K the multiplication instead of addition is used, although the disadvantage of the calculation technique is clear—a very wide interval of K-values counted for countries of diverse electromobility incentives history.

Clearly, other indicators such as the average income or average vehicle prices should have been considered in order to find out to which extent selected countries are prepared for the widespread utilization of electric cars, as well. However, here we focus mainly on the diffusion of electric vehicles in reference of the charging infrastructure, what is reflected in the definition of the infrastructural country electromobility coefficient K. We understand its utilization being helpful especially for customers that are considering buying an EV and wonder whether the local environment and country infrastructure is satisfactory for its daily use. It is expected that the distance range, safety, as well as economic, legal and other aspects of purchasing an EV would be considered by customers separately.

On the basis of carried out review, calculations, and objectively critical analysis of research results we have posted our conclusions regarding the research questions. The schema of processes that preceded the writing of this article is recorded at Figure 3 (own structurogram based on the realized processes).

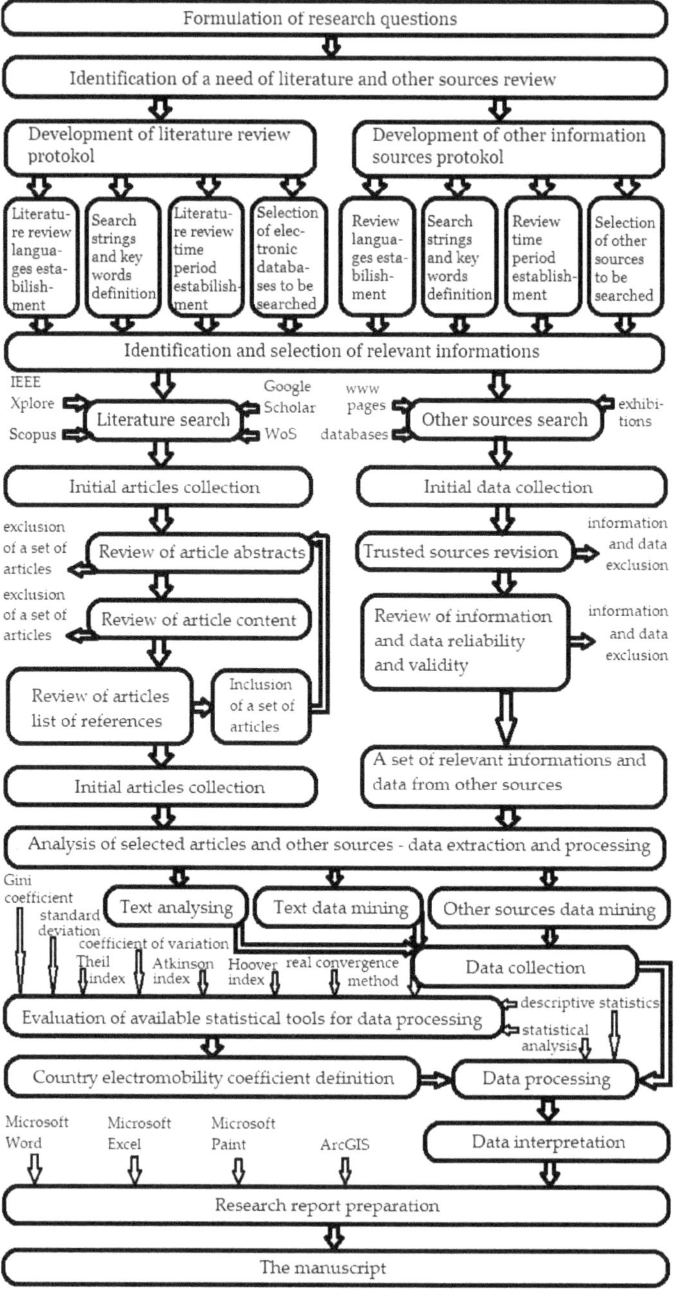

Figure 3. Structurogram of scientific approach and scientific work of the paper.

3.1. The Choice of Reference Countries

In our study we focused on the situation in V4 countries. The countries of Visegrád Group belong to Central European region and closely cooperate in several areas of common interest within the framework of pan-European integration. All V4 countries achieved the goal of being a part of the EU at the same time (1 May 2004), but not only their EU history has had the same features. These countries have always been a part of the same civilization based on the same cultural and intellectual values and common roots of traditions. The list of their mutual cooperation projects is long, including the projects relating to electromobility. In order to have the top and standard reference values for V4 countries comparison considering the electromobility preparation, we decided to analyze the same types of data for V4 countries as for European electromobility leaders and compare them with the EU average. EU is considered as the union of 28 states here, hence, inclusive United Kingdom, as all the data subject to the EU are from the range of years 2014–2019.

According to Electric Vehicle World Sales Database [17] the European leader in the share of electric vehicles within new sold cars referred to 2019 is Norway. It clearly belongs to electromobility leaders of Europe. Although it is not a member of the EU, we decided to add it into our set of selected countries. Among other European electromobility leaders we have included into our countries of selection EU countries Denmark, Germany, and Netherlands.

Norway, Germany, Denmark and Netherlands have high potential within electromobility. All these countries have high level of economic power, good infrastructure and there are a lot of government incentives towards sustainable technology, green energy, and reductions of greenhouse gas emissions in these countries, as well. For example, Norway has agreed upon a ban against emission cars in 2025, while Denmark has a 100% renewable energy target towards 2050 [20]. Netherlands uses renewable energy from many different sources (biofuels, waste, wind, sun, geothermal and hydro sources) and has one of the most ambitious targets for climate-change mitigation—the 49% reduction in carbon emissions by 2030 [51]. Germany has also ambitious plans towards green energy. Five years ago, Germany's share of renewable energy was 28% of its gross electricity consumption, with a goal of achieving a 40% share by the end of 2020 [20]. Germany is also known for its strong history in car manufacturing and for its ambitions being the front-runner country in mobility and electromobility. Therefore, these countries have been chosen into our set.

3.2. Electric Vehicles-Cars

According to the degree that electricity is used in cars as their energy source there are three main types of electric vehicles (EVs): battery electric vehicles (BEVs), plug-in hybrid electric vehicles (PHEVs), and hybrid electric vehicles (HEVs): HEVs are powered by both electricity and gasoline. PHEVs can recharge the battery through either external source of electrical power or regenerative braking. The fully-electric vehicles with rechargeable batteries and no gasoline engine are those of BEV category. Their high-capacity battery packs allow using their electric motor and all onboard electronics without any harmful emissions, what is the point towards global warming mitigation [52]. BEVs have to be charged from an external source (for details on these sources see [53]).

3.3. The Data on Electric Vehicles-Cars

Actually, there are more than one million EVs in Europe. After the industry geared up to meet the 95 g CO_2/km target for 2020/2021, over 30 BEV and PHEV models were introduced or improved in 2019. The EV sales in Europe noticed 44% growth in 2019 and pushed the global BEV and PHEV share to 2.5%. The share leader for 2019 with 56% of new car from the PHEV class was Norway, followed by Iceland (24.5%) and the Netherlands (15%). The largest volume growth contributors were Germany, Netherlands and Norway (see Table 2). Among other markets with over 1 million total sales in 2019, China lead with a plug-in share of 5.2%, followed by United Kingdom (3.2%), Germany (2.9%), France

(2.8%), and Canada (2.7%). By these facts within the year 2019 the Europe's share in global BEV and PHEV sales increased from 20% to 26% [17].

This trend is clearly notable from the time development of the total number of passenger cars, BEVs and PHEVs in the last years. Tables 1 and 2 show these data for selected European countries, EU average and V4 countries for time range 2015–2017 and 2018–2019, respectively. These data were processed on the base of [16,25].

Table 1. The total number of passenger cars, BEVs (battery electric vehicles) and PHEVs (plug-in hybrid electric vehicles) in 2015–2017, (source: elaborated by authors based on [16,25]).

	BEV			PHEV			The Total Number of Passenger Carsat the Turn of the Years		
	2015	2016	2017	2015	2016	2017	2015/2016	2016/2017	2017/2018
EU (incl. UK)	118,044	172,768	247,337	128,458	196,219	272,152	252,075,544	257,540,942	262,947,936
V4	1578	2041	3565	729	1131	2251	31,111,843	32,477,515	33,792,296
Czech Republic	1018	1090	1118	204	340	568	5,158,516	5,368,660	5,592,738
Hungary	204	405	1153	99	210	522	3,192,132	3,308,495	3,467,861
Poland	219	348	896	270	425	816	20,723,423	21,675,388	22,503,579
Slovakia	137	198	398	156	156	345	2,037,772	2,124,972	2,228,118
Denmark	7491	8686	9432	588	770	1223	2,392,175	2,465,934	2,529,960
Germany	28,268	41,857	59,672	17,439	32,049	58,312	45,071,209	45,803,560	46,484,594
Netherlands	9368	13,105	21,115	78,163	98,903	98,217	8,336,414	8,439,318	8,594,600
Norway	61,393	97,615	130,532	10,164	30,828	59,368	2,592,324	2,639,245	2,693,021

Table 2. The total number of passenger cars, BEVs and PHEVs in 2018–2019, (source: elaborated by authors based on [16,25]).

	BEV		PHEV		The Total Number of Passenger Cars at the Turn of the Years
	2018	2019	2018	2019	2018/2019
EU (incl. UK)	389,863	626,264	385,409	517,172	267,834,417
V4	6772	10,391	4339	6655	35,196,697
Czech Republic	2030	2837	859	1326	5,802,520
Hungary	2460	3696	1268	2122	3,638,374
Poland	1487	2902	1592	2381	23,429,016
Slovakia	795	956	620	826	2,326,787
Denmark	10,898	16,331	4526	8412	2,593,568
Germany	101,477	152,886	92,214	136,509	47,095,784
Netherlands	44,984	107,536	97,702	95,885	8,787,283
Norway	162,525	222,796	86,518	105,535	2,720,013

3.4. The Data on Charging Stations

The millionth EV sale reached in 2018 was important milestone on the road to electrification and meeting emission targets, as well as a clear signal of consumer intent. However, crucial for capturing this momentum is the access to low-cost, well-located charging stations (Viktor Irle and Matt Allen for The Guardian [54]).

Tables 3 and 4 show the development of the alternative fuel infrastructure for BEVs and PHEVs in selected European countries, the average of European Union and the average of Visegrád Group alliance (processed based on charts published at [25]). The data reflect remarkable growth of infrastructure between 2014 and 2019.

Table 3. The total number of charging stations at the turn of the years, (source: elaborated by authors based on [25]).

	2014/2015	2015/2016	2016/2017	2017/2018	2018/2019	2019/2020
EU (incl. UK)	26,193	48,375	75,240	106,825	119,543	166,723
V4	452	868	1234	1739	2347	3244
Czech Republic	160	308	408	559	572	1049
Hungary	123	178	197	257	572	692
Poland	119	298	319	507	714	919
Slovakia	50	84	310	416	489	584
Denmark	923	1300	2397	2469	2541	2678
Germany	2864	5328	15,379	24,422	26,196	39,922
Netherlands	11,981	18,044	25,453	33,387	36,789	50,466
Norway	5385	5763	7779	9547	11,041	12,473

Table 4. The yearly growth of infrastructure in % at the turn of the years, (source: elaborated by authors based on [25]).

	2014/2015	2015/2016	2016/2017	2017/2018	2018/2019
EU (incl. UK)	85	56	42	12	40
V4	89	80	40	44	40
Czech Republic	93	32	37	2	83
Hungary	45	11	31	123	21
Poland	150	7	59	41	29
Slovakia	68	269	34	10	27
Denmark	41	84	3	3	6
Germany	86	189	59	7	52
Netherlands	51	41	31	10	37
Norway	7	35	23	16	13

As Table 3 reflects only the time development of the number of charging stations, it does not reflect their utilization. One can expect that charging stations of some country would be used mostly by inhabitants of this country and only rarely by foreigners. Hence, the total number of charging points has to be considered together with the total number of BEVs and PHEVs registered in the country. Therefore, Table 5 shows the rounded values of plug-in electric vehicles per public charging point (processed based on available material at [25]). In this case the small numbers are considered as the good ones. The higher numbers refer to the missing infrastructure.

Table 5. The number of BEVs and PHEVs per public charging point, (source: elaborated by authors based on [25]).

			BEV					PHEV		
	2015	2016	2017	2018	2019	2015	2016	2017	2018	2019
EU (incl. UK)	2	2	2	3	4	3	3	3	4	3
V4	2	2	2	3	3	1	1	1	2	2
Czech Republic	4	3	2	3	2	1	1	1	1	2
Hungary	1	2	4	4	5	1	1	2	2	3
Poland	1	1	2	2	3	1	1	1	2	2
Slovakia	1	1	1	2	2	2	1	1	1	1
Denmark	6	4	4	4	6	0	0	0	2	3
Germany	5	3	3	4	4	4	2	2	3	3
Netherlands	1	1	1	1	2	4	4	3	3	2
Norway	10	12	12	14	16	2	4	6	7	8

Debating the public charging points, the highest importance has those of at least 22 kW, so called fast charging points (see [53]). These are usually situated along highways.

Their amount per 100 km of highway for countries of selection shows Table 6 processed based on [25]. In this case the greater number means the better developed infrastructure along highways, what is the main requested effect.

Table 6. The number of fast (the power is at least 22 kW) public charging points per 100 km of highway at the turn of the years, (source: elaborated by authors based on [25]).

	2015/2016	2016/2017	2017/2018	2018/2019	2019/2020
EU (incl. UK)	5	7	12	15	23
V4	3	5	8	15	28
Czech Republic	3	6	11	22	38
Hungary	1	2	3	4	7
Poland	0	1	4	11	15
Slovakia	8	12	15	24	50
Denmark	21	24	28	30	35
Germany	6	10	19	26	47
Netherlands	12	18	21	28	35
Norway	133	201	381	519	655

3.5. Combined Index Calculation

As some of the countries have higher density of highways or longer highways roads, the data in Table 6 have to be discussed in wider context reflecting the highway net and area of the country. Therefore, it is worth to use the data from Table 7 that was processed based on [25,51,55–58].

Table 7. The data on population development, area and highways length of selected countries and alliances, (source: elaborated by authors based on [25,51,55–58]).

	Area	Highways	Population of the Country at the End of the Years				
	km × km	Length in km	2015/2016	2016/2017	2017/2018	2018/2019	2019/2020
EU (incl. UK)	4,475,757	77,573	511,218,529	512,191,098	513,213,363	513,480,000	514,292,912
V4	533,617	6207	63,803,055	63,771,937	63,747,782	63,718,669	63,675,585
Czech Republic	78,866	1240	10,618,857	10,641,034	10,665,677	10,689,209	10,708,981
Hungary	93,030	1936	9,752,975	9,729,823	9,707,499	9,684,679	9,660,351
Poland	312,685	2549	37,989,220	37,953,180	37,921,592	37,887,768	37,846,611
Slovakia	49,036	482	5,442,003	5,447,900	5,453,014	5,457,013	5,459,642
Denmark	43,094	1308	5,711,349	5,732,274	5,752,126	5,771,876	5,792,202
Germany	357,121	13,009	82,193,768	82,658,409	83,124,418	83,517,045	83,783,942
Netherlands	41,543	3055	16,981,295	17,021,347	17,059,560	17,097,130	17,134,872
Norway	323,802	523	5,250,949	5,296,326	5,337,962	5,378,857	5,421,241

The data presented in Tables 1–7 have been used in order to derive the base parameters that help to describe the countries' development towards electromobility: the share of BEVs and PHEVs in passenger cars for the given country or alliance of countries, yearly or time horizon growth of this share, percentage yearly jump in number of BEVs and PHEVs, average yearly growth of infrastructure ... But these data have been also used in order to count the infrastructural country electromobility coefficient K obtained as a product of E_p, N_k, N_e, and N_p, where

$$E_p = N_{B+P} \times P^{-1} \times 1000, \tag{3}$$

$$N_k = N \times A^{-1}, \tag{4}$$

$$N_e = N \times {N_{B+P}}^{-1}, \tag{5}$$

$$N_p = N \times P^{-1} \times 1000, \tag{6}$$

where N_{B+P} is the total number of battery electric vehicles and plug-in hybrid electric vehicles in the country, P is the total number of inhabitants of the respective country,

N is the total number of public charging points, and A is the total area of the country of selection.

Hence, we have:

$$K = E_p \times N_k \times N_e \times N_p = N^3 \times P^{-2} \times A^{-1} \times 10^6 \quad (7)$$

4. Results and Discussion

The total number of passenger cars, BEVs and PHEVs grew fast between 2015 and 2019. While at the end of 2015 there were 252,075,544 passenger cars on European Union roads out of which 0.047% were BEVs and 0.051% were PHEVs (Table 1), at the beginning of 2019 there were 267,834,417 passenger cars out of which 0.146% were BEVs and 0.144% were PHEVs (Table 2). These average fractions of BEVs and PHEVs are small, as some countries, such as e.g., countries of Visegrád Group alliance, only very recently moved towards electromobility. At the end of 2015 there were 31,111,843 passenger cars in V4 countries out of which 0.005% were BEVs and 0.002% were PHEVs (Table 1), at the beginning of 2019 there were 35,196,697 passenger cars out of which 0.019% were BEVs and 0.012% were PHEVs (Table 2). Among V4 countries the fastest growth considering the BEVs and PHEVs share reached Hungary that between years 2015–2019 increased the BEVs share from 0.006% to 0.068% and PHEVs share from 0.003% to 0.035%, hence, in both cases, more than 11 times. BEVs' shares of other V4 countries grew 1.75–6 times: the growth from 0.001% to 0.006% is associated with Poland, the growth from 0.007% to 0.034% is associated with Slovak Republic and the lowest growth from 0.020% to 0.035% is associated with Czech Republic. PHEVs' shares of other V4 countries grew 3.4–7 times: the growth from 0.001% to 0.007% is associated with Poland, the growth from 0.004% to 0.015% is associated with Czech Republic and the lowest growth from 0.008% to 0.027% is associated with Slovak Republic. At the other hand, Poland steadily has the lowest both BEVs and PHEVs share among V4 countries, but all V4 countries are far below EU average and in a much worse position compared to countries like Germany, Netherlands or Norway. Considering the total numbers of both BEVs and PHEVs, except of the year 2014, when there were 156 registered PHEVs in Slovakia and only 99 in Hungary, Slovakia takes always the last position among V4 countries within the referenced time period. It points out to the fact that incentives towards electromobility in Slovakia came only in the recent years. However, they seem to be effective, as the amount of BEVs in Slovakia jumped by 20.3% (from 795 to 956) and the amount of PHEVs jumped by 33.2% within the last year (Table 2).

Clearly, the total amount of vehicles, and hence, EVs, partially depends on the number of inhabitants of the country. Therefore, is worth to count the total number of BEVs and PHEVs per 1000 inhabitants of the country (E_p) and compare it with the total number of passenger cars per 1000 inhabitants (C_p)—see Table 8. During the reference time period Hungary followed by Slovakia had the least number of passenger cars per 1000 inhabitants. Except of Denmark, there was roughly 1 car per 2 persons, if not less, in other countries. At the other side, Denmark together with Netherlands and Germany belong to those countries in the EU, where rough 1–12 people out of 1000 own an electric car, and in a case of non-European Union country Norway tens of people out of 1000 own it. The discussed ratio is very much smaller in the case of V4 countries, although, during the last 5 years it grew more than 19 times in a case of Hungary, more than 10 times in a case of Poland, more than 6 times in a case of Slovakia, and more than 3 times in a case of Czech Republic. This shows an increasing interest in buying an electric car in V4 countries. The growth in interest in Slovakia is similar to the growth in Germany.

Table 8. The total number of plug-in electromobiles per 1000 inhabitants (E_p) vs. the total number of passenger cars per 1000 inhabitants (C_p) at the turn of the years.

	C_p				E_p				
	2015/2016	2016/2017	2017/2018	2018/2019	2015/2016	2016/2017	2017/2018	2018/2019	2019/2020
EU (incl. UK)	493.087	502.822	512.356	521.606	0.482	0.720	1.012	1.510	2.223
V4	487.623	509.276	530.094	552.377	0.036	0.050	0.091	0.174	0.268
Czech Republic	485.788	504.524	524.368	542.839	0.115	0.134	0.158	0.270	0.389
Hungary	327.298	340.037	357.235	375.683	0.031	0.063	0.173	0.385	0.602
Poland	545.507	571.109	593.424	618.379	0.013	0.020	0.045	0.081	0.140
Slovakia	374.452	390.053	408.603	426.385	0.054	0.065	0.136	0.259	0.326
Denmark	418.846	430.184	439.830	449.346	1.415	1.650	1.852	2.672	4.272
Germany	548.353	554.131	559.217	563.906	0.556	0.894	1.419	2.319	3.454
Netherlands	490.917	495.808	503.800	513.962	5.155	6.580	6.995	8.346	11.872
Norway	493.687	498.316	504.504	505.686	13.627	24.251	35.575	46.300	60.564

For the better visualization Figure 4 (own chart based on obtained data) shows the development of the total number of BEVs and PHEVs per 1000 inhabitants of countries of selection.

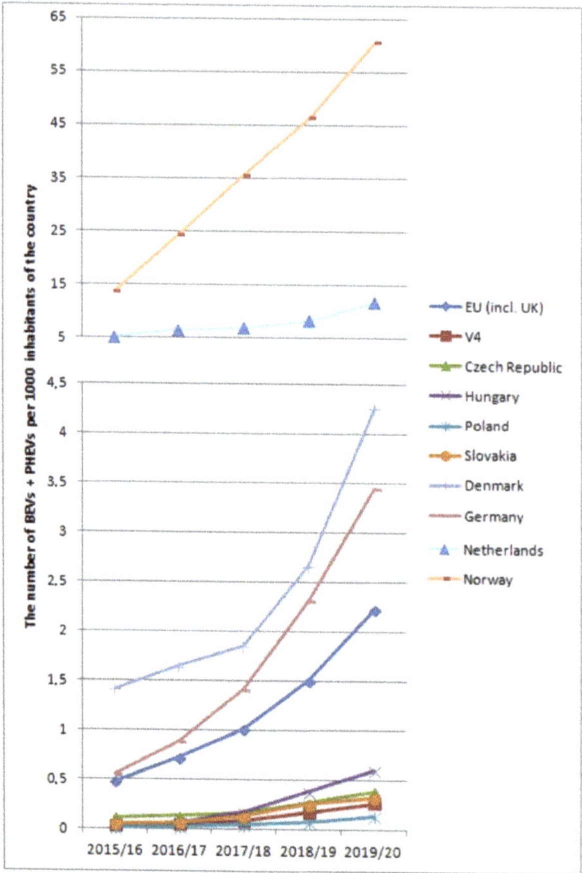

Figure 4. The development of the parameter E_p (the total number of battery electric vehicles plus plug-in hybrid electric vehicles per 1000 inhabitants of the country).

Debating Figure 4, one can see that there is a big difference between Norway and other European countries of selection. The very low share of electric cars in V4 countries reflects in the flat shape of the corresponding lines at Figure 4. The similar is not true debating the parameter C_p–the total number of passenger cars per 1000 inhabitants of the country. In this case V4 countries are not behind the EU countries of selection and even Norway, see Table 8. The total number of charging stations in V4 countries jumped from several hundred in 2014 to several thousands in 2019 (Table 3). Depending on the actual incentives, the yearly growth of infrastructure (Table 4) in the countries of V4 varied from 2% (Czech Republic, 2017–2018) to 269% (Slovakia, 2015–2016). In average, the yearly growth of infrastructure reached 58.6% during the spotted time horizon. The highest was in Slovakia (81.6%), followed by Poland (57.2%), Czech Republic (49.4%) and Hungary (46.2%). Except of Hungary, this growth is higher like EU average (47.0%), what points out to the fast development of these countries regarding the electromobility infrastructure. However, it has to be remarked, that countries like Netherlands, Germany and Norway have nowadays decades of thousands charging stations (Table 3), while the whole V4 alliance only 3244 (2019). Similarly, between 2016–2017 and 2017–2018 the yearly growth of electromobility infrastructure in Denmark reached only 3% (Table 4), but it has to be noticed, that the total number of public charging stations in Denmark only, with the area of 43,094 km^2, has been each year from the range 2014–2018 higher than the total number of public charging stations in Visegrád Group alliance of the area 533,617 km^2. Only in 2019 the V4 countries overtake Denmark considering the cumulative number of public charging points.

The above example of Denmark shows that it is important to consider the alternative fuel charging infrastructure depending on the area of considered country. Table 9 shows the number of public charging stations per km^2 of the country.

Table 9. The number of public charging stations per km^2 of the country (N_k) at the turn of the years.

	2015/2016	2016/2017	N_k 2017/2018	2018/2019	2019/2020
EU (incl. UK)	0.01081	0.01681	0.02387	0.02671	0.03725
V4	0.00212	0.00366	0.00499	0.00641	0.00890
Czech Republic	0.00391	0.00517	0.00709	0.00725	0.01330
Hungary	0.00191	0.00212	0.00276	0.00615	0.00744
Poland	0.00095	0.00102	0.00162	0.00228	0.00294
Slovakia	0.00171	0.00632	0.00848	0.00997	0.01191
Denmark	0.03017	0.05562	0.05729	0.05896	0.06214
Germany	0.01492	0.04306	0.06839	0.07335	0.11179
Netherlands	0.43435	0.61269	0.80367	0.88556	1.21479
Norway	0.01780	0.02402	0.02948	0.03410	0.03852

While there was roughly 1 public charging point per 2 km^2 of Netherlands at the end of 2015, there was only a very small fraction (0.00171) of public charging point per 1 km^2 of Slovakia—see Figure 5 (own chart based on obtained data), what was the second least number among V4 countries. Nowadays, when there is statistically more than 1 public charging point per 1 km^2 of Netherlands, Slovakia reached the value 0.01191. With a similar value as Czech Republic, it is 3 times lesser than in the case of European Union average and this average is approximately the same as the density of public charging stations in Norway. Although Slovakia is far behind the EU's electromobility leaders and even behind the EU average considering this parameter, within V4 countries it is in a good position.

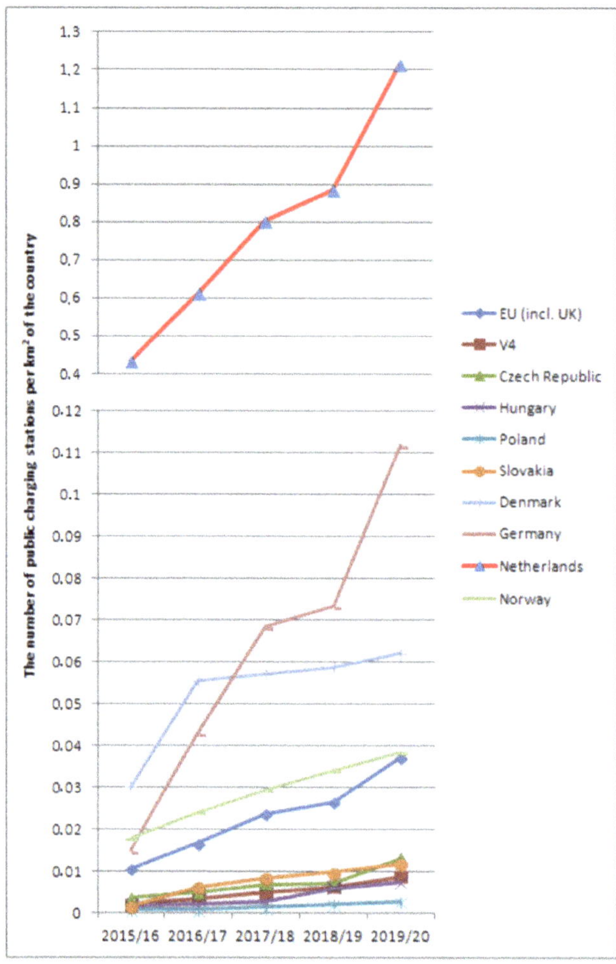

Figure 5. The development of the parameter N_k (the number of public charging stations per km^2 of the country).

The lesser is the number of BEVs and PHEVs per one public charging point, the more satisfactory is the electromobility infrastructure in the country. As all the values in Table 5 are rounded, there are several 0s in the line associated to Denmark. This means that there are lesser than 0.49 PHEVs for one public charging point. Among selected countries, the highest values in Table 5 are associated with Norway. Although there were several thousands of public charging points during 2014–2017, and nowadays, there are more than 12,000 of them (Table 3), due to the large volume of EVs in Norway, it is not satisfactory. According to the suggestions of the European Commission [5] that qualify the sufficiency of the public charging stations, it is recommended that states should be at the ratio not more than 10 electric cars per one public charging point. Hence, it can be said that the percentage yearly growth of infrastructure here (see Table 4) was insufficient, especially considering BEVs.

Thanks to the relatively small amount of EVs in Czech Republic, Hungary, Poland and Slovakia, the number of public charging points was almost sufficient during 2015–2018 in V4 countries and it only slightly moved up in the last two years—see Table 5. However, the values are more favorable than in the EU average. Only Hungary as a part of V4 countries in the recent years has the discussed rate slightly above the EU average in the BEVs case.

From values of Table 5 one can observe, that due to the relatively small number of BEVs and PHEVs on Slovak roads there are steadily 1–2 EVs per 1 public charging point during the recent 5 years' time horizon. As the number of PHEVs and BEVs grow every year, the above fact points out to the accurately fast growth of electromobility infrastructure—see Table 3. The asymmetric percentage yearly growth of Slovak electromobility infrastructure (see Table 4) seems to be appropriate. On the basis of suggestions, the European Commission [5] somebody should argue that 1–2 EVs per 1 public charging point reflect a very favorable position of Slovakia, not only the accurately fast growth of electromobile infrastructure. However, similarly like in the previous paragraph, we should not use too optimistic words here, as in both cases one has to take into account also the total number of EVs and charging points in the respective country. In the case when there are only 2 EVs in some country and singe 1 charging point, the statistics says the same, but, hardly it can be said that this is very good unless the area of the state itself is not larger than area of some city. According to pure statistics, Slovakia might make an impression to be very well prepared for electromobility from the infrastructural point of view. However, the above data do not reflect the spatial coverage of Slovakia by public charging stations.

Some partial information about the spatial coverage of Slovakia by public charging stations can be obtained from Table 6. The existence of fast charging stations along highways is crucial in order to ensure electromobility. The bigger is their density, the better is country prepared for using EVs for large distance travels. The rapid growth of electromobility infrastructure in the last five years had a consequence that nowadays there are 50 charging points per 100 km of Slovak highway, what is 6.25 times more than in 2015. Hence, Slovakia is nowadays in a more than double better position than the average of EU. It was above the EU average during the whole monitored time horizon. In the last two years also Czech Republic among V4 countries overcame the EU average. By the recent values Slovakia and Czech Republic overtake countries like Denmark and Netherlands, and in a case of Slovakia also Germany.

However, one can complain by discussing the amount of highways in each country. Slovakia with 482 km highways on area of 49,036 km^2 is nowadays similarly equipped by highways' fast charging points like Germany with 13,009 km highways on area of 357,121 km^2 (Table 7). Norway with roughly the same amount of highways as Slovakia has 655 fast charging points per 100 km of highway and during the whole considered time horizon the number of fast charging stations along highways was counted in hundreds. Hence, it is worth to discuss the location of Slovak fast charging stations more in detail.

According to statistic referred to middle of 2018 [53] there were 104 fast charging stations and 347 public charging places of lower power in Slovakia. In the same reference point Norway had 2267 fast charging stations and 8617 public charging places of lower power and Netherlands had 811 fast charging stations and 34,021 public charging points of lower power. Hence, there were 91 vehicles for 1 fast charging station and 19 vehicles for one charging place without differentiation of its power in Norway, and 160 vehicles for 1 fast charging station but only three vehicles for one charging place without differentiation of its power in Netherlands. It is known that Norway and Netherlands are the leading European countries considering the EVs penetration—see Figure 4, but, within this comparison, Slovakia with 13 vehicles per one public fast charging place and three vehicles per one charging place without differentiation of its power was in some sense in the better position.

The growth of reliable charging infrastructure in big cities, along highways, and country roads is the key stimulus for sustainable traffic development. The position of fast charging stations (usually more than one at each location) in the middle of 2018 is depicted at Figure 6, which we have created based on the data published in [53].

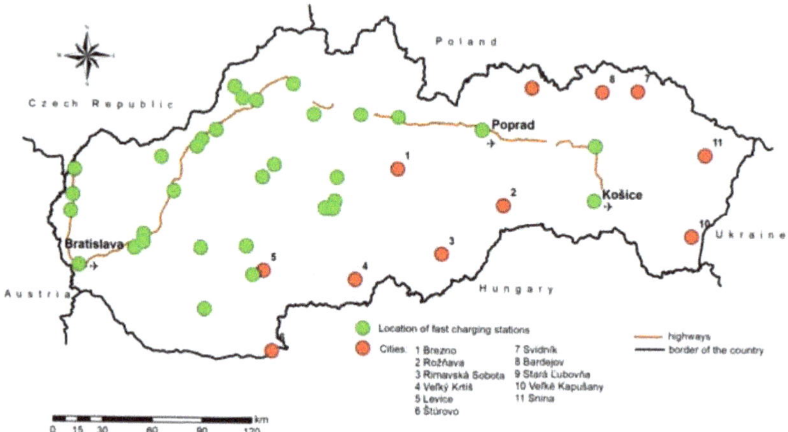

Figure 6. Localization of public fast charging stations in Slovakia in 2018.

One can see that the covering along highways was really good already in 2018. The problem was in the north-east, south-east, and outermost east of the country, what are exactly the areas without highways. It was more than uncomfortable to travel the distance e.g., between Košice and Brezno through Rožňava by affordable BEV of small range that is unable to travel this distance without charging the battery. The even worse situation was in a case of travelling from Košice (the second largest city of Slovakia, city with international airport) to Bratislava (the capital of Slovak Republic, the largest city in Slovakia, city with international airport) by so called southern route via Rožňava, Rimavská Sobota, Veľký Krtíš and Levice. Similar situation was between, let's say, Štúrovo and Košice, Svidník and Poprad (city with international airport) trough Bardejov and Stará Ľubovňa, or from Veľké Kapušany (through Snina) to Svidník—see Figure 6. In this case our results correspond to the findings of Daňo and Rehák [38] that in 2018 point out to the lack of charging stations at some places of the Slovak road network.

Actual position of the fast charging stations (usually more than one at each place) is depicted at Figure 7. We have drawn the situation based on the union of data obtained from [59–61] in March 2020.

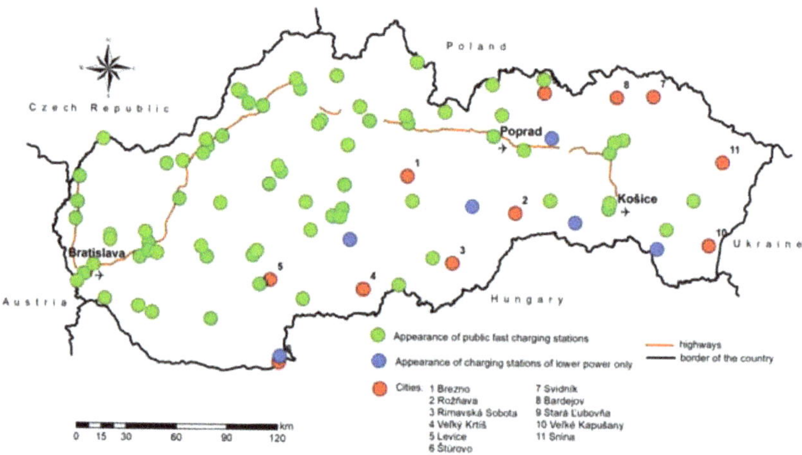

Figure 7. Localization of public fast charging stations in Slovakia in 2020.

As it is clear from Figure 7 there is better electromobile infrastructure along Slovak highways in 2020 as it was two years before. Some of the fast charging stations appeared also on those, so called, problematic places. Moreover, for example, in a case of travelling from Košice to Bratislava by southern route, in necessity, one can also use public charging stations of lower power—their position is depicted by blue circles at Figure 7. That means that Slovakia is fast moving forward country in building electromobile infrastructure. However, the bad situation considering the stage of preparation for electromobility is still in the north-east and outermost east of the country—see the missing green or blue circles nearby e.g., Bardejov, Svidník, Snina or Veľké Kapušany.

At the other side, it has to be noted, that owning public charging stations is a good commercial move for store chains, hotels and restaurants [62]. Therefore, a lot of them offer their clients the possibility to charge their EVs during resting, shopping, eating or stay. Not all of these charging stations are depicted at Figure 6 or Figure 7, as, officially, charging of EV at some of these charging points might be conditioned by utilization of another provider's service, hence, in this case we are not speaking about public charging stations in the true sense of the word.

Table 10, which complements Table 5, shows the rates of public charging stations per BEVs and PHEVs (cumulative) registered in countries and alliances of selection. These data supported by the chart at Figure 8 (own chart based on obtained data) show that demands on infrastructure grew by number of EVs registered in countries. Therefore, EU is now in the worse situation like V4 countries considering this comparative criterion.

Table 10. Number of public charging stations per plug-in electromobile (N_e) at the turn of the years.

	2015/2016	2016/2017	N_e 2017/2018	2018/2019	2019/2020
EU (incl. UK)	0.19625	0.20391	0.20563	0.15419	0.14581
V4	0.43390	0.47351	0.33526	0.23223	0.21815
Czech Republic	0.25205	0.28531	0.33155	0.19799	0.25198
Hungary	0.58746	0.32032	0.15343	0.15343	0.11894
Poland	0.60941	0.41268	0.29614	0.23189	0.17395
Slovakia	0.28669	0.87571	0.55989	0.34558	0.32772
Denmark	0.16091	0.25349	0.23172	0.16474	0.10823
Germany	0.11657	0.20809	0.20699	0.13525	0.13795
Netherlands	0.20614	0.22724	0.27978	0.25783	0.24809
Norway	0.08054	0.06056	0.05027	0.04433	0.03799

While between 2015 and 2018 there were roughly five vehicles registered in the EU per one public charging point, the share of charging point for one EV is 0.14581 now. At the other side, in average, lesser than five vehicles share one charging point in V4 countries also nowadays and this ratio was more favorable here several years ago.

The values of the infrastructural country electromobility coefficient K can be found in Table 11. The wide range of K's values reflects the diverse gauge of infrastructural electromobility preparation in the countries of selection. Clearly, countries of Visegrád Group alliance are far beyond the European electromobility leaders such as Netherlands, Denmark or Norway. However, example of Germany, which five years ago had K's value roughly double as big as Hungary has today, gives a hope, that with appropriate positive stimuli from the government it is possible to magnify the infrastructural country electromobility coefficient even several hundred times. Germany by multiplying its infrastructural country electromobility coefficient from the turn of the years 2015/2016 by almost 405 times at the turn of the years 2019/2020 overcame the value of infrastructural country electromobility coefficient of Denmark. Among V4 countries the similar trend follows Slovakia which multiplied its infrastructural country electromobility coefficient from the turn of the years 2015/2016 by almost 334 times at the turn of the years 2019/2020. The multiplying constant of other selected countries is in at most decades.

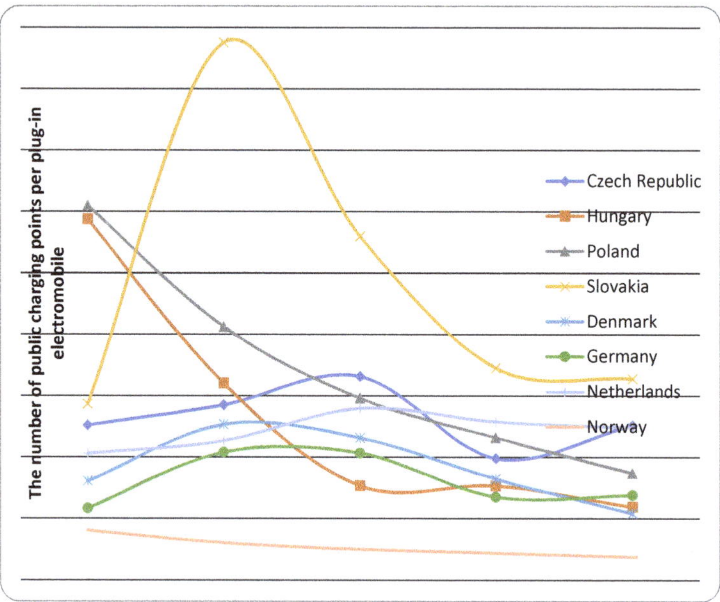

Figure 8. The development of the parameter N_e (the number of public charging points per plug-in electromobile).

Table 11. Values of the infrastructural country electromobility coefficient (K) for selected European countries at the turn of the years.

	2015/2016	2016/2017	K 2017/2018	2018/2019	2019/2020
Czech Republic	3.286×10^{-6}	7.605×10^{-6}	19.470×10^{-6}	20.769×10^{-6}	127.626×10^{-6}
Hungary	0.637×10^{-6}	0.868×10^{-6}	1.936×10^{-6}	21.448×10^{-6}	38.169×10^{-6}
Poland	0.059×10^{-6}	0.072×10^{-6}	0.290×10^{-6}	0.811×10^{-6}	1.733×10^{-6}
Slovakia	0.408×10^{-6}	20.470×10^{-6}	49.373×10^{-6}	80.076×10^{-6}	136.268×10^{-6}
Denmark	1562.917×10^{-6}	9725.981×10^{-6}	$10{,}555.764 \times 10^{-6}$	$11{,}427.815 \times 10^{-6}$	$13{,}283.960 \times 10^{-6}$
Germany	62.690×10^{-6}	1490.716×10^{-6}	5902.977×10^{-6}	7216.711×10^{-6}	$25{,}380.484 \times 10^{-6}$
Netherlands	$490{,}409.626 \times 10^{-6}$	$1{,}370{,}033.624 \times 10^{-6}$	$3{,}078{,}213.117 \times 10^{-6}$	$4{,}100{,}243.658 \times 10^{-6}$	$10{,}537{,}487.330 \times 10^{-6}$
Norway	$21{,}438.319 \times 10^{-6}$	$51{,}825.365 \times 10^{-6}$	$94{,}312.723 \times 10^{-6}$	$143{,}669.872 \times 10^{-6}$	$203{,}908.726 \times 10^{-6}$

Extreme accession of Slovakia towards electromobility among V4 countries exposes Figure 9. It is clear from the chart that Slovakia steadily holds the rapid growth of its infrastructural country electromobility coefficient which is far above the V4 average. Only in the last year the infrastructural country electromobility coefficient of Czech Republic jumped up to a similar value as the Slovak one. It can be expected that governments' benefits towards electromobility development of both countries will secure this trend in the future, as well.

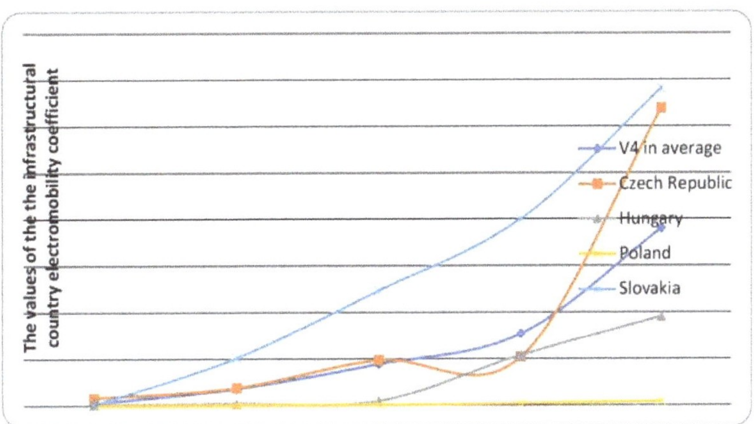

Figure 9. The progress of the infrastructural country electromobility coefficient (K) of V4 countries.

In order to better cross-country comparison of the infrastructural country electromobility coefficient values, the K's values can be normalized. For example, it can be done in such a way that the values associated to Netherlands—the country that reach the highest K's values within the whole reference time period, will be changed to 100 and the values associated to other countries will be proportionally recalculated so that all the rescaled values of the parameter reduce to fit between 0 and 100. These way normalized values of the infrastructural country electromobility coefficient can be found in Table 12. Although again we have a wide range of parameter's values, due to the normalization, the obtained K_n's values might be possibly easier evaluated.

Table 12. Normalized values of the infrastructural country electromobility coefficient (K_n) for selected European countries at the turn of the years.

	K_n				
	2015/2016	**2016/2017**	**2017/2018**	**2018/2019**	**2019/2020**
Czech Republic	0.00067	0.00056	0.00063	0.00051	0.00121
Hungary	0.00013	0.00006	0.00006	0.00052	0.00036
Poland	0.00001	0.00001	0.00001	0.00002	0.00002
Slovakia	0.00008	0.00149	0.00160	0.00195	0.00129
Denmark	0.31870	0.70991	0.34292	0.27871	0.12606
Germany	0.01278	0.10881	0.19177	0.17601	0.24086
Netherlands	100.00000	100.00000	100.00000	100.00000	100.00000
Norway	4.37151	3.78278	3.06388	3.50393	1.93508

Except of Netherlands, only Norway achieved within the whole reference time period the K_n's values greater than 1. However, the descending sequence of K_n's values points out to the fact that due to the increased interest of purchasing an electromobile in Norway the growth of electromobile infrastructure was not as satisfactory as in the case of the trendsetter Netherlands. Similar can be observed in the case of Denmark. The contrary to that fact is the ascending tendency of the sequence of normalized K's values in the case of Germany.

Again, we can observe that considering K_n the V4 countries are far beyond the European electromobility leaders, especially Netherlands. The best position considering the normalized values of the infrastructural country electromobility coefficient within V4 countries belongs to Slovakia, while Poland is in the worst position considering the parameter's values. However, in the case of both countries the positive is the non-descending tendency in the K_n's values between the years 2015–2019 (2015–2020 in the case of Poland, respectively). The lesser value of the normalized K's value associated to Slovakia at the turn of the years 2018/2019 compared to the turn of the years 2019/2020 can be explained by the government's incentives towards the increase of the EV's share on Slovak roads and proportionally not the same fast speed of infrastructural growth compared to the trendsetter.

Debating Slovakia, at the beginning of 2020, Slovak government in the first round approved support for the construction of charging stations for 72 towns and municipalities in the Slovak Republic, one third of which is situated in the least developed regions considering electromobile infrastructure. Another 60 charging stations, among which more than one half should be the fast charging stations of 100 kW and remarkable amount of them the ultra-fast charging stations of 350 kW, should be placed in Slovakia by one European charging stations provider until the end of 2020 with the aim of supporting the electromobile infrastructure in least developed regions considering the electromobility [58]. New charging stations could be expected from the side of other European providers, as well. Government's steps such as endowments for EVs purchasing will surely lead to the greater amount of EVs in Slovakia. At the beginning of 2020 in the first round Slovak government approved 110 grant applications for purchasing electric vehicles, in the second round another 115 grant applications [63]. As a result of these and similar government measures the significant growth of the infrastructural country electromobility coefficient of Slovakia can be expected in the forthcoming years.

On the base of our results and the values of the infrastructural country electromobility coefficient, we disagree with Knapčíková, according to who Slovak Republic does not develop sufficiently in electromobility as it lags behind the surrounding countries [64]. However, we agree with her that owners of electric vehicles have no extra benefits, what is together with missing infrastructure contra-productive considering the motivation towards electromobility. We also hope that the stimuli of the Ministry of Economy of the Slovak Republic in order to support electromobility narrowing to the promotion of the sale of vehicles, favor parking, or forgiveness of road tax and toll roads would lead to the development of electromobility in Slovakia, the lower greenhouse gas emissions, and the reduction of local pollution equal to sample of efficient and environmentally friendly smart cities [65].

5. Conclusions

In this paper we aimed to find out in which position Visegrád Group countries are compared to EU average and the electromobility leaders of Europe. In order to fulfill our goal, we asked and answered four questions of research. Some of the contributions of the paper can be found in data collection, their processing, analyzing and visualization. As among V4 countries the position of Slovakia has been investigated more in detail, the drawings of the maps showing the development of charging stations in Slovakia help to understand the local trends of recent years, what is added value of the paper. The introduction and computation of the infrastructural country electromobility coefficient and its components help to decide whether the respective country itself is infrastructural prepared for (car) electromobility, what should be helpful, as well. The advantage of parameter's utilization is in quick cross-country comparisons of its values. The greater values reflect the higher stage of electromobility preparedness from the infrastructural point of view. As the parameter itself is for the given set of countries within a wide interval of values, its normalized values rescaled on 0–100 interval might make the cross-country comparison easier. This is done in the paper, as well.

Regarding the questions *Q1* and *Q2* one can sum up, that all the countries of Visegrád Group alliance only very recently made their first steps and legislative arrangements in order to support electromobility. Therefore, in a lot of discussed indicators they are far away from the European electromobility leaders and even far away from EU average. However, important is that one can see the progress in the matter, what is clearly to be seen from the chart at Figure 9 depicting the yearly growth of the infrastructural country electromobility coefficient of V4 countries. According to our opinion this parameter is worth for describing the extent to which is particular country prepared for car electromobility from the infrastructural point of view. Assuming the future, one can expect that thanks to the government intentions, the established trend will be preserved also in the forthcoming years. Restrictions connected with the 2020 coronavirus pandemic and the economic crisis after its beat might temporarily affect it. However, as the whole world is changing towards sustainability, lower emissions, waste reduction, renewable sources of energies, and environmentally friendly materials (see e.g., [66]) one can await that the growth towards electromobility will be kept, as well.

Regarding the questions *Q3* and *Q4*, based on the data obtained, we can sum up that although Slovakia within the referenced time period 2014–2019 had the least total number of registered BEVs + PHEVs and second least number of BEVs + PHEVs per 1000 inhabitants among V4 countries and all the countries of selection, its stage of preparation for electromobility is rather good. Very rapid yearly growth of infrastructure, especially along highways, pushed it with 50 fast charging stations per 100 km of highway to the double as good position as is European Union average. Moreover, by recent values it overtook countries like Denmark, Netherlands and Germany. Debating Slovakia, the problem of missing charging stations is in the north-east and outermost east, and partially in the south-east of the country, although, in average there are only 1–2 EVs per 1 public charging point here. As the number of PHEVs and BEVs grows every year, without reflecting the location, one can say that Slovakia has accurately rapid growth of electromobility infrastructure reverberative its needs. The rate of its moving forward in building electromobile infrastructure and overall preparation for electromobility can be discussed via its infrastructural country electromobility coefficient. Slovakia by multiplying its infrastructural country electromobility coefficient by 334 in the last five years is the fastest forward moving country towards electromobility from the V4 countries and has high potential of reduction of environmental impacts from the transport in the forthcoming years. Similar steep growth of K values is associated with Germany that is nowadays one of the EU electromobility leaders.

Clearly, the utilization of BEVs and PHEVs is possible with certain limits e.g., range, long-distance facilities, sufficient initial capital and similar. Other problems are associated with the production of energy itself. A large part of it is produced by so called "dirty process", what is not sustainable. In order to support the sale of BEVs and PHEVs, future research should focus on the use of renewable energy sources and "clean energy" production.

Author Contributions: Conceptualization: E.F.S. and E.G.; methodology: E.F.S. and M.I.; software: E.F.S. and M.I.; validation: A.R., M.S. and V.F.; formal analysis: E.F.S., M.I. and V.F.; investigation and resources: E.F.S., M.I., A.R., E.G., M.S. and V.F.; data curation: E.F.S., M.I. and E.G.; writing—original draft preparation: E.F.S.; writing—review and editing: M.I., A.R., E.G., M.S. and V.F.; visualization: E.F.S. and M.I.; supervision: A.R., E.G., M.S. and V.F.; project administration and funding acquisition: E.F.S., A.R. and M.S. All authors have read and agreed to the published version of the manuscript.

Funding: This work is supported by the Scientific Grant Agency of the Ministry of Education, Science, Research, and Sport of the Slovak Republic and the Slovak Academy Sciences as part of the research project VEGA 1/0797/20: "Quantification of Environmental Burden Impacts of the Slovak Regions on Health, Social and Economic System of the Slovak Republic", project VEGA 1/0264/21: "Application of modern methods in the analysis and modeling of technological and other processes used in the acquisition and processing of earth resources in order to optimize them", project VEGA 1/0588/21: "The research and development of new methods based on the principles of modeling, logistics and simulation in managing the interaction of mining and backfilling processes with regard to economic efficiency and the safety of raw materials extraction", and is supported by the project of the Cultural and Educational Grant Agency of the Ministry of Education, Science, Research and Sport of the Slovak Republic and the Slovak Academy of Sciences project No. KEGA 006TUKE-4/2019. This contribution is the result of the implementation of the project Historical Mining-tracing and learning from ancient materials and mining technologies, no. 18111, EIT/RAW MATERIALS/SGA2019/1 supported by EIT-the European Institute of Innovation and Technology, a body of the European Union, under the Horizon 2020, the EU Framework Programme for Research and Innovation.

Institutional Review Board Statement: Not applicable.

Informed Consent Statement: Not applicable.

Data Availability Statement: The data presented in this study are available on request from the corresponding author.

Acknowledgments: The authors would like to thank anonymous referees for their valuable comments that improved the quality of the manuscript.

Conflicts of Interest: The authors declare no conflict of interest.

References

1. Coppola, P.; Arsenio, E. Driving societal changes towards an electromobility future. *Eur. Transp. Res. Rev.* **2015**, *7*, 37. [CrossRef]
2. European Environment Agency. *Focusing on Environmental Pressures from Long-Distance Transport*; Report 7; EEA: Luxembourg, 2014.
3. Donada, C.; Perez, Y. Electromobility at the crossroads. *Int. J. Automot. Technol. Manag.* **2016**, *16*, 1–15.
4. IEA. Tracking Transport 2019. Available online: https://www.iea.org/reports/tracking-transport-2019/electric-vehicles (accessed on 23 February 2020).
5. European Commission. *Towards an Integrated Strategic Energy Technology (SET) Plan: Accelerating the European Energy System Transformation*; Communication from the European Commission, COM Report 6317; European Commission: Brussels, Belgium, 2015.
6. Straka, M.; Chovan, T.; Bindzár, P.; Žatkovič, E.; Hricová, R. Possibilities and limitations of electromobiles utilization. *Appl. Mech. Mater.* **2015**, *708*, 159–164. [CrossRef]
7. Shafiei, E.; Davidsdottir, B.; Stefansson, H.; Asgeirsson, E.I.; Fazeli, R.; Gestsson, M.H.; Leaver, J. Simulation-based appraisal of tax-induced electro-mobility promotion in Iceland and prospects for energy-economic development. *Energy Policy* **2019**, *133*, 110894. [CrossRef]
8. Lin, X.; Sovacool, B.K. Inter-niche competition on ice? Socio-technical drivers, benefits and barriers of the electric vehicle transition in Iceland. *Environ. Innov. Soc. Trans.* **2020**, *35*, 1–20. [CrossRef]
9. Ryghaug, M.; Skjølsvold, T.M. Nurturing a regime shift toward electro-mobility in Norway. In *The Governance of Smart Transportation Systems*; Springer: Cham, Switzerland, 2019; pp. 147–165.
10. Deuten, S.; Vilchez, J.J.G.; Thiel, C. Analysis and testing of electric car incentive scenarios in the Netherlands and Norway. *Technol. Forecast. Soc. Change* **2020**, *151*, 119847. [CrossRef]
11. Hovi, I.B.; Pinchasik, D.R.; Figenbaum, E.; Thorne, R.J. Experiences from battery-electric truck users in Norway. *World Electr. Veh. J.* **2020**, *11*, 5. [CrossRef]
12. Figenbaum, E.; Assum, T.; Kolbenstvedt, M. Electromobility in Norway: Experiences and opportunities. *Res. Transp. Econ.* **2015**, *50*, 29–38. [CrossRef]
13. Figenbaum, E.; Kolbenstvedt, M. *Pathways to Electromobility: Perspectives Based on Norwegian Experiences*; Transportøkonomisk institutt (Institute of Transport Economics), Norwegian Centre for Transport Research: Oslo, Norway, 2015; pp. 1–65.
14. Norsk Elbilforening. Available online: https://elbil.no/english/norwegian-ev-policy/ (accessed on 11 January 2020).
15. Slovak Investment and Trade Development Agency. Available online: https://www.sario.sk/sites/default/files/files/Ak%C4%8Dn%C3%BD%20pl%C3%A1n%20rozvoja%20elektormobility%20v%20SR.pdf (accessed on 11 January 2020).
16. European Automobile Manufacturers' Association. Available online: https://www.acea.be/uploads/press_releases_files/20180201_AFV_Q4_2017_FINAL.PDF (accessed on 11 January 2020).
17. Electric Vehicle World Sales Database. Available online: http://www.ev-volumes.com/country/total-world-plug-in-vehicle-volumes/ (accessed on 23 February 2020).
18. Cansino, J.M.; Sánchez-Braza, A.; Sanz-Díaz, T. Policy instruments to promote electromobility in the EU28: A comprehensive review. *Sustainability* **2018**, *10*, 2507. [CrossRef]
19. Rehák, R. Electromobility in European Union. *Ekonomika Cestovného Ruchu a Podnikanie* **2018**, *10*, 53–63.
20. Johansen, S.K. E-Mobility Maturity Model: Measuring E-Mobility Readiness of Countries. Master's Thesis, Mannheim University, Department of Information Systems II, Mannheim, Germany, 16 January 2018.
21. Schickram, S.; Gleyzes, D.; Lienkamp, M. Evaluation of the electromobility potential index and results for 46 major cities. In Proceedings of the EVS27 International Battery, Hybrid and Fuel Cell Electric Vehicle Symposium, Barcelona, Spain, 17–20 November 2013.
22. Schickram, S.; Weber, M.; Lienkamp, M. Electromobility potential index. Evaluating the potential for sustainable success of electric vehicles in megacities. In Proceedings of the 2013 Eight International Conference and Exhibition on Ecological Vehicles and renewable Energies (EVER), Monte Carlo, Monaco, 27–30 March 2013.
23. Higueras-Castillo, E.; Liébana-Cabanillas, F.J.; Muñoz-Leiva, F.; García-Maroto, I. Evaluating consumer attitudes toward electromobility and the moderating effect of perceived consumer effectiveness. *J. Retail. Consum. Serv.* **2019**, *51*, 387–398. [CrossRef]
24. Nanaki, E.A.; Kiartzis, S.; Xydis, G.A. Are only demand-based policy incentives enough to deploy electromobility? *Policy Stud.* **2020**, 1–17. Available online: https://rsa.tandfonline.com/doi/abs/10.1080/01442872.2020.1718072#.YA2bFBY1WUl (accessed on 23 February 2020).

25. European Alternative Fuels Observatory. Available online: https://www.eafo.eu/countries/eu-uk%20efta%20turkey/23682/summary (accessed on 8 January 2020).
26. Bühne, J.; Gruschwitz, D.; Hölscher, J.; Klötzke, M.; Kugler, U.; Schimeczek, C. How to promote electromobility for European car drivers? Obstacles to overcome for a broad market penetration. *Eur. Transp. Res. Rev.* **2015**, *7*, 1–9. [CrossRef]
27. Tucki, K.; Orynycz, O.; Świć, A.; Mitoraj-Wojtanek, M. The development of electromobility in Poland and EU states as a tool for management of CO_2 emissions. *Energies* **2019**, *12*, 2942. [CrossRef]
28. Drożdż, W.; Starzyński, P. Economic conditions of the development of electromobility in Poland at the background of selected countries. *Eur. J. Serv. Manag.* **2018**, *28*, 133–140. [CrossRef]
29. Witaszek, M.; Witaszek, K. Emission of selected toxic exhaust fume components from various vehicles. *Polish Zeszyty Naukowe Politechniki Slaskiej Transport* **2015**, *29*, 105–112.
30. Sendek-Matysiak, E. Analysis of the electromobility performance in Poland and proposed incentives for its development. In *Automotive Safety, Proceedings of the 2018 XI International Science-Technical Conference, Častá–Papiernička, Slovakia, 18–20 April 2018*; IEEE: New York, NY, USA, 2018; pp. 1–7.
31. Kupczyk, A.; Mączyńska, J.; Redlarski, G.; Tucki, K.; Bączyk, A.; Rutkowski, D. Selected aspects of biofuels market and the electromobility development in Poland: Current trends and forecasting changes. *Appl. Sci.* **2019**, *9*, 254. [CrossRef]
32. Kłos, M.; Marchel, P.; Paska, J.; Bielas, R.; Błędzińska, M.; Michalski, Ł.; Wróblewski, K.; Zagrajek, K. Forecast and impact of electromobility development on the Polish electric power system. In *E3S Web of Conferences*; EDP Sciences: Les Ulis, France, 2019; p. 01005.
33. Drożdż, W. The development of electromobility in Poland. *Virtual Econ.* **2019**, *2*, 61–69. [CrossRef]
34. Skrúcaný, T.; Kendra, M.; Stopka, O.; Milojević, S.; Figlus, T.; Csiszár, C. Impact of the electric mobility implementation on the greenhouse gases production in central European countries. *Sustainability* **2019**, *11*, 4948. [CrossRef]
35. Jóźwiak, V.; Wąsiński, M. Hungary's climate policy. *PISM Bull.* **2019**, *160*, 1–3.
36. Jursová, S.; Burchart-Korol, D.; Pustějovská, P. Low-carbon transport in Czech-Polish cross border area: Attitude of Czech and Polish public to electromobility. *New Trends Product. Eng.* **2019**, *2*, 79–88. [CrossRef]
37. Igliński, H. Comparative analysis of electromobility development level in Central and Eastern Europe countries. *Problemy Transportu Logistyki* **2018**, *44*, 15–23. [CrossRef]
38. Daňo, F.; Rehák, R. Electromobility in the European Union and in the Slovakia and its development opportunities. *Int. J. Multidiscipl. Bus. Sci.* **2018**, *4*, 74–83.
39. Rehák, R. Electromobility in the European Union and in the Slovakia and its development opportunities. In *Central and Eastern Europe in the Changing Business Environment, Proceedings of 16th International Joint Conference, Prague, Czech Republic and Bratislava, Slovakia, 27 May 2016*; Oeconomica Publishing House, University of Economics: Prague, Czech Republic, 2016; pp. 316–326.
40. Hrudkay, K.; Jaroš, J. Conceptual development of electromobility in conditions of Slovak municipalities. *Acta Logist.* **2019**, *6*, 147–154. [CrossRef]
41. Černohorský, J.; Jandura, P.; Kuprová, K. Sustainable electromobility in the Liberec Region and in the middle Europe in general. In Proceedings of the 2019 International Conference on Electrical Drives & Power Electronics (EDPE), The High Tatras, Slovakia, 24–29 September 2019; IEEE: New York, NY, USA, 2019; pp. 194–200.
42. Svetlík, J. Elektro drive to the implementation of the bike. *Transfer Inovácií* **2015**, *31*, 135–138.
43. Okorie, O.; Salonitis, K.; Charnley, F.; Moreno, M.; Turner, C.; Tiwari, A. Digitisation and the circular economy: A review of current research and future trends. *Energies* **2018**, *11*, 3009. [CrossRef]
44. Mohamad, F.; Teh, J. Impacts of energy storage system on power system reliability: A systematic review. *Energies* **2018**, *11*, 1749. [CrossRef]
45. Denyer, D.; Tranfield, D. Producing a systematic review. In *The Sage Handbook of Organizational Research Methods*; Buchanan, D.A., Bryman, A., Eds.; Sage Publications Ltd., MPG Books Group: Bodmin, UK, 2009; pp. 671–689.
46. Moher, D.; Shamseer, L.; Clarke, M.; Ghersi, D.; Liberati, A.; Petticrew, M.; Shekelle, P.; Stewart, L.A.; PRISMA-P Group. Preferred reporting items for systematic review and meta-analysis protocols (PRISMA-P) 2015 statement. *Syst. Rev.* **2015**, *4*, 1. [CrossRef]
47. Bai, X.; Liu, D.; Tan, J.; Yang, H.; Zheng, H. Dynamic identification of critical nodes and regions in power grid based on spatio-temporal attribute fusion of voltage trajectory. *Energies* **2019**, *12*, 780. [CrossRef]
48. Klamár, R.; Madziková, A.; Rosič, M. Theoretical concept of regional inequalities in the context of creating university textbook–Regional development–factors, inequalities and cross-border cooperation. In *Useful Geography: Transfer from Research to Practice, Proceedings of 25th Central European Conference, Brno, Czech Republic, 12–13 October 2017*; Svobodová, H., Ed.; Masaryk University: Brno, Czech Republic, 2018; pp. 1–16.
49. Matlovič, R.; Klamár, R.; Kozoň, J.; Ivanová, M.; Michalko, M. Spatial polarity and spatial polarization in the context of supranational and national scales: Regions of Visegrad countries after their accession to the EU. *Bull. Geogr. Soc. Econ. Series* **2018**, *41*, 59–78. [CrossRef]
50. Fecková Škrabuľáková, E.; Grešová, E.; Khouri, S. Similarity assessment of economic indicators of selected countries by graph theory means. *Transform. Bus. Econ.* **2019**, *18*, 333–347.
51. Eurostat. Available online: https://ec.europa.eu/eurostat (accessed on 1 March 2020).
52. EVgo Services LLC. Available online: https://www.evgo.com/why-evs/types-of-electric-vehicles/ (accessed on 2 February 2020).
53. E-Car. Available online: https://www.e-car.sk/flipbook/elektromobilita/mobile/index.html (accessed on 23 February 2020).

54. The Guardian. Available online: https://www.theguardian.com/environment/2018/aug/26/electric-cars-exceed-1m-in-europe-as-sales-soar-by-more-than-40-per-cent (accessed on 25 February 2020).
55. World Bank Group. Available online: https://data.worldbank.org/indicator/SP.POP.TOTL?end=2018&locations=EU&start=2014 (accessed on 1 March 2020).
56. Vienna Institute of Demography (VID); International Institute for Applied Systems Analysis (IIASA). European Demographic Datasheet 2018. 2018. Available online: http://www.populationeurope.org/ (accessed on 1 March 2020).
57. Statista, Inc. Available online: https://www.statista.com/statistics/253372/total-population-of-the-european-union-eu/ (accessed on 1 March 2020).
58. Worldometer. Available online: https://www.worldometers.info/population/europe/ (accessed on 1 March 2020).
59. Chargemap. Available online: https://chargemap.com/map (accessed on 13 March 2020).
60. Evmap. Available online: https://www.evmapa.cz/ (accessed on 13 March 2020).
61. GreenWay Infrastructure. Available online: https://greenway.sk/nase-stanice/ (accessed on 13 March 2020).
62. Slovak Electric Car Salon 2020, 24–26 January 2020, Bratislava, Slovakia. Available online: https://autozurnal.com/salon-elektromobilov-2020-bratislava/ (accessed on 26 January 2020).
63. MôjElektromobil (My Electric Car). Available online: https://www.mojelektromobil.sk/ (accessed on 20 March 2020).
64. Knapčíková, L. Electromobility in the Slovak Republic: A green approach. *Acta Logist.* **2019**, *6*, 29–33. [CrossRef]
65. Csiszár, C.; Földes, D. Analysis and modelling methods of urban integrated information system of transportation. In Proceedings of the 2015 Smart Cities Symposium Prague (SCSP), Prague, Czech Republic, 24–25 June 2015.
66. Kostúr, K.; Laciak, M.; Durdán, M. Some influences of underground coal gasification on the environment. *Sustainability* **2018**, *10*, 1512. [CrossRef]

Article

CFD Performance Analysis of a Dish-Stirling System for Microgeneration

Davide Papurello [1,2,*], Davide Bertino [3] and Massimo Santarelli [1,2]

1 Department of Energy (DENERG), Politecnico di Torino, Corso Duca degli Abruzzi, 24, 10129 Turin, Italy; Massimo.santarelli@polito.it
2 Energy Center, Politecnico di Torino, Via Paolo Borsellino, 38/16, 10138 Turin, Italy
3 Department of Applied Science and Technology (DISAT), Politecnico di Torino, Corso Duca degli Abruzzi, 24, 10129 Turin, Italy; bertino.davide1@gmail.com
* Correspondence: davide.papurello@polito.it; Tel.: +39-34-0235-1692

Abstract: The sustainable transition towards renewable energy sources has become increasingly important nowadays. In this work, a microgeneration energy system was investigated. The system is composed of a solar concentrator system coupled with an alpha-type Stirling engine. The aim was to maximize the production of electrical energy. By imposing a mean value of the direct irradiance on the system, the model developed can obtain the temperature of the fluid contained inside the Stirling engine. The heat exchanger of the microgenerator system was analyzed, focusing on the solar coupling with the engine, with a multiphysical approach (COMSOL v5.3). A real Stirling cycle was implemented using two methods for comparison: the first-order empirical Beale equation and the Schmidt isothermal method. Results demonstrated that a concentrator of 2.4 m in diameter can generate, starting from 800 W/m^2 of mean irradiance, a value of electrical energy equal to 0.99 kWe.

Keywords: CFD; COMSOL multiphysics; solid works; dish-stirling; renewable energy; stirling engine; CSP

Citation: Papurello, D.; Bertino, D.; Santarelli, M. CFD Performance Analysis of a Dish-Stirling System for Microgeneration. *Processes* **2021**, *9*, 1142. https://doi.org/10.3390/pr9071142

Academic Editors: Sergey Zhironkin and Radim Rybar

Received: 28 May 2021
Accepted: 28 June 2021
Published: 30 June 2021

1. Introduction

The depletion of fossil sources and the environmental crisis has worsened over the years and has led to an inevitable transition towards the exploitation of renewable energy sources. Figure 1 demonstrates a six-fold increase of global investment in renewable energies from 2004–2014 [1].

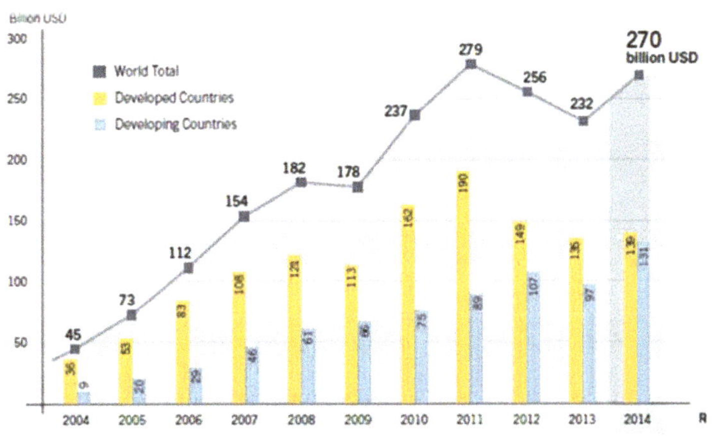

Figure 1. Global investment in renewable energies in billion dollars: period 2004–2014 [2].

The most important renewable energy source comes from the sun, which can be used to produce energy through various technologies which are constantly improved. The development of solar-based systems is important for greater security and independence, and the environmental impacts are limited [3]. The exploitation of concentrating systems is more advantageous than other solar technologies for energy generation [4]. Recent updates involve the coupling of concentrated solar power systems with Micro Gas Turbines (MGT) for a technology linked to new small high-temperature solar receivers [5], as opposed to the Stirling technology which is already considered an established technology. Solar energy is not only useful to produce electrical energy but also for recent technologies such as "photovoltaic powered reverse osmosis" (PV-RO) and "solar thermal powered reverse osmosis" (ST-RO) which can sustainably perform desalination [6].

The solar source is not constant, it is a function of the position of the solar concentrator in terms of latitude, longitude, and the consequent "Direct Normal Irradiance" [7]. This value can be measured with a pyrheliometer mounted on the solar concentrator [8]. Figure 2 shows the different techniques to exploit the solar source which have been developed recently [9]. The integration of systems such as photovoltaic panels and CSP systems into power grids represents a remarkable application that could significantly increase the ability to generate sustainable electrical power [9,10].

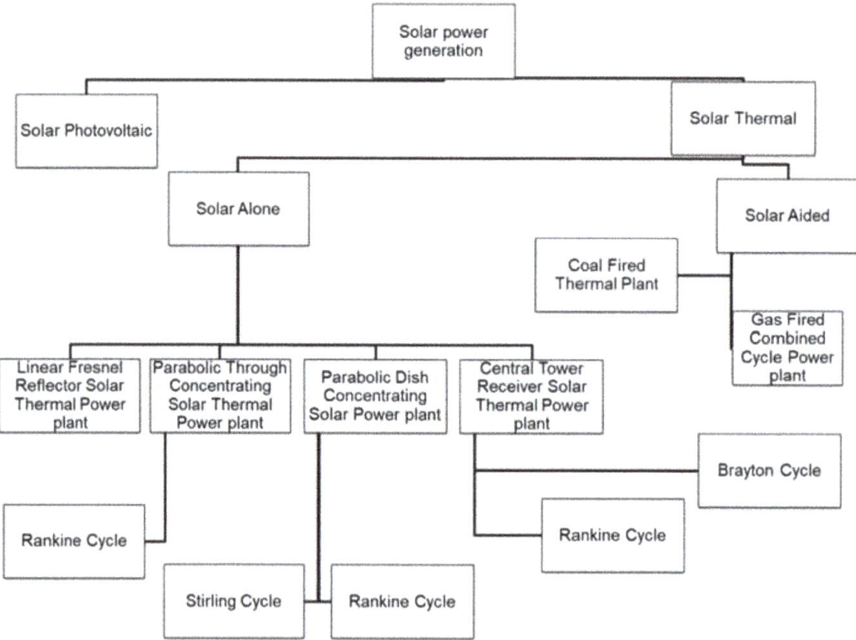

Figure 2. Solar power technologies.

Most of the systems presented refer to indirect systems using thermodynamic cycles.

Islam et al., (2018) investigated solar technologies and found that the most promising choice considers a thermodynamic cycle coupled with an electrical generation system [11]. Molten salts are the most commonly used thermal fluids [12,13]. An example of large-scale applied solar technologies is the "Solana Generating Station", completed in 2013 in Arizona (USA) [14]. The plant includes a 280 MW parabolic trough solar plant that can provide energy for up to 70,000 households in the Arizona area [15].

Air Brayton cycles and supercritical CO_2 Brayton cycles were investigated in pilot plant case studies [16,17].

An example of Stirling cycle technology is the EuroDISH, a system that can produce up to 10 kW of electrical energy with a concentrator of 7.5 m in diameter. The system is shown in Figure 3 and can be found at the "Plataforma Solar de Almería" in Almería, Spain [18,19].

Figure 3. EuroDISH system, Almería.

Another dish-Stirling cycle was built in Jodhpur (India), with a 50 MW CSP system. This station is composed of 2000 units of the parabolic dish-Stirling, each having 25 kWe capacity [20]. This aspect highlights the modularity of such technologies.

Another current example of valid application of the dish-Stirling is for the absorption and compression of cooling systems. Current works [21] show that compression systems are able to produce up to 53% of cooling power more than the amount of heat supplied.

The second main component of the dish-Stirling is the Stirling engine, an external combustion engine with two separate cylinders on the expansion side and the compression side. The space between the hot and cold side of the engine is called "regenerator" and acts as a heat recuperator to reduce the work of the compression and expansion of the pistons [22].

Since this type of engine is an external combustion engine, it can be fed by various kinds of external heat sources: fuel, solar, geothermal, or waste heat [23]. The heat source hits the walls of the hot side heat exchanger, inside which the heat carrier fluid reaches a sufficient temperature level to start the Stirling cycle. The continuous heating and refrigerating of the heat carrier inside the two cylinders are crucial for the continuation of the cycle [24].

This project focuses on the Stirling cycle fed by solar energy through a solar dish concentrator. The solar concentrator focuses the "Direct Normal Irradiance", DNI, on a plane called the "focal plane". On this plane, an α-Stirling engine uses this energy to heat its fluid carrier contained inside the engine and initiates the Stirling cycle to generate electrical power.

This article aims to evaluate the electrical production of renewable energy thanks to dish-Stirling technology. The project will consider two different configurations, while the highest performances will be chosen. The starting point for the simulation is the Direct Normal Irradiance (DNI), values obtained from the Politecnico di Torino tracking system. This average value is typically measured in hotter seasons (spring and summer).

Following the same principle, the most suitable material in terms of temperature reached by the heat carrier fluid inside the pipe of the exchanger was evaluated experimentally. After deciding the most suitable material between the two materials presented, the temperature of the fluid was used to simulate a Stirling cycle, with constant electrical power production. In the end, this value will be compared with lower values of DNI, typical of colder seasons such as winter and autumn.

2. Geometry and Design Parameters of the System

The solar concentrator used in the project is mounted on the roof of Politecnico di Torino (Energy Center). The solar concentrator was built by El.Ma. Electronic Machining srl (TN, Italy). The geometrical data used for the simulation of geometric optics are reported in Table 1 presented below:

Table 1. Parameters of the solar dish concentrator.

Name	Value	Description
f	0.92 m	Focal length
phi	0.9028 rad	Rim angle
d	2.4 m	Dish diameter
A	4.48 m^2	Dish projected surface area
psim	0.00465 rad	Maximum solar disc angle

These parameters can be implemented in the COMSOL Multiphysics 5.3 library, in which it is possible to create the geometry of a parabolic concentrator with negligible thickness for the Ray Tracing analysis. The geometry of the concentrator is shown in Figure 4 [19], the rim angle can be calculated from focal length and dish diameter:

$$d = 4 \cdot f \cdot [\csc(\varphi) - \cot(\varphi)] = 2.4 \text{ m} \quad (1)$$

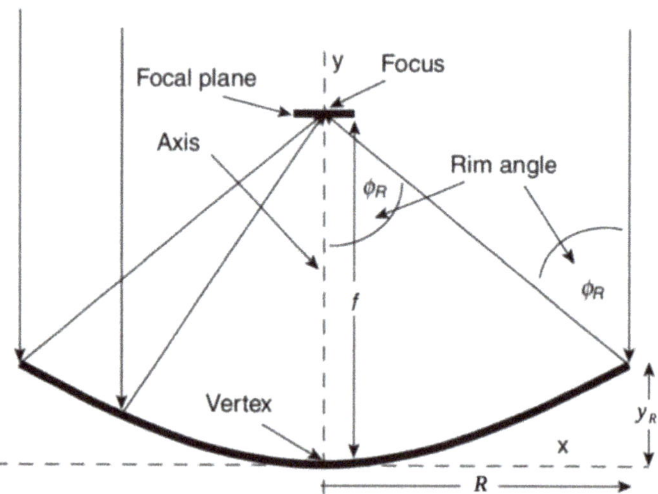

Figure 4. Geometry of a parabolic concentrator.

As far as the mean irradiation value is concerned, a constant initial value was chosen for the simulation. This parameter can be measured by the concentrator through a pyrheliometer mounted on the concentrator. The average value defined for the simulation is 800 W/m^2. This value is mainly measured during hotter seasons of the year, while for autumn and winter, the values tend to decrease slightly.

The hot side heat exchanger is the area of the engine which receives the concentrated energy and, for this project, the geometry for the heat exchanger was developed using CAD software SolidWorks 2019, whereas the CFD simulations were carried out using COMSOL Multiphysics 5.3. The presented work follows the following steps: the analysis of the two chosen configurations for the engine between the single-cylinder and the twin-cylinder.

Next, another CFD simulation was implemented to choose the most suitable material for the hot heat exchanger between AISI 310 and Inconel 625, both of which show high resistance at high temperatures. Lastly, the third and final simulation allows to find the desired value of temperature reached by the fluid inside the engine. This is an important value used to simulate a real Stirling cycle with two methods: the first-order empirical method known as "Beale equation" and the second-order method known as "Schmidt analysis" to evaluate the heat power produced which, by coupling the system with a permanent magnet generator gives the final result: the electrical power output of the system.

The Stirling engine chosen for the project is designed by an Italian Stirling engine manufacturer "Genoastirling s.r.l.". An example of this engine is shown in Figure 5 in both the single-cylinder (ML1000) and twin-cylinder configuration (ML3000).

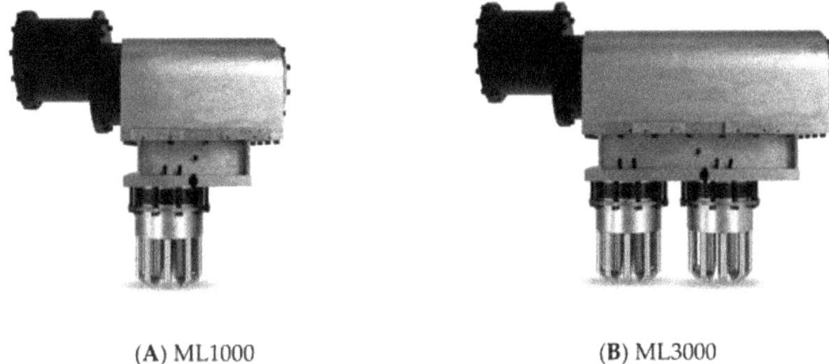

(**A**) ML1000 (**B**) ML3000

Figure 5. Genoastirling engines.

The engine designed by Genoastirling is an alpha-type engine that uses air as a heat carrier throughout the cycle. The cold temperature for the Stirling cycle is 40 °C and the objective of this article is to evaluate the hot temperature reached by the fluid for the analysis of power production. In Table 2, the main geometric and operative parameters of the engine can be found.

Table 2. Parameters of the engine.

Parameters	Value	Unit of Measurement
	Geometric Parameters	
Connecting rod length	210	mm
Diameter of the cylinder	110	mm
Cylinder stroke	55.2	mm
Offset compression-expansion cycle	90	sexagesimal degree °
Compression cylinder displacement	524.6	cm^3
Expansion cylinder displacement	524.6	cm^3
Compression cylinder dead volume	153.3	cm^3
Expansion cylinder dead volume	153.3	cm^3
	Operative Parameters	
Rotational speed	600	rpm
Working fluid	Air	-
Average pressure indicated	15	bar
Cooling fluid	Water	-
Hot side starting temperature	700	°C

The engine's hot side heat exchanger is placed on the focal plane and receives the energy from the concentrator on its external surface. The receiver is therefore represented by the entire surface of the hot side heat exchanger and has to be designed to be implemented in COMSOL Multiphysics 5.3 through the function "LiveLink for SolidWorks". The heat exchanger for the hot side is called "Nazgul" and a model of the exchanger created using SolidWorks version 2019 is shown in Figure 6 with the technical parameters defined in Table 3.

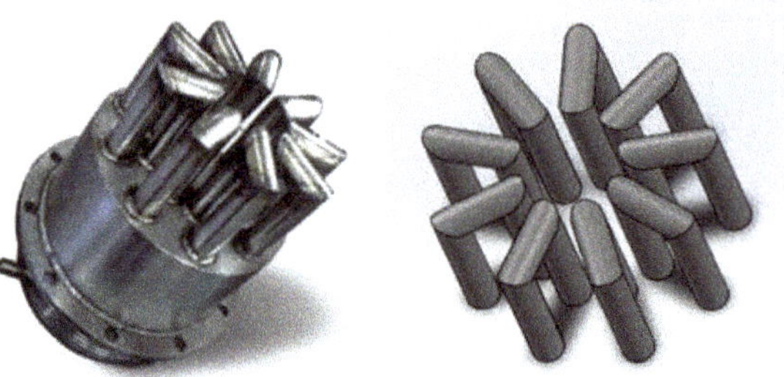

Figure 6. Hot side heat exchanger.

Table 3. Characteristics of the Nazgul heat exchanger.

Technical Data for the Exchanger	
Number of pipes	9
External diameter (mm)	20
Internal diameter (mm)	14
Pipe thickness (mm)	3
Max pipe height (mm)	120
Twin-cylinder exchangers' center distance (mm)	274

By knowing all the technical details of the heat exchanger, the model of the system was developed and implemented in COMSOL Multiphysics for the Ray Tracing simulation to decide the most suitable configuration between the single and the twin-cylinder. The resulting geometries for the two distinct simulations are presented in Figure 7.

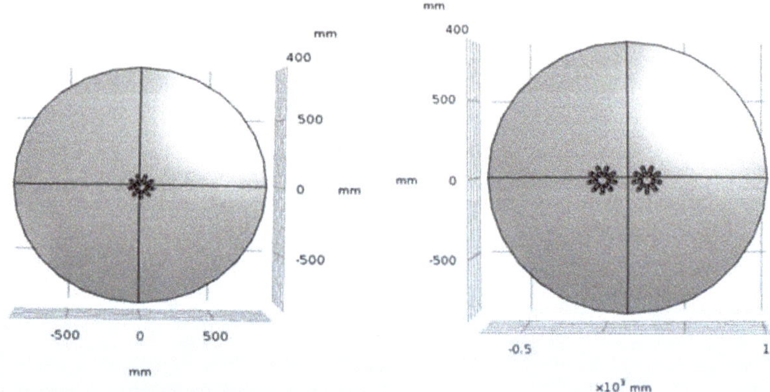

Figure 7. Configurations for the CFD simulation.

By implementing the various deviations from the ideal concentration such as the absorption coefficient equal to 0.15 and the optical limitations of the mirror with a "limb darkening" model coupled with a surface roughness of 1.75 microns, the simulation has been successfully run. The analysis for the results is based on the Stefan–Boltzmann law, where σ is the Stefan–Boltzmann coefficient equal to 5.67×10^{-8} W/(m^{2}K^4):

$$T = \left(\frac{q}{\sigma}\right)^{0.25} \tag{2}$$

The parameter defined in Equation (2) as "q" can be defined in COMSOL 5.3 as the heat source at the wall (W/m^2), and the results obtained for the two temperature profiles with the single-cylinder and the twin-cylinder engine configurations are shown respectively in Figure 8a,b.

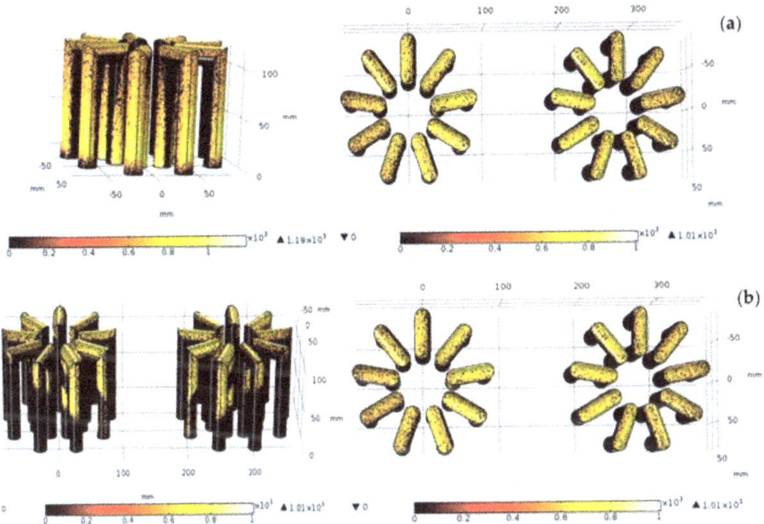

Figure 8. (**a**) Single-cylinder temperature profile (in K). (**b**) Twin-cylinder temperature profile (in K).

Figure 8 show that the higher encumbrance of the twin-cylinder engine results in a poorer temperature distribution along the external surface of the heat exchanger and a lower overall average temperature. The maximum reachable temperature in the single-cylinder configuration reaches a value of around 900 °C, this value drops down to 740 °C for the twin-cylinder. For these reasons, the choice for the continuation of the project is the single-cylinder one given the higher performances obtained by such configuration. The simulation also allows to obtain the value of "average heat source at the wall" which, for this specific case, is equal to 5990 W/m^2.

3. Materials

Having obtained the configuration for the system, it is important to evaluate the two proposed materials for the hot side heat exchanger. The materials taken into consideration for the simulation are AISI 310 and Inconel 625, both of which are proven to be highly resistant at high temperatures. These materials are resistant to temperatures above 1000 °C, therefore, are both suitable for the application since the system does not reach temperatures higher than 920 °C. The characteristics of the two materials at high temperature are defined in Table 4.

Table 4. Physical parameters for the materials.

Physical Properties	AISI 310	Inconel 625
Density (kg/m^3)	7550	8440
Heat capacity at constant pressure (J/kg/K)	610	560
Thermal conductivity (W/m/K)	23.7	20.3

The second simulation is performed to choose the most suitable material in terms of reached temperature on the internal pipe, which represents the solid–fluid interface for the Stirling engine.

The external tube's temperature is equal to the external temperature (20 °C), and the simulation is a transitory heat exchange in solids. The objective is to comparatively evaluate the material that can guarantee the highest temperature with an equal time. The geometry inevitably considers the width of the pipe equal to 3 cm and the model used for the simulation is shown in Figure 9.

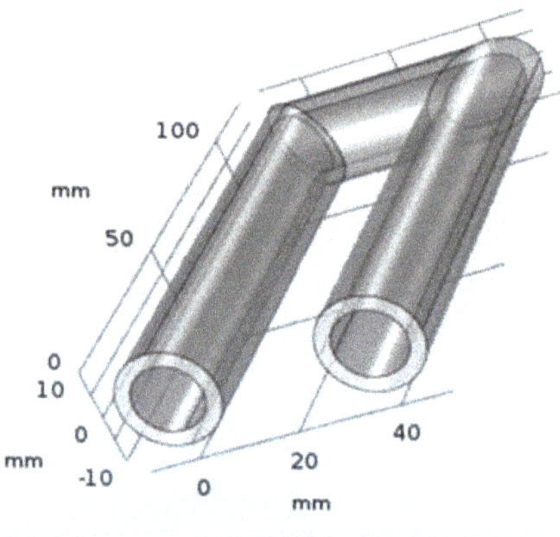

Figure 9. Geometry of the exchanger tube.

4. Results

4.1. Temperature Distribution

The parameter chosen as input for the simulation is the average heat source at the wall equal to 5990 W/m^2 previously calculated. The results are presented in Figure 10 and show AISI 310 and Inconel 625 and their temperature profile at a constant time value of 28 min. A fixed time value was chosen to evaluate which material can guarantee a higher temperature on the internal side of the pipe. The value settled for both simulations with AISI 310 and Inconel 625 is 28 min. Figure 11 shows the difference in the temperature profile between the two materials at the same fixed time value. This was decided to choose the most suitable material for the exchanger. The higher the reached temperature, the higher will be the production of thermal energy. This is why it is important to verify which material reaches the highest exchange temperature with the fluid.

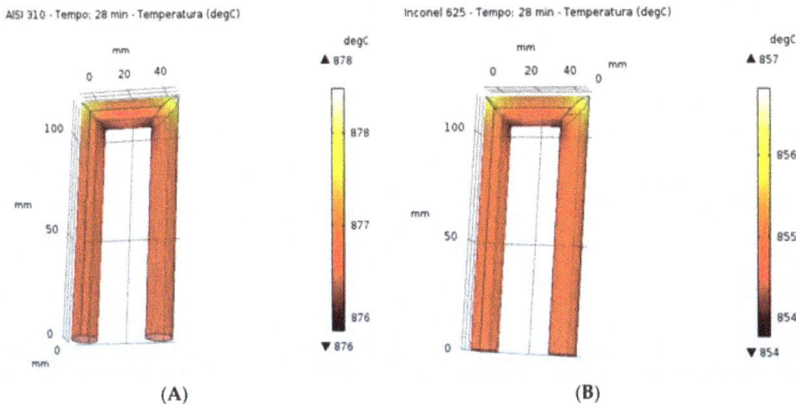

Figure 10. Temperature profiles in the internal tube. (**A**) AISI 310, (**B**) Inconel 625.

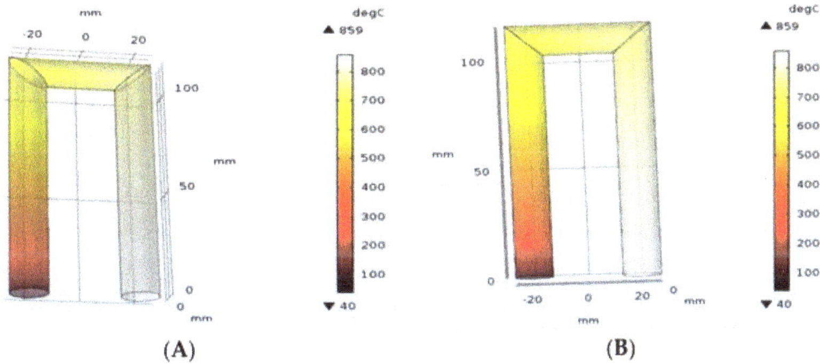

Figure 11. Hot air temperature profile. (**A**) k-ε, (**B**) k-Ω.

The results presented above easily demonstrate how AISI 310 allows to reach a temperature of about 876 °C, whereas, in the same amount of time, 28 min, Inconel 625's temperature profile enables the system to reach a temperature equal to 856 °C. Therefore, AISI 310 was chosen as the design material for the heat exchanger. By imposing the same time on the two different simulations, it is clear that the most suitable material for this application is AISI 310 because it allows the system to reach a higher interface temperature than Inconel 625. These two materials are used by Genoastirling as materials for the exchanger section.

The third and final simulation sets itself the objective to simulate the fluid flow inside the pipe for the evaluation of the temperature reached by air, the heat carrier for the engine, inside the hot side of the heat exchanger. This step has to be preceded by the fluid regime evaluation. It is very important to evaluate the fluid flow also to set the final simulation on COMSOL Multiphysics 5.3. In addition, by calculating the Reynolds and Prandtl number, it will also be possible to calculate the Nusselt number. The Nusselt number is useful to calculate the heat transfer coefficient, a value that has to be implemented in the simulation to calculate the final temperature achieved by the fluid inside the pipe.

Through Aspen Plus V9, the main physical properties of the system have been calculated at an average temperature of 500 °C and at the indicated average pressure of the cycle equal to 15 bar. Table 5 reports the main parameters for the calculation of the Reynolds and Prandtl number.

Table 5. Properties of the system.

Property	Value
Density (kg/m^3)	6.7
Dynamic viscosity (Pa·s)	3.64×10^{-5}
Thermal conductivity (W/m/K)	0.056
Heat capacity at constant pressure (J/kg/K)	1102.02
Average velocity (m/s)	15
Hydraulic diameter (m)	0.014

The results shown in Table 5 provide the following values for Reynolds and Prandtl numbers:

$$Re = \frac{\rho \cdot v \cdot D_h}{\mu} = 38,654 \quad Pr = \frac{C_p \cdot \mu}{k} = 0.716 \tag{3}$$

where:

- ρ is the fluid density, kg/m^3;
- v is the average velocity of the fluid, m/s;
- Dh is the hydraulic diameter, m;
- μ is the dynamic viscosity of the fluid, Pa·s;
- Cp is the specific heat at constant pressure of the fluid, J/kg/K;
- k is the thermal conductivity of the fluid, W/m/K.

The Reynolds number calculated confirms that the flow of the fluid inside the pipe is turbulent and it is possible to calculate the Nusselt number by using the Dittus–Boelter correlation. The Nusselt number is in turn used to calculate the heat transfer coefficient.

$$Nu = 0.023 \cdot Re^{0.8} \cdot Pr^{0.4} = 94 \tag{4}$$

$$h_e = \frac{Nu \cdot k}{D_h} = 376 \left(\frac{\text{W}}{\text{m}^2 \cdot \text{K}} \right) \tag{5}$$

The multiphysics simulation can now be implemented to study the temperature profile of the fluid inside the heat exchanger. The two physics chosen for the simulation are: "heat transfer in fluids" and "turbulent flow". As far as the latter is concerned, two methods for turbulent flow study have been compared: k-ε and k-Ω, with an internal pressure at the inlet of the pipe equal to 15 bar with a constant fluid velocity of 15 m/s. The former, instead, can simulate the external conditions of 890 °C of external temperature and can also be used to insert the convective heat transfer coefficient previously calculated. The inlet temperature of the gas is equal to 40 °C for the Stirling cycle. The results are shown in Figure 11.

The results shown in Figure 11 provide an equal final temperature with both turbulent methods: 859 °C. This value guarantees a value of pressure equal to 21 bar, comparable with the literature results [25].

This temperature value is high enough to start the Stirling cycle (700 °C) for the Genoastirling engine. This value of temperature can be used to simulate the Stirling cycle for the energy output of this system.

4.2. Real Stirling Cycle Analysis

The results obtained in the previous paragraphs are extremely useful to analyze a real Stirling cycle. The approach used in this section follows a two-step methodology: the first method used to calculate the thermal power output is the empirical Beale equation, a first-order method that provides an initial value for the power generated. This value has to be compared to a more accurate method: the second-order Schmidt analysis, which can provide a more detailed model for the value of thermal power [26].

The first-order empirical analysis is based on the Beale equation:

$$P = Be \cdot p \cdot f \cdot V_0 \tag{6}$$

where

- P is the thermal power, expressed in kW;
- p is the average pressure in the cycle, bar;
- f is the frequency of the cycle, Hz;
- V_0 is the displacement of the power piston, expressed in cm^3;
- Be is the "Beale Number", a dimensionless parameter.

The information required to perform this calculation is reported in Table 6.

Table 6. First-order power calculation.

Beale Number	-	**0.015**
Average pressure	bar	15
Cycle frequency	Hz	10
Piston displacement	cm^3	524.6
Thermal Power	W	1180

The Beale equation provides an empirical simplified estimation of thermal power. This method could be used to verify the validity of this equation and add more effectiveness to the simulation.

The second-order method is known as the "Schmidt analysis" and uses as simplification the isothermal transformation of both the expansion and the compression side. The temperature profile of the regenerator is constant as well and calculated as the logarithmic mean of the first two. The three sections of the engine are simplified as shown in Figure 12 [27,28].

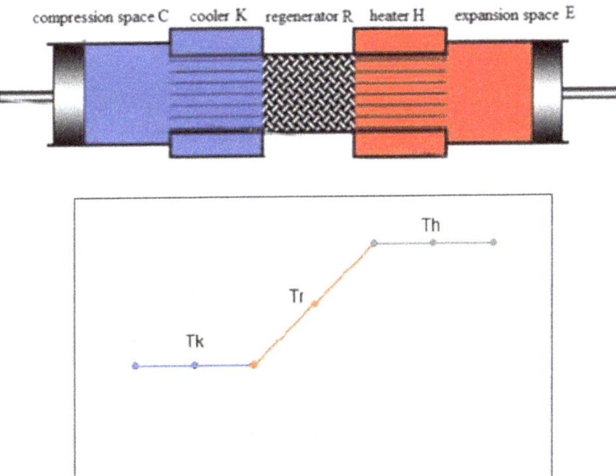

Figure 12. Temperature profile of the engine.

This method is a simplified method as well and for an even more thorough analysis, the adiabatic model must be implemented. Therefore, discrepancies are expected between the calculated values with this method and the values measured experimentally.

Since the α-Stirling engine is a two-piston configuration, the model can be simplified using the equation presented by G. Walker in the 1980 "Stirling engines" manual [29], whose equation is presented as follows:

$$W = P \cdot V_T \cdot \frac{\pi \cdot (1 - AU)}{K + 1} \cdot \left(\frac{1 - DL}{1 + DL}\right)^{1/2} \cdot \frac{DL \cdot sin(ET)}{1 + (1 - DL^2)^{1/2}} \quad (7)$$

The output of the equation, in this case, is the work per cycle of the Stirling cycle. This equation is also a function of three parameters:

$$S = \frac{2 \cdot RV \cdot AU}{AU + 1} \quad (8)$$

$$DL = \frac{\left(AU^2 + 2 \cdot AU \cdot K \cdot cos(AL) + K^2\right)^{1/2}}{(AU + K + 2 \cdot S)} \quad (9)$$

$$ET = tan^{-1}\left(\frac{K \cdot sin(AL)}{AU + K \cdot cos(AL)}\right) \quad (10)$$

These parameters have been calculated starting from the geometric and operative parameters of the engine. In Table 7, the list of parameters and their respective values used for the calculation are reported. The results of such calculation can be highlighted in the bottom part of the table.

Table 7. Second-order analysis.

Definition	Symbol	U.O.M.	Value
Max pressure	P	Mpa	2.1
Total swept volume	VT	cm^3	1049.2
Expansion swept volume	VL	cm^3	657.94
Compression swept volume	VK	cm^3	569.05
Swept volume ratio	K	-	1,16
Cold side temperature	TC	K	313
Hot side temperature	TH	K	1132
Temperature ratio	AU	-	0.28
Regenerator temperature	TR	K	637.09
Phase angle	AL	°	90
Total dead volume	VD	cm^3	459.9
Dead volume ratio	RV	-	0.70
Work per cycle	W	J/cycle	372.98
Thermal power	P_{th}	W	3729.8
Cycle efficiency	%	-	30
Actual thermal power	P_{eff}	W	1118.93

The result obtained through the Walker equation is comparable to the value obtained with the Beale equation. The two results are comparable: 1.18 kW for the Beale equation and 1.12 kW for the Walker equation.

As previously stated, these two methods provide estimated values of thermal output, but since the two values are very close, it is possible to confirm that the expected experimental values measured with this system will not have significant discrepancies with the calculated ones.

The cycle efficiency has been set at 30%. This value comes directly from Walker's Stirling Engine Design Manual. In this work, the worst-case scenario presented by Walker has been evaluated and therefore the most pessimistic value.

This claim is supported by data found in literature which prove that efficiencies in Stirling engines range between high-20's up to 42% [30].

By using these two methods, we can conclude that the value of thermal power that the system is able to produce in these conditions is 1.12 kW.

The last section of the analysis is the coupling of the system with a permanent magnet generator to verify the electrical output of the dish-Stirling. For a complete system, the Mecc Alte Eogen [31] generator can be coupled for the generation of electrical energy. The efficiency of generation for the chosen generator is 88%. In conclusion, the result of the entire analysis is presented in Table 8.

Table 8. Results of the simulation.

Parameter	U.O.M.	Value
Average irradiance on the concentrator	W/m^2	800
Hot side temperature reached	°C	859
Generated thermal power	kW	1.12
Efficiency of the electrical generator	%	88
Generated electrical power	kW	0.99

As shown in Table 8, an irradiance value of 800 W/m^2 allows the system to produce 996 W of electrical power.

The result of the simulation shows that a single-cylinder configuration of the engine with AISI 310 as coating material can allow the heat carrier fluid to reach a temperature of 859 °C.

This value can be used for the thermal energy production analysis, with an approximate resulting thermal power produced of 1.12 kW. Being the efficiency of the permanent magnet generator for this specific case equal to 88%, the resulting production of electrical energy is equal to 996 W. The result is in line with the expected results gained from the literature [32,33].

Furthermore, the specifics of the Genoastirling ML1000 single-cylinder Stirling engine set the maximum electrical power produced by the engine at 1.1 kW. This experimental value is close to the result obtained under similar conditions [34]. Therefore, it is possible to surmise that the electrical power produced by this technology, taking into consideration all the losses caused by the many components present in the system, is around 1 kW for a solar parabolic concentrator with a diameter of 2.4 m.

This result was compared with a much bigger CSP system with a production of 25 kW of electrical power simulated in similar conditions [35]. Obviously, since the diameter of the concentrator taken as comparison is much bigger (12.5 m^2 of diameter), the value taken for comparison is the power to area ratio to verify the effectiveness of the simulation. The modularity of the CSP technology should allow to obtain similar data from different sizes of concentrators.

In fact, the comparison from the simulation from literature allows to obtain a power-to-surface ratio equal to 0.204 kW/m^2. The current study simulated from Politecnico di Torino instead enables to reach a power-to-surface ratio equal to 0.205 kW/m^2. The results based on G. Walker's analysis [29] showed that a CSP system is useful for micro-generation purposes. The simulation is in line with both data from similar size concentrators and with scaled concentrators with bigger sizes.

Table 9 [36] compares the real efficiency of the Stirling engine mounted in various cities. This shows that the efficiency of the engine for small-scale microgeneration applications is stable between the high 20′s and 40%, with fluctuations during the different months of the year. The solar irradiance has also been compared to the same microgeneration systems to verify the validity of the simulation. Figure 13 shows irradiance values, which are coherent with the value recorded from this article.

Table 9. Actual efficiency of the Stirling engine installed in the selected cities.

City	January	February	March	April	May	June	July	August	September	October	November	December
Teheran	0.29	0.32	0.33	0.34	0.34	0.34	0.34	0.33	0.32	0.31	0.28	0.27
Tabriz	0.28	0.32	0.33	0.34	0.34	0.34	0.34	0.33	0.32	0.3	0.28	0.26
Bandar Abbas	0.31	0.33	0.34	0.34	0.34	0.34	0.34	0.34	0.33	0.32	0.3	0.3
Rasht	0.28	0.31	0.33	0.34	0.34	0.34	0.34	0.33	0.32	0.3	0.27	0.26
Yazd	0.3	0.33	0.34	0.34	0.34	0.34	0.34	0.34	0.33	0.32	0.3	0.29

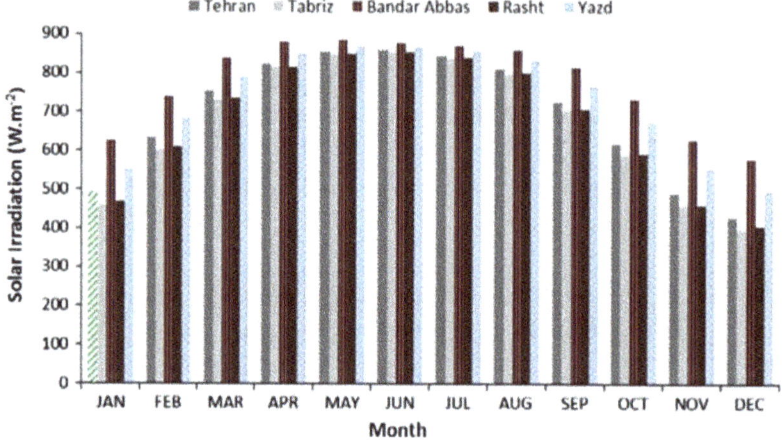

Figure 13. Monthly solar irradiance in the selected cities.

5. Conclusions

The obtained result is interesting since the solar concentrator used for the project has a relatively small size. Therefore, by increasing the size of the concentrator, it is foreseeable to increase the energy production. However, the downside of this technology is represented by two main factors: the installation cost of such technologies, and the inconsistency of the mean solar irradiance. The latter can be analyzed in further detail by operating the previous simulations with a starting irradiance value of 400 W/m^2. The resulting electrical power output will drop to 810 W. This value was obtained by following the same steps of the analysis, starting from the temperature of the fluid in the internal pipe followed by the Stirling cycle analysis to produce electrical energy.

The drop in power output is not excessive but has to be taken into consideration for a potential productivity analysis on an annual basis. This is a disadvantage of any solar technology, which depends on the position of the Sun and its irradiance. However, the sun tracker placed on the concentrator enables to minimize these losses by following the movement of the Sun across the sky, compared to fixed systems such as photovoltaic panels for the small-scale generation which are not mounted with a tracking system and are therefore more subjected to these fluctuations.

The results show that this technology has various positive aspects, but also showed a margin of improvement to reach a clear and sustainable expansion which can eventually facilitate the highly required transition to "clean" energy production methods.

Author Contributions: Conceptualization, D.P.; Data curation, D.B.; Formal analysis, D.P.; Methodology, D.B.; Software, D.B.; Supervision, D.P. and M.S.; Writing—original draft, D.B.; Writing—review

& editing, D.P.; Funding, M.S. All authors have read and agreed to the published version of the manuscript.

Funding: This research received no external funding.

Institutional Review Board Statement: Not applicable.

Informed Consent Statement: Informed consent was obtained from all subjects involved in the study.

Data Availability Statement: Open access.

Conflicts of Interest: The authors declare no conflict of interest.

Nomenclature

°C	Degrees Celsius
Be	is the "Beale Number", [-]
CAD	Computer-Aided Drafting
CFD	Computational Fluid Dynamics;
Cp	is the specific heat at constant pressure of the fluid, [J/kg/K]
CSP	Concentrated Solar Power
D_h	is the hydraulic diameter, [m]
DNI	Direct Normal Irradiance (W/m^2)
f	is the frequency of the cycle, [Hz]
J	Joule
k	is the thermal conductivity of the fluid, [W/m/K]
kW_{el}	Kilowatts of electrical power
Nu	Nusselt number
p	is the average pressure in the cycle, [bar]
P	is the thermal power, [kW]
Pr	Prandtl number
PV-RO	photovoltaic powered reverse osmosis
rad	Radians
Re	Reynolds number
rpm	Revolutions per minute
ST-RO	solar thermal powered reverse osmosis
UOM	Unit Of Measurement
Vo	is the displacement of the power piston, [cm^3]
W	Watt
Greek Symbols	
σ	Stefan–Boltzmann coefficient, [$W/(m^2 \cdot K^4)$]
ϕ	rim angle, [rad]
v	is the average velocity of the fluid, [m/s]
μ	is the dynamic viscosity of the fluid, [Pa·s]
ρ	is the fluid density, [kg/m^3]

References

1. Shahbaz, M.; Raghutla, C.; Chittedi, K.R.; Jiao, Z.; Vo, X.V. The effect of renewable energy consumption on economic growth: Evidence from the renewable energy country attractive index. *Energy* **2020**, *207*, 118162. [CrossRef]
2. Lohani, B.; Tulej, P.; Dixon, R.K.; Frankl, P.; Amin, A.Z.; Alers, M.; Radka, M.; Monga, P.; George, A.M.; Giner-Reichl, I.; et al. *An Agreement for the Establishment of the Caribbean Commission*; International Organisations; University of Wisconsin Press: Madison, WI, USA, 1947; Volume 1, pp. 251–256.
3. Devabhaktuni, V.; Alam, M.; Shekara Sreenadh Reddy Depuru, S.; Green, R.C.; Nims, D.; Near, C. Solar energy: Trends and enabling technologies. *Renew. Sustain. Energy Rev.* **2013**, *19*, 555–564. [CrossRef]
4. Kabir, E.; Kumar, P.; Kumar, S.; Adelodun, A.A.; Kim, K.-H. Solar energy: Potential and future prospects. *Renew. Sustain. Energy Rev.* **2018**, *82*, 894–900. [CrossRef]
5. Giovannelli, A. State of the Art on Small-Scale Concentrated Solar Power Plants. *Energy Procedia* **2015**, *82*, 607–614. [CrossRef]
6. Ahmed, F.E.; Hashaikeh, R.; Hilal, N. Solar powered desalination–Technology, energy and future outlook. *Desalination* **2019**, *453*, 54–76. [CrossRef]

7. Carrillo Caballero, G.E.; Mendoza, L.S.; Martinez, A.M.; Silva, E.E.; Melian, V.R.; Venturini, O.J.; del Olmo, O.A. Optimization of a Dish Stirling system working with DIR-type receiver using multi-objective techniques. *Appl. Energy* **2017**, *204*, 271–286. [CrossRef]
8. Roth, P.; Georgiev, A.; Boudinov, H. Cheap two axis sun following device. *Energy Convers. Manag.* **2005**, *46*, 1179–1192. [CrossRef]
9. Siva Reddy, V.; Kaushik, S.C.; Ranjan, K.R.; Tyagi, S.K. State-of-the-art of solar thermal power plants—A review. *Renew. Sustain. Energy Rev.* **2013**, *27*, 258–273. [CrossRef]
10. Sobri, S.; Koohi-Kamali, S.; Rahim, N.A. Solar photovoltaic generation forecasting methods: A review. *Energy Convers. Manag.* **2018**, *156*, 459–497. [CrossRef]
11. Islam, M.T.; Huda, N.; Abdullah, A.B.; Saidur, R. A comprehensive review of state-of-the-art concentrating solar power (CSP) technologies: Current status and research trends. *Renew. Sustain. Energy Rev.* **2018**, *91*, 987–1018. [CrossRef]
12. Ummadisingu, A.; Soni, M.S. Concentrating solar power–Technology, potential and policy in India. *Renew. Sustain. Energy Rev.* **2011**, *15*, 5169–5175. [CrossRef]
13. Braun, F.G.; Hooper, E.; Wand, R.; Zloczysti, P. Holding a candle to innovation in concentrating solar power technologies: A study drawing on patent data. *Energy Policy* **2011**, *39*, 2441–2456. [CrossRef]
14. Suresh, N.S.; Thirumalai, N.C.; Rao, B.S.; Ramaswamy, M.A. Methodology for sizing the solar field for parabolic trough technology with thermal storage and hybridization. *Sol. Energy* **2014**, *110*, 247–259. [CrossRef]
15. Solana Solar Power Generating Station, Arizona, US-Power Technology | Energy News and Market Analysis. Available online: https://www.power-technology.com/projects/solana-solar-power-generating-arizona-us/ (accessed on 24 June 2021).
16. Coco-Enríquez, L.; Muñoz-Antón, J.; Martínez-Val, J.M. Integration between direct steam generation in linear solar collectors and supercritical carbon dioxide Brayton power cycles. *Int. J. Hydrog. Energy* **2015**, *40*, 15284–15300. [CrossRef]
17. Elsafi, A.M. On thermo-hydraulic modeling of direct steam generation. *Sol. Energy* **2015**, *120*, 636–650. [CrossRef]
18. Mancini, T.; Heller, P.; Butler, B.; Osborn, B.; Schiel, W.; Goldberg, V.; Buck, R.; Diver, R.; Andraka, C.; Moreno, J. Dish-Stirling Systems: An Overview of Development and Status. *J. Sol. Energy Eng.* **2003**, *125*, 135–151. [CrossRef]
19. Barlev, D.; Vidu, R.; Stroeve, P. Innovation in concentrated solar power. *Sol. Energy Mater. Sol. Cells* **2011**, *95*, 2703–2725. [CrossRef]
20. Zayed, M.E.; Zhao, J.; Elsheikh, A.H.; Zhao, Z.; Zhong, S.; Kabeel, A.E. Comprehensive parametric analysis, design and performance assessment of a solar dish/Stirling system. *Process. Saf. Environ. Prot.* **2021**, *146*, 276–291. [CrossRef]
21. Romage, G.; Jiménez, C.; de Jesús Reyes, J.; Zacarías, A.; Carvajal, I.; Jiménez, J.A.; Pineda, J.; Venegas, M. Modeling and Simulation of a Hybrid Compression/Absorption Chiller Driven by Stirling Engine and Solar Dish Collector. *Appl. Sci.* **2020**, *10*, 9018. [CrossRef]
22. Nielsen, A.S.; York, B.T.; MacDonald, B.D. Stirling engine regenerators: How to attain over 95% regenerator effectiveness with sub-regenerators and thermal mass ratios. *Appl. Energy* **2019**, *253*, 113557. [CrossRef]
23. Barreto, G.; Canhoto, P. Modelling of a Stirling engine with parabolic dish for thermal to electric conversion of solar energy. *Energy Convers. Manag.* **2017**, *132*, 119–135. [CrossRef]
24. Thombare, D.G.; Verma, S.K. Technological development in the Stirling cycle engines. *Renew. Sustain. Energy Rev.* **2008**, *12*, 1–38. [CrossRef]
25. Torres García, M.; Carvajal Trujillo, E.; Vélez Godiño, J.; Sánchez Martínez, D. Thermodynamic Model for Performance Analysis of a Stirling Engine Prototype. *Energies* **2018**, *11*, 2655. [CrossRef]
26. Hirve, N.S. Thermodynamic Analysis of a Stirling Engine Using Second Order Isothermal and Adiabatic Models for Application in Micropower Generation System. Master's Thesis, University of Washington, Seattle, WA, USA, 2015.
27. Organ, A.J. Thermodynamic Analysis of the Stirling Cycle Machine—A Review of the Literature. *Proc. Inst. Mech. Eng. Part. C J. Mech. Eng. Sci.* **1987**, *201*, 381–402. [CrossRef]
28. Urieli, I.; Berchowitz, D.M. *Stirling Cycle Engine Analysis*; Adam Hilger Ltd.: London, UK, 1984.
29. Walker, G. *Stirling Engines*; Clarendon Press: Oxford, UK, 1980; ISBN 978-0-19-856209-2.
30. Zayed, M.E.; Zhao, J.; Elsheikh, A.H.; Li, W.; Sadek, S.; Aboelmaaref, M.M. A comprehensive review on Dish/Stirling concentrated solar power systems: Design, optical and geometrical analyses, thermal performance assessment, and applications. *J. Clean. Prod.* **2021**, *283*, 124664. [CrossRef]
31. Magneti Permanenti | Mecc Alte. Available online: https://www.meccalte.com/it/prodotti/alternatori/magneti-permanenti (accessed on 24 June 2021).
32. Zhang, C.; Xu, Q.; Zhang, Y.; Arauzo, I.; Zou, C. Performance analysis of different arrangements of a new layout dish-Stirling system. *Energy Rep.* **2021**, *7*, 1798–1807. [CrossRef]
33. Hafez, A.Z.; Soliman, A.; El-Metwally, K.A.; Ismail, I.M. Solar parabolic dish Stirling engine system design, simulation, and thermal analysis. *Energy Convers. Manag.* **2016**, *126*, 60–75. [CrossRef]
34. Motore ML1000 | I Nostri Motori | Genoastirling. Available online: https://genoastirling.it/ita/motore-ml1000.php (accessed on 24 June 2021).
35. Lashari, A.A.; Shaikh, P.H.; Leghari, Z.H.; Soomro, M.I.; Memon, Z.A.; Uqaili, M.A. The performance prediction and techno-economic analyses of a stand-alone parabolic solar dish/stirling system, for Jamshoro, Pakistan. *Clean. Eng. Technol.* **2021**, *2*, 100064.
36. Moghadam, R.S.; Sayyaadi, H.; Hosseinzade, H. Sizing a solar dish Stirling micro-CHP system for residential application in diverse climatic conditions based on 3E analysis. *Energy Convers. Manag.* **2013**, *75*, 348–365. [CrossRef]

MDPI

Article

Evaluation of the Performance of Mining Processes after the Strategic Innovation for Sustainable Development

Katarína Teplická *, Samer Khouri, Martin Beer and Jana Rybárová

Institute of Earth Resources, Technical University of Košice, 04200 Košice, Slovakia; samer.khouri@tuke.sk (S.K.); martin.beer@tuke.sk (M.B.); jana.rybarova@tuke.sk (J.R.)
* Correspondence: katarina.teplicka@tuke.sk; Tel.: +421-556022997

Abstract: The article summarizes the arguments within the scientific discussion about performance management in mining companies and their significance for obtaining competitiveness in the market of mining companies in the direction of sustainable development and economic growth. The main goal of the paper is to evaluate the performance indicators of mining processes after the implementation of strategic innovation—a new layout of the mining area focused on a combination of stationary and mobile mining equipment and their influence on the environment in a selected mining company in Slovakia in area of mining of limestone. Methods of research were focused on using economic indicators for the valuation of the efficiency and functionality of the mining processes. We used Pareto analysis for evaluation that points to critical mining processes and their significance in the financial area with orientation to costs, revenues. This research was used economic analysis with direction to efficiency, an indicator of cost and profit. Those indicators create a base for effective business in the mining area. The research empirically confirms that the new innovation of layout of mining place brings improvement of mining processes and indicators point to effective (over limit 0.70) and functional (over limit 0.90) mining processes in the year 2020. Pareto analysis showed the best processes (mining, expedition, transport, sorting) for financial benefits, the volume of production, demand, the satisfaction of customers, the cover needs of industries but at the same they are processes with high costs. Strategic innovation brought improvement too in the area of the environment. The results of the research can be useful for other mining companies in performance management and achievement mining market position.

Keywords: mining processes; evaluation; efficiency; functionality; performance

Citation: Teplická, K.; Khouri, S.; Beer, M.; Rybárová, J. Evaluation of the Performance of Mining Processes after the Strategic Innovation for Sustainable Development. *Processes* **2021**, *9*, 1374. https://doi.org/10.3390/pr9081374

Academic Editor: Chunjiang An

Received: 24 May 2021
Accepted: 4 August 2021
Published: 6 August 2021

1. Introduction

Management of mining processes and monitoring of their performance is a basic prerequisite for continual improvement, elimination of waste, the introduction of innovations in mining, gaining competitive advantages, and application in the competitive environment of the mining industry by the view of sustainable development and economic growth. Jiskani et al. (2020) support the argument that improving the competitiveness of the mining industry aids in the promotion of its sustainable development by the SWOT analysis for the mining industry. Jiskani et al. (2021) [1] comment that the performance indicators are evaluated in smart mining, but it is important to the importance of performance indicators framework should be highlighted. Cehlár et al. (2019) [2] stated that the new mining machines mean improving mining operations, more efficiently boosting productivity and employee safety, and opportunities for future use of earth resources and new technologies in the mining industry [3]. Optimization of the mining area and its layout means competitiveness in the market of mining industries (quality management). Jiskani et al. (2020) described the importance of mine optimization, safety, and green mining strategy as an essential pathway to achieving sustainable mining. The role of mining safety and green mining for sustainable mining is the base idea for future mining [4]. Krupnik et al. (2020) stated that process and operation management and mining process improvement represent an approach that enables to ensure the maximum performance of mining processes, to fill in the strategic goals of the mining companies, and to ensure the competitiveness on the mining industry market [5]. Sutoová et al. (2018) stated that the achievement of the performance in mining processes means focusing on innovation, developing knowledge, improving internal processes, eliminating waste, minimizing costs, improving technology, and using production factors effectively [6]. Zhou et al. (2020) stated the concept of green mining is to improve the mining industry in a holistic way so that it is safe, efficient, and environmentally sustainable. They created an evaluation index system of green surface mining based on the theory of green grades. The evaluation model is comprised of three attributes (safety, efficiency, and environment), nine criteria, and 35 indicators [7]. Chen et al. (2020) presented the idea of green mining was proposed as a practical approach to making the mining industry more sustainable than before. Green mining is a contemporary mining model centered on the sustainability of resources, environment, and socio-economic benefits. Its purpose is to develop and apply technologies and processes that increase environmental performance, while maintaining competitiveness throughout the entire mining cycle from exploration to post-closure [8]. The main goal of the paper was to evaluate the performance indicators of mining processes after the implementation of strategic innovation—a new layout of the mining area focused on a combination of stationary and mobile mining equipment. Mining processes relate to negative environmental impacts (environmental management) and high energy consumption (energy management). Mirmozaffari et al. (2020) stated that it is important continually to evaluate ecological efficiency affected by CO2 consumption that is one part of the performance of mining processes [9]. Negative impacts of mining processes to the adjacent territories create a base for evaluation. The contamination of territories is developed by active seepage of liquid waste of mining (waste management). Menshikova et al. (2020) commented particularly important is to evaluate water balance in the mining processes and a significant amount of wastewater (water management). The results of performance indicators are to substantiate the need to manage the seepage discharge process by means of enclosing dams to ensure environmentally safe operation of the tailings dump [10]. The growing availability of information and data analysis capabilities provides new opportunities for improving the performance of mining operations. By using real-time measurements and artificial intelligence, it is possible to respond faster to changing conditions, while accurate and timely information permits rapid evaluation of changes and fast business improvement decisions (operation management). The base of the blueprint is a performance framework (Figure 1) with a system of key performance indicators for mining control.

The synergy of all various forms of management creates the base of improvement in mining companies. Visser et al. (2020) commented that the introduction of the blueprint for operations management enables mines to leverage the developments in information and analysis capabilities to improve the performance of operations and prevention of risks (risk management) [11]. An important part of performance evaluation in mining companies is effective risk management in mining. Jiskani et al. (2020) stated that the significance of risk identification, risk analysis, and risk management for sustainable development in the mining industry must be highlighted. Sustainability has always been a core concern in the mining industry because the exploitation of mineral resources poses adverse impacts on the environment and society. As a result, the industry is seeking spectacular progress to foster sustainable mining practices that can enable it to be financially viable, environmentally friendly, and socially responsible. Since mining is inevitably endemic with risks, risk assessment could enable the industry to deal with the risks that have significant impacts on its sustainable development [12]. Risk needs to be properly identified before it may be estimated and later responded to adequately. Tworek et al. (2018) stated that prerequisite is an effective system of risk management in mining [13]. One part of risk management in the mining area is the health and safety management system. Health risk management depends on mining technologies, mining areas, and mining processes (Risk management) [14]. The best important part of management is financial and cost management in mining companies.

Typical responses by mining companies are to cut costs or increase production what is presented as financial indicators—profit, revenues, and costs. Hall et al. (2003) stated that economic risk is important to know in mining companies. Most mineral deposits respond to increased production rates. Many mine failures can be prevented by a close examination of the tonnage grade curve and an understanding of how margins, net cash flows, and resulting business risk change with cut-off grades and rates of production. The synergy effect of cost and financial management is a base for improvement in mining companies [15]. The economic situation of mining companies entails a strategic approach to cost reduction planning. The slowdown of the economy forces the business sector to restructure and reduction of cost, proceeding to rationalization processes which leads to optimization of mining processes, increasing of machine utilization, the use of materials, and people. Domaracká et al. (2013) commented that for the right decision-making it is necessary to have information about the financial situation and about critical areas in the mining process during risk management [16]. Puzder et al. (2017) commented that one of the risks in mining companies is an economic indicator—cost ratio. The cost ratio indicator is the fundamental performance indicator of the mining companies and it is important to create a cost model for its evaluation orientated to minimize mining costs [17]. The next problem for evaluation of the performance of mining is sustainable development and is it attracting new investment. Solving it requires access to international capital markets and preparing financial statements with international requirements based on the data generated by the accounting system. Tyuleneneva (2017) stated that the framework of international financial reporting standards is important to base for new investment in mining companies [18]. Evaluation of the performance of mining must be introduced through the life cycle assessment (LCA). The mining industry is a potential field where sustainability and LCA can be implemented due to intense energy requirements and equipment utilization [19]. LCA is beginning with the mining process. The rationalization of the transportation of raw material in a mining company creates a base for other processes for example sorting, milling, and crushing. The application of generally applicable logistics principles may result in the increased efficiency of the transportation process. The main input of the rationalization proposal is the analysis of technical parameters of belt conveyors and following their optimization [20]. All mining processes would have to be managed effectively. Strategic development in mining companies is orientated to vertical integration. It means using outsourcing for the specific needs of the basic mining company during the IPO chain. The integration process may concern mining, processing, or mining–processing–metallurgical operations [21]. Bye (2007) stated that in the frame of strategic development is important to evaluate mining areas not only for grade and tonnage predictions but also for predictions of rock mass quality. The development of 3D multi-parametric models facilitates the provision of resource information well in advance of the mining. This information can be used for overall mine planning and evaluation, costing, mining optimization, and slope design. This allows the full range of mining activities to be interconnected, thereby lowering costs and improving efficiencies [22]. Mining companies using simulation modeling as an integral component in their development to determine which combination of infra-structure options, operating performance, and operating rules best achieve the goals of the mining process [23]. Except for simulation, models are used, namely, the technical–economic model for calculating a suitable mining method with accepted technical and economic factors [24]. The conventional discontinuity survey process in the mining industry to be a time-consuming one and it is technically challenging due to the limited accessibility of fresh rock exposures. A rapid and robust rock mass property quantification system is desirable for rock structure design during mining operations. An image-based and fully automatic rock mass geological strength index (GSI) rating system is used in praxis. The GSI system includes both structure rating (SR) and joint condition digital imaging (JCDI) to represent the bulk rock and discontinuity surface conditions of the rock mass [25]. Information management is today part of mining processes. The raw material policy is focused on the performance of all mining processes. Performance indicators of mining processes show to efficiency and functionality of mining processes. In developed countries, support tools for process optimization are increasingly used, which ultimately affects the quality of the final product [26]. Process management can be provided by managerial approaches (Figure 2), which are focused on special areas of management.

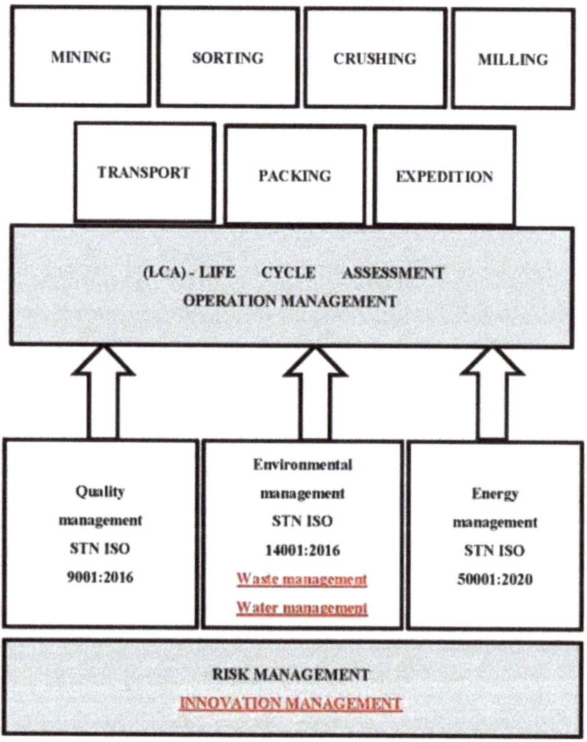

Figure 1. Performance framework for evaluation mining processes. Source: own source.

In this paper, we used instruments of various forms of management and we showed how important a synergy of forms of management is in the mining companies. The main goal of the paper was to evaluate the performance indicators of mining processes after the implementation of strategic innovation—a new layout of the mining area focused on a combination of stationary and mobile mining equipment and their influence on the environment in a selected mining company in Slovakia in area of mining for limestone.

The most frequently implemented management approaches in mining companies include: facility management (FM), activity-based management (ABM), project management (PM), and human resource management (HRM), alongside other forms of management that are presented in Figures 1 and 2. These managerial approaches focus on mining processes to shorten the time of product implementation, utilization of input resources in a particular process, sensitivity to the interconnection of activities in the process, risk documentation, and consistency and completeness in filling reports, records, and processed documentation. The orientation of process management is focused on measuring key performance indicators of mining processes, which are in the synergy of other management approaches.

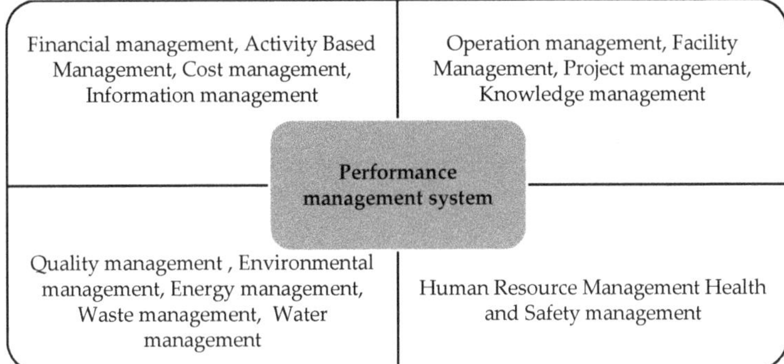

Figure 2. The synergy of various forms of management as a base for the performance management system. Source: own source.

2. Materials and Methods

In this article, we evaluated mining processes after the implementation of strategic innovation—a new layout of machines in the mining area. The complex process of research was done by a research algorithm (Figure 3). The object of research was the chosen mining company in Slovakia that deals with the mining of raw material, limestone.

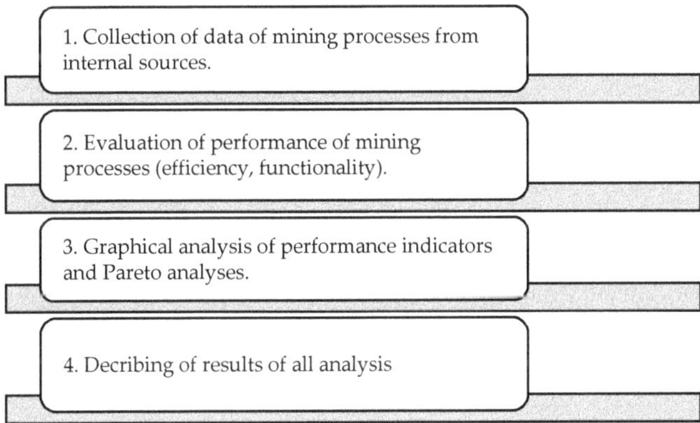

Figure 3. Algorithm of research in the mining company. Source: own source.

We evaluated performance indicators of processes in mining. The base processes in the IPO chain (input, process, output) (Figure 4) are processes in the mining company such as mining, crushing, sorting, grinding, packing, expedition, and transport. These processes create a base of the IPO chain and are the main processes in mining. These processes were evaluated. In the first step, we collected data of mining processes from internal documents of the mining company in the financial accounting and financial statements. We collected data through a personal visit to the mining company, and at the economic department we obtained outputs from the company's internal databases and double-entry bookkeeping. Data named "plan" are information of budgeting which was prepared last time period. The budgeting used plan method was based on percentual increasing of data. Data named "reality" are information of accounting statement which was prepared at the end of the year. Information creates data from two years—one year before implementation of the new layout, and after the implementation of the new layout.

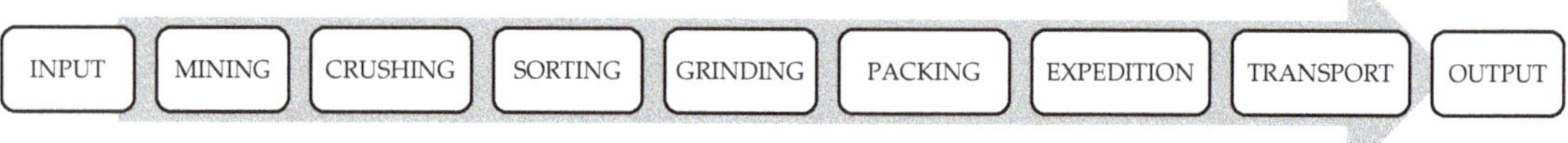

Figure 4. IPO chain of mining processes in the mining company.

We collected data for economic analyses for performance indicators in mining processes in the selected mining company in Slovakia. Table 1 contains data of various industry areas where the mining company sells raw material and products of raw material -limestone. Data are expressed volume unit (tone) for two the period years 2019 and 2020. Data are introduced for reality (current period) and plan (budgeting) of production of raw material.

Table 1. Production volume of raw material—limestone for various industries.

Type of Industry (Tone)	2019 (Plan)	2019 (Reality)	2020 (Plan)	2020 (Reality)
Construction (0–8)	32,000	35,300	34,000	30,500
Construction (8–32)	4500	5200	14,600	15,100
Construction (32–63)	52,360	54,600	65,820	87,650
Steel industry	7900	8200	18,500	20,400
Steel industry	15,600	17,500	90,000	156,000
Chemical industry	18,600	20,700	32,000	30,200
Glass industry	1400	1500	1650	1700
Agriculture	2500	2100	2100	1900

Source: Internal document of the mining.

Table 2 contains data of mining processes in the mining company. Data were expressed as volume unit (ton) for two period years, 2019 and 2020. Data were introduced for reality (current period) and plan (budgeting) of production of raw material.

Table 2. Production volume of raw material, limestone, for mining processes.

Mining Processes (Tone)	2019 (Plan)	2019 (Reality)	2020 (Plan)	2020 (Reality)
Mining	134,860	145,100	258,270	343,450
Transport	125,600	78,500	215,600	200,700
Crushing	75,860	74,650	154,600	158,750
Sorting	128,900	131,200	185,600	189,600
Grinding	45,600	43,500	74,650	74,800
Packing	33,450	31,530	48,500	49,650
Expedition	115,600	120,630	225,450	256,700

Source: Internal document of the mining.

In the second step, we analyzed performance indicators—coefficient of efficiency, index of functionality—for each process because these are two basic indicators of process evaluation according to ISO standard 9001 in the field of process management. We evaluated each area of industry where the mining company sold raw material, limestone.

Coefficient of efficiency (K) expresses the degree of fulfillment of the plan of the given process. It can be calculated by various parameters (production, revenue, failures, number of contracts, costs). Results of the coefficient evaluation are compared by limit values (Table 3). Formula for coefficient of efficiency (K), where (Xs) = value for reality, (Xp) = value for plan. This formula is based on the essence of process management:

$$K = \frac{Xs}{Xp} \quad (1)$$

Table 3. Limit of effective process [27].

Coefficient of Efficiency (K)	Limit
Effective	$K \geq 0.85$
Mostly effective	$0.85 > K \geq 0.70$
Ineffective	$K < 0.70$

The index of functionality (I) expresses the degree of trend (development) of the given process through the coefficient of efficiency for two continual time's period. It can be calculated by various parameters (production, revenue, failures, number of contracts, costs). Results of the index of functionality evaluation were compared by limit values (Table 4). The formula index of functionality (I), where $K1$ = value of coefficient of efficiency for current period, $K0$ = value of coefficient of efficiency for base period:

$$I = \frac{K1}{K0} \quad (2)$$

Table 4. Limit of functionality of process [27].

Index of Functionality	Limit
Functional	$I \geq 1$
Mostly functional	$1 > I \geq 0.90$
Nonfunctional	$I < 0.90$

The limit of the effective process and limit of the functionality of process are values established by long-term theories from the quality guru Deming and are used today in the quality management system. For those indicators, the ISO norm 10,014: economic of quality exists. The coefficient of efficiency and index of functionality are important indicators for assessing the business environment and the functioning of mining processes in mining companies around the world. These indicators significantly affect the functioning of mining processes, the demand for extracted raw material, interest in processed raw materials, the use of the machinery park, and complex influence on the financial situation in the mining area [27].

We also used Pareto analysis to research the performance of mining processes. Pareto analysis is named after the Italian economist Vilfredo Pareto, who at the end of the 19th century found that 80% of the wealth was owned by 20% of people. The Pareto rule also applies in business processes, e.g., 80% of the company's revenues come from 20% of customers, 20% of products generate 80% of profit, and 20% of possible causes generate 80% of problem situations in production.

The Pareto analysis procedure is employed for the following reasons:

1. To identify the causes in the monitored process (complaints, errors, costs, injuries, failures);
2. To arrange the causes in descending order (MAX-MIN), from the largest to the smallest values;
3. To determine the relative abundance (%);
4. To determine the cumulative relative abundance (%);
5. To construct a bar graph of causes from the largest value to the smallest value;
6. To construct a Lorenz curve (line graph) of cumulative frequencies of observed causes by means of a secondary axis in the graph;

7. To draw the ratio 80/20—80% on the cumulative number of the perpendicular to the Lorenz curve, from the point of intersection of the Lorenz curve perpendicular to the x-axis (20%). From point 0 to point 20%, the area of causes is decisive for the implementation of changes, after 20%, the area of causes is insignificant.

Relative abundance (%) expresses the share of the part in the whole in percentage expression, where Xi = economic value (production, profit, revenue, costs, number of employees) and KUM Ra means cumulating of percentage expression:

$$Ra = \frac{Xi}{SUM\ Xi} * 100(\%) \tag{3}$$

The Pareto analysis creates one of the models for a production system that describes failures in the mining machines area. This model is based on filling business strategy goals in the area of machines park of a mining company [28].

Process performance indicators also include financial indicators. In this area, we evaluated the indicator of economic efficiency (e), economic result (P), and cost ratio indicator (n).

The modern way of evaluating performance is based on the assumption that a company is efficient if it is able to achieve defined strategic goals. The efficiency of the company is a prerequisite for the company's competitiveness. The Global Competitiveness Report (BCI) assesses business-level competitiveness on the basis of indicators of performance. For economic analyses, we needed data of financial accounting as revenues, costs. Table 5 contains data of revenue for various industry areas (construction, steel, chemical, glass industry, agriculture). Data were expressed value units (euro) for two period years, 2019 and 2020. Data were present for actuality.

Table 5. Revenue of raw material, limestone.

Type of Industry	2019 (Reality) (€)	2020 (Reality) (€)
Construction	546,600	1,034,600
Steel industry	175,800	1,920,500
Chemical industry	150,000	210,560
Glass industry	97,000	95,000
Agriculture	75,200	87,400

Source: Internal document of the mining.

Table 6 contains data of costs for various industry areas (construction, steel, chemical, glass industry, agriculture). Data were expressed value units (euro) for two period years, 2019 and 2020. Data were present for actuality.

Table 6. Costs of raw material, limestone.

Type of Industry	2019 (Reality) (€)	2020 (Reality) (€)
Construction	339,120	1,025,700
Steel industry	125,600	1,758,000
Chemical industry	149,800	207,560
Glass industry	85,600	85,700
Agriculture	68,520	65,700

Source: Internal document of the mining.

Economic indicators present those formulas:

Formula efficiency (e) is where X = value of revenue (€), Y = value of cost (€):

$$e = \frac{X}{Y} \tag{4}$$

Efficiency (e) expresses the index between revenue and costs. The value of the coefficient (e) should be above level $e > 1$.

Formula profit (P) is where X = value of revenue (€), Y = value of cost (€):

$$P = X - Y \ (€) \tag{5}$$

Profit (P) expresses the difference between revenue and costs. The value of the profit (P) should be positive.

Formula cost ratio (n) is where X = value of revenue (€), Y = value of cost (€):

$$n = \frac{Y}{X} \tag{6}$$

Cost ratio (n) expresses the index between costs and revenue. The value of the cost ratio (n) should be level $n < 1$.

Financial indicators create the base of performance. The dissatisfaction with financial indicators led to a focus on areas of performance measurement such as a balanced scorecard, environmental indicators, quality indicators, and technic indicators. Moreover, many recent studies have focused on the sustainability concept and performance measurement interconnection. This approach is important for mining companies and evaluation mining processes [29].

3. Results

The research was orientated to process evaluation in mining. The object of research was chosen mining company in Slovakia that deals with the mining of raw material, limestone. In the selected mining company, we evaluated all processes in the IPO chain. Processes were evaluated after the implementation of strategic innovation, a new layout of machines in the mining area (Figure 5). The mining machines were placed in the workplace in another place. This change was done for the efficiency of the mining process.

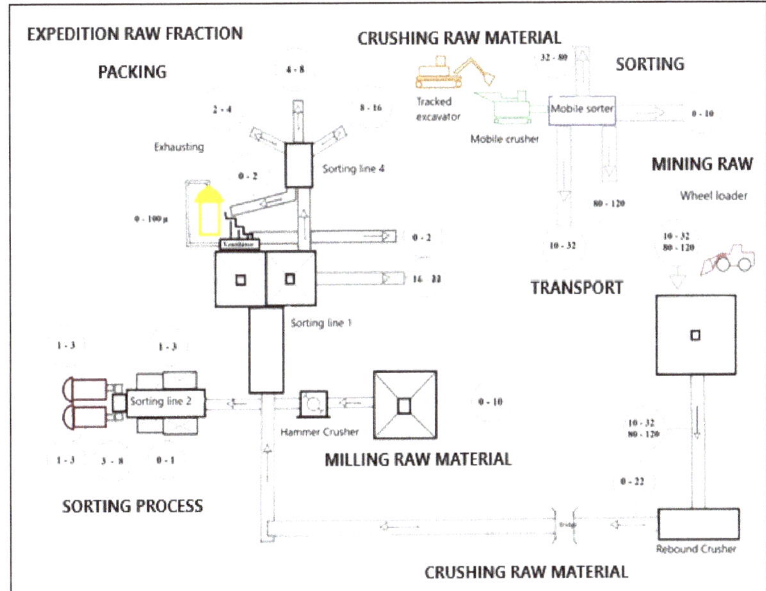

Figure 5. Layout in mining company after implementation new strategic innovation. Source: [30].

After the new layout of mining machines in the mining area, the mining processes were optimized. Based on the change of the working space, we evaluated the performance of the processes before the change and after the change of the mining layout area. The results of the evaluation of the efficiency indicator (Table 7) and functionality indicator (Table 8) of mining processes are presented in tables. The coefficient and index were calculated by formula 1 and formula 2. The results of the evaluation of the efficiency of mining processes point to critical mining process—transport in a mining company in the year 2019 at the level under limit e < 0.70. This means that this process is ineffective. The other processes are higher as limit level. The main goal of this indicator is to plan new changes in the area of internal transport in a mining company. This problem was in the year 2020 solved by implementing a new layout. The coefficient of efficiency was changed on the over-limit level on value K = 0.93.

Table 7. Coefficient of efficiency (K).

Mining Processes (Tone)	K(0) 2019	K(1) 2020
Mining	1.075	1.32
Transport	**0.62**	0.93
Crushing	0.98	1.03
Sorting	1.01	1.02
Grinding	0.95	1.002
Packing	0.94	1.02
Expedition	1.04	1.13

Source: own calculation.

Table 8. Index of functionality (I).

Mining Processes (Tone)	I
Mining	1.22
Transport	**1.5**
Crushing	1.05
Sorting	1.009
Grinding	1.05
Packing	1.08
Expedition	1.09

Source: own calculation.

The positive benefits we evaluated in this research through the coefficient of efficiency and index of functionality were recorded in the year 2020 because the coefficient of efficiency for transport was increased over a limit level K > 0.70 and the index of functionality for the year 2020 was recorded value over a limit level I >1 because its value was I = 1.5. All mining processes were effective. It means that each mining process was plan filling, and the mining, crushing, and grinding increased from the year 2019 after the new innovation strategy layout. For the mining company, the new innovation means new opportunities, meeting customer requirements, process efficiency, cost reduction, and downtimes.

The results of the evaluation of the efficiency (Table 9) and functionality (Table 10) orientated on the industry area where the mining company sells raw material and fractions of raw material we evaluated. In both years, mining processes were effective for all industry areas because the value of coefficient of efficiency in both 2019 and 2020 achieved over-limit level K < 0.70—for ineffective process. The lower results were recorded by area of construction (0–8), chemical industry, and agriculture.

In all areas, all purchasers recorded positive indicators of efficiency over-limit levels. The mining company fills the requirements of customers in the various areas—construction, chemical industry, glass industry, steel industry, and agriculture. This state is very important to the sustainability of mining market.

Index of functionality (Table 10) recorded nonfunctional mining processes in two areas—construction (0–8) and construction (8–32). These areas did not fulfill the requirements of customers. The other areas were fulfilled. This means that the trend during the two years recorded caused some problems in the mining processes as internal transport and downtimes at mining place. This problem was solved by a new layout at the mining place. In the new layout new mobile machines and other mobile crushers were used. The change of machines park brings the efficiency of mining processes, including crushing, grinding, and transport. This fact was described in values of functionality over the limit level. Managing and improving business processes increases business performance. There are several perspectives on managing and improving business processes such as productivity, efficiency, performance, time, cost, accuracy, flexibility, and output quality. Performance is the ability of a company to achieve the set goals and bring effect to all stakeholders. This represents a high probability of success in competing with other companies [31].

Table 9. Coefficient of efficiency (K) for type of industry.

Type of Industry (Tone)	K(0) 2019	K(1) 2020
Construction (0–8)	1.10	**0.89**
Construction (8–32)	1.15	1.03
Construction (32–63)	1.04	1.33
Steel industry	1.03	1.10
Steel industry	1.12	1.73
Chemical industry	0.99	**0.94**
Glass industry	1.07	1.03
Agriculture	0.84	**0.90**

Source: own calculation.

Table 10. Index of functionality (I) for type of industry.

Type of Industry (Tone)	I
Construction (0–8)	**0.80**
Construction (8–32)	**0.89**
Construction (32–63)	1.27
Steel industry	1.067
Steel industry	1.54
Chemical industry	0.94
Glass industry	0.96
Agriculture	1.07

Source: own calculation.

In the second step of algorithm of research, we prepared the Pareto analysis (Table 11). In the second step of the algorithm of research, we prepared a Pareto analysis. This analysis stated which processes were critical. Rules 20/80 mean that 20% mining processes create 80% operational costs (Figure 6). This analysis points to minimize costs in mining processes, including mining, expedition, transport, and sorting. Transport and sorting were solved by the new layout of the mining space.

Table 11. Pareto analyses for mining processes.

Mining Processes	Production (Tone)	Ra (%)	KUM Ra (%)
Mining	343,450	27	27
Expedition	256,700	20	47
Transport	200,700	16	63
Sorting	189,600	15	78
Crushing	158,750	12	90
Grinding	74,800	6	96
Packing	49,650	4	100

Source: own source.

Results of Pareto analysis point to critical processes in the mining company but at the same to processes that bring high revenues to satisfy the demand of raw material of various industry sections. Lorenz curve explains relation 20/80. It means only 20% of mining processes (mining, expedition, transport, sorting) create high production 80%. At the same time, this Lorenz curve explains 20% of mining processes create 80% operational costs. At last Lorenz curve explains 20% of mining processes create 80% revenues, which are important for the financial stability of the mining company. The process of mining contains a lot of various processes, but only some processes create a high level of costs. Those processes are important to solve and optimize costs.

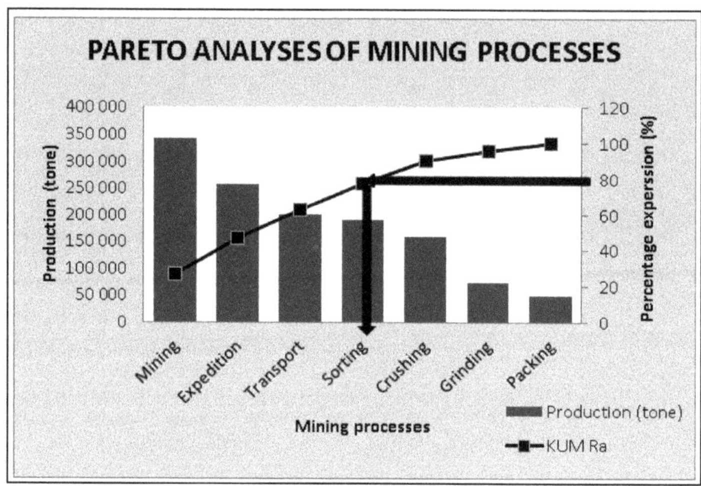

Figure 6. Pareto analyses and Lorenz curve. Source: own source.

We investigated the financial stability of the mining company through economic indicators (Table 12) as economic efficiency (e), cost ratio (n), and economic result (P). For economic analysis, we used data from the years 2019 and 2020 and we calculated economic indicators by formula 4, 5, 6. Indicator of efficiency (e) expresses the index between revenue and costs. The value of the coefficient (e) should be above level e > 1. In the mining company, this indicator is higher as (1) in the year 2019, 2020. It means that the mining company fulfills the goals of production and demand, and it uses all production factors optimal. The indicator economic result (P) in the mining company recorded high value in the years 2019 and 2020 means profit, not loss, which is important for the financial stability and for new investment. The high profit was recorded in the area for construction in the year 2019, and for the steel industry in the year 2020.

Table 12. Economic analyses.

Type of Industry	2019 (e)	2020 (e)	(P) (€) 2019	(P) (€) 2020	2019 (n)	2020 (n)
Construction	1.6	1.01	**207,480**	8900	**0.6**	0.99
Steel industry	1.4	1.1	50,200	**162,500**	0.7	0.9
Chemical industry	1.001	1.01	200	3000	0.999	0.99
Glass industry	1.1	1.1	11,400	9300	0.9	0.9
Agriculture	1.1	1.3	6680	21,700	0.9	**0.8**

Source: own source.

The cost ratio indicator (n) expresses the index between costs and revenue. The value of the cost ratio (n) should be level n < 1. This indicator informs the mining company about operational costs to one unit of revenue. In the year 2019 was this indicator the lowest in the construction area of industry and in the year 2020 was this indicator the lowest in the agriculture area. The structure of this indicator runs from 0.6 to 0.99. It means the risk for mining companies because costs directly to the value of revenue, which means a loss in the future period. Comparison analysis for economic efficiency and cost ratio (Figure 7) express through a limit level—value (1). Efficiency must be over this limit level and cost ratio must be under this limit level.

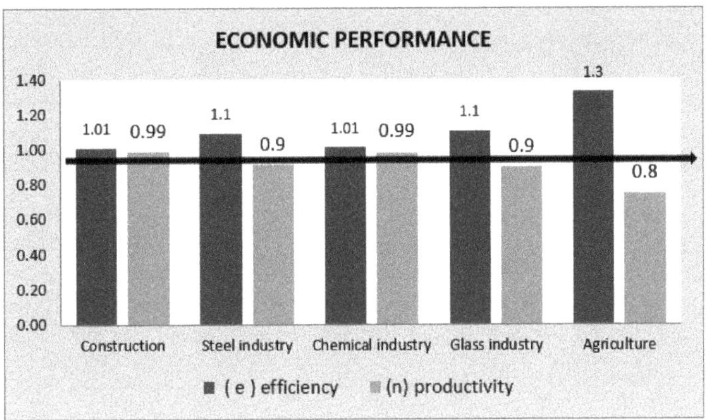

Figure 7. Economic analyses. Source: own source.

Comparison analysis points to economic efficiency over-limit level (1), where the cost ration indicator is under limit level (1). Both indicators showed a positive trend, but in the future they will be important to monitor.

4. Discussion

In this paper, we dealt with strategic innovation and its impact on performance indicators in mining companies for all mining processes. It is important to state that various innovations change performance indicators and bring improvement and increasing of profit of the mining company. Green mining is based on safety, environment, employees, results of the company, requirements of customers, and the end to achieve market position. The evaluation of the performance (Figure 8) of mining processes is connected with the strategic innovations that mining companies plan on the basis of the achieved results in the area of process efficiency and functionality.

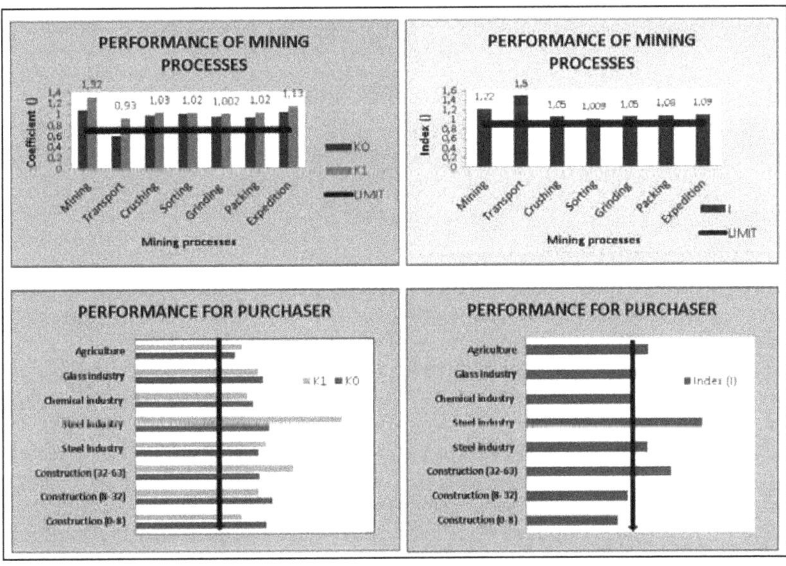

Figure 8. Dashboarding of performance of mining processes. Source: own source.

Innovations are also part of the goals of the business strategy, which must be achieved and fulfilled in the set time [32]. Based on the implemented strategic innovation in the mining company, the performance indicators improved. The processes were evaluated as functional and efficient. The change in layout caused a reduction in time downtime by transporting the raw material to the crushing and grinding process. Mobile crushers were introduced and belt conveyors were moved. Sorting lines were added for each shredder, as well as mobile sorting lines. The change in in-house transport has reduced the operating costs of equipment, transport costs, and raw material storage costs. The use of mining equipment was in line with the technical capacity of the machines at the level of 85%. All these measures brought the mining company better results in the performance of mining processes. In praxis managerial instruments were used such as TQM, Kaizen, Six Sigma, Controlling, 5S, Kanban, and JIT, which could ensure that the customers' needs are met and thus contribute to the higher performance of the mining companies [33]. Using quality management instruments and methods, businesses can increase their productivity and efficiency, decrease risks, reduces the unwanted variability in the processes, and associated non-productive costs for increasing the production quality and customer satisfaction [34]. Potkany et al. (2020) presented the results of the research that indicate dependence between the business size, capital structure, and use of quality management instruments. The enterprises applying at least one of the quality management instruments achieved higher performance measured by indicator ROE (above 7.5%) [35]. Using a new approach as outsourcing or facility management is one direction for improvement in mining companies. The search for the potential for cost savings, gaining more time for the core business in mining companies, but also increasing the quality of outsourced activities is offered through coordinated management of facility management support processes. Facility management is a vital part of successfully operating companies because joins people, processes, the building, and technology and brings benefits for companies [36]. All the instruments of various management areas (quality, energy, environmental, financial, cost, human resource, risks, information, and operation management) brought improvements for the mining processes and for mining companies. The significant change was the layout of the mining area, using alternative energy—solar energy, using mobile machines, and others.

5. Conclusions

The universal tool for measuring performance in mining companies is the subject of research in many countries around the world. Achieving a competitive advantage and gaining a foothold in the mining companies' market provides a basic impetus for evaluating the performance of mining processes in the direction of sustainable development with orientation on green mining. Achieving performance in mining companies means focusing on innovation, developing knowledge, improving mining processes, using alternative resources, and protecting the environment. The right strategic move for mining companies is to focus on innovative strategies. The mining company introduced innovation in the form of the layout of the mining area and thus improved the mining processes and achieved positive results in the financial area, thus ensuring the financial stability of the mining company and satisfies customers in several of the industry. This approach brought improvements in the environment and the use of alternative sources for the achievement of energy. The research empirically confirms that the new innovation of layout of mining place brings improvement of mining processes and indicators point to effective (over limit 0.70) and functional (over limit 0.90) mining processes in the year 2020. The Pareto analysis showed the best processes (mining, expedition, transport, sorting) for financial benefits, the volume of production, demand, the satisfaction of customers, and the cover needs of industries, but at the same they are processes with high costs. The results of the research can be useful for other mining companies in performance management and achievement mining market position. The introduction of new modern tools for measuring performance is a prerequisite for building new performance management models in direction of sustainable development.

Author Contributions: Conceptualization, K.T.; methodology, software, validation, formal analysis, investigation, resources, data curation, writing—original draft preparation, writing—review and editing, visualization, supervision, project administration, funding acquisition, K.T., S.K., M.B. and J.R. All authors have read and agreed to the published version of the manuscript.

Funding: This research was funded by VEGA 1/0317/19, "Research and development of new smart solutions based on the principles of Industry 4.0, logistic, 3D modelling and simulation for streamlining production in the mining and building industry" and VEGA 1/0797/20, "Quantification of Environmental burden impacts of the Slovak regions on health, social and economic system of the Slovak republic.

Conflicts of Interest: The authors declare no conflict of interest.

References

1. Jiskani, I.M.; Shah, S.A.A.; Qingxiang, C.; Zhou, W.; Lu, X. A multi-criteria based SWOT analysis of sustainable planning for mining and mineral industry in Pakistan. *Arab. J. Geosci.* **2020**, *13*, 1–16. [CrossRef]
2. Jiskani, I.M.; Cai, Q.; Zhou, W.; Shah, S.A.A. Green and climate-smart mining: A framework to analyze open-pit mines for cleaner mineral production. *Resour. Policy* **2021**, *71*, 102007. [CrossRef]
3. Cehlár, M.; Čulková, K.; Pavolová, H.; Khouri, S. Sustainability of Business with Earth Sources in V4. In *Proceedings of the 4th International Innovative Mining Symposium;* E3S Web of Conferences; Edition Diffusion Presse Sciences: Bristol, UK, 2019; Volume 105, pp. 1–7. [CrossRef]
4. Jiskani, I.M.; Cai, Q.; Zhou, W.; Chang, Z.; Chalgri, S.R.; Manda, E.; Lu, X. Distinctive model of mine safety for sustainable mining in Pakistan. *Min. Metall. Explor.* **2020**, *37*, 1023–1037. [CrossRef]
5. Krupnik, L.; Yelemessov, K.; Beisenov, B.; Baskanbayeva, D. Substantiation and process design to manufacture polymer-concrete transfer cases for mining machines. *Min. Miner. Deposit.* **2020**, *14*, 103–109. [CrossRef]
6. Sütőová, A.; Zgodavová, K.; Lajczyková, M. Quality and effectiveness evaluation of the geological services using CEDAC method. *Acta Montanistica. Slovaca* **2018**, *23*, 18–25.
7. Zhou, Y.; Zhou, W.; Lu, X.; Jiskani, I.M.; Cai, Q.; Liu, P.; Li, L. Evaluation index system of green surface mining in China. *Min. Metall. Explor.* **2020**, *37*, 1093–1103. [CrossRef]
8. Chen, J.; Jiskani, I.M.; Jinliang, C.; Yan, H. Evaluation and future framework of green mine construction in China based on the DPSIR model. *Sustain. Environ. Res.* **2020**, *30*, 1–10. [CrossRef]
9. Mirmozaffari, M.; Shadkam, E.; Khalili, S.M.; Kabirifar, K.; Yazdani, R.; Gashteroodkhani, T.A. A novel artificial intelligent approach: Comparison of machine learning tools and algorithms based on optimization DEA Malmquist productivity index for eco-efficiency evaluation. *Int. J. Energy Sect. Manag.* **2020**, *2*, 1–15.
10. Menshikova, E.; Fetisov, V.; Karavaeva, T.; Blinov, S.; Belkon, P.; Vaganov, S. Reducing the negative technogenic impact of the mining enterprise on the environment through management of the water balance. *Minerals* **2020**, *10*, 1145. [CrossRef]
11. Visser, W.F. A blueprint for Performance—Driven Operations Management. *Min. Metall. Explor.* **2020**, *37*, 823–831. [CrossRef]
12. Jiskani, I.M.; Cai, Q.; Zhou, W.; Lu, X. Assessment of risks impeding sustainable mining in Pakistan using fuzzy synthetic evaluation. *Resour. Policy* **2020**, *69*, 101820. [CrossRef]
13. Tworek, P.; Tchorzewski, S.; Valouch, P. Risk Management in coal mines methodical proposal for Polish and Czech hard coal mining industry. *Acta Montan. Slovaca* **2018**, *23*, 72–80.
14. Hancock, M.G. Evaluation of risk management systems for non-occupational disease and illness in the workforce of Papua New Guinean mining operations. *Australas. Inst. Min. Metall. Publ. Ser.* **2007**, *2*, 111–118.
15. Hall, B.E.; Vries, J.C. Quantifying the economic risk of suboptimal mine plans and strategies. *Min. Risk Manag. Conf.* **2003**, *15*, 59–69.
16. Domaracká, L.; Muchová, M.; Gonos, J. Evaluation of innovative aspects in mining company. In Proceedings of the 13th International Geoconference on Science and Technologiesin Geology—SGEM, Albena, Bulgaria, 16–22 June 2013; pp. 463–468.
17. Puzder, M.; Pavlik, T.; Molokač, M.; Hlavňová, B.; Vaverčák, N.; Samaneh, I.B.A. Cost ratio model proposal and consequential evaluation of model solutions of manufacturing process in mining company. *Acta Montan. Slovaca* **2017**, *22*, 270–277.
18. Tyuleneva, T. The prospects of accounting at mining enterprises as a factor of ensuring their sustainable development. In Proceedings of the 2nd International Innovative Mining Symposium, E3S Web of Conferences, Russia, Kemerovo, Russia, 20–22 November 2017; Volume 21, pp. 4009–4015. [CrossRef]
19. Demirel, N.; Erkayaoglu, M. Sustainability comparison of mining industries by life cycle assessment for Turkey and European Union. In Proceedings of the SWEMP 2016, 16th International Symposium on Environmental Issues and Waste Management in Energy and Mineral Production, Istanbul, Turkey, 5–7 October 2016.
20. Ambrisko, L.; Marasová, D.; Grendel, P.; Lukáč, S. Application of logistics principles when designing the process of transportation of raw materials. *Acta Montan. Slovaca* **2015**, *20*, 141–147.

21. Kudelko, J. Economic evaluation of backward vertical integration in mining industry. In Proceedings of the 12th International Multidisciplinary Scientific Geoconference, SGEM, Albena, Bulgaria, 17–23 June 2012; Volume 1, pp. 547–554.
22. Bye, A. The application of multi-parametric block models to the mining process. *J. South. Afr. Inst. Min. Metall.* **2007**, *107*, 51–58.
23. Hoare, R. The role of simulation modelling in project evaluation. *Australas. Inst. Min. Metall. Publ. Ser.* **2007**, *4*, 121–126.
24. Aghababaei, S.; Jalalifar, H.; Hosseini, A. Applyiing a technical-economic approach to calculate a suitable panel width for longwall mining method. *J. Min. Environ.* **2021**, *12*, 113–126.
25. Yang, S.; Liu, S.M.; Zhang, N.; Li, G.C.; Zhang, J. A fully automatic image based approach to quantifying the geological strength index of underground rock mass. *Int. J. Rock Mech. Min. Sci.* **2021**, *140*, 104585. [CrossRef]
26. Pavolová, H.; Šimková, Z.; Seňová, A.; Wittenberger, G. Macroeconomic indicators of raw material policy in Slovakia. In Proceedings of the First Interregional Conference Sustainable Development of Eurasian Mining Regions, Kemerovo, Russia, 25–27 November 2019; Edition Diffusion Presser Sciences: London, UK, 2019; pp. 1–12.
27. Dvorský, J.; Gavurová, B.; Čepel, M.; Červinka, M. Impact of selected economic factors on the business environment: The case of selected East European Countries. *Pol. J. Manag. Stud.* **2020**, *22*, 96–110. [CrossRef]
28. Gomes, J.G.C.; Okano, M.T.; Otola, I. Creation of indicators for classification of business models and business strategies in production systems. *Pol. J. Manag. Stud.* **2020**, *22*, 142–157. [CrossRef]
29. Rajnoha, R.; Lesníková, P.; Krajčík, V. Influence of business performance measurement systems and corporate sustainability concept to overall business performance: Save the planet and keep your performance. *Econ. Manag.* **2017**, *1*, 111–128. [CrossRef]
30. Teplicka, K.; Straka, M. Sustainability of extraction of raw material by a combination of mobile and stationary mining machines and optimization of machine life cycle. *Sustainability* **2020**, *12*, 24. [CrossRef]
31. Ruzekova, V.; Kittová, Z.; Steinhauser, D. Export Performance as a Measurement of Competitiveness. *J. Compet.* **2020**, *12*, 145–160. [CrossRef]
32. Stojanovic, A.; Milosevic, I.; Arsic, S.; Urosevic, S.; Mihaljovic, I. Corporate Social Responsibility as a Determinant of Employee Loyalty and Business Performance. *J. Compet.* **2020**, *12*, 149–166. [CrossRef]
33. Alglawe, A.; Schiffauerova, A.; Kuzgunkaya, O. Analysing the cost of quality within a supply chain using system dynamics approach. *Total. Qual. Manag. Bus. Excell.* **2019**, *30*, 1630–1653. [CrossRef]
34. Pattanayak, A.K.; Prakash, A.; Mohanty, R.P. Risk analysis of estimates for cost of quality in supply chain: A case study. *Prod. Plan. Control.* **2019**, *30*, 299–314. [CrossRef]
35. Potkany, M.; Gejdos, P.; Lesnikova, P.; Schmidtova, J. Influence of quality management practices on the Business performance of Slovak manufacturiong eterprises. *Acta Polytech. Hung.* **2020**, *17*, 161–180. [CrossRef]
36. Kamodyova, P.; Potkany, M.; Kajanova, J. Facility management—trend for management of supporting business processes and increasing of competitiveness. *AD Alta J. Interdiscip. Res.* **2020**, *10*, 122–127.

 processes

Article

Introducing Oxygen Vacancies in $Li_4Ti_5O_{12}$ via Hydrogen Reduction for High-Power Lithium-Ion Batteries

Yiguang Zhou, Shuhao Xiao, Zhenzhe Li, Xinyan Li, Jintao Liu, Rui Wu * and Junsong Chen *

School of Materials and Energy, University of Electronic Science and Technology of China, Chengdu 610054, China; 202021030225@std.uestc.edu.cn (Y.Z.); 18653460879@163.com (S.X.); uestclizhenzhe@163.com (Z.L.); 18896552001@163.com (X.L.); liujt1989@163.com (J.L.)

* Correspondence: ruiwu0904@uestc.edu.cn (R.W.); jschen@uestc.edu.cn (J.C.)

Abstract: $Li_4Ti_5O_{12}$ (LTO), known as a zero-strain material, is widely studied as the anode material for lithium-ion batteries owing to its high safety and long cycling stability. However, its low electronic conductivity and Li diffusion coefficient significantly deteriorate its high-rate performance. In this work, we proposed a facile approach to introduce oxygen vacancies into the commercialized LTO via thermal treatment under Ar/H_2 (5%). The oxygen vacancy-containing LTO demonstrates much better performance than the sample before H_2 treatment, especially at high current rates. Density functional theory calculation results suggest that increasing oxygen vacancy concentration could enhance the electronic conductivity and lower the diffusion barrier of Li^+, giving rise to a fast electrochemical kinetic process and thus improved high-rate performance.

Keywords: $Li_4Ti_5O_{12}$; oxygen vacancy; high-rate performance; electronic conductivity; density functional theory

Citation: Zhou, Y.; Xiao, S.; Li, Z.; Li, X.; Liu, J.; Wu, R.; Chen, J. Introducing Oxygen Vacancies in $Li_4Ti_5O_{12}$ via Hydrogen Reduction for High-Power Lithium-Ion Batteries. *Processes* **2021**, *9*, 1655. https://doi.org/10.3390/pr9091655

Academic Editor: Sergey Zhironkin

Received: 21 July 2021
Accepted: 8 September 2021
Published: 14 September 2021

1. Introduction

Lithium-ion batteries (LIBs), as the dominant energy storage device, have been widely applied in portable electronic devices and electric vehicles [1–4]. Even though graphite could be considered as the most successful anode material for LIBs [5–7], it still suffers from large volume expansion, poor rate capability arising from its low Li^+ diffusion coefficient, and also dendrite formation which would cause severe safety problems [8,9]. As a result, graphite may not be suitable for applications where safety and low-frequency maintenance are the primary concerns, such as batteries for buses or large-scale power plants.

Recently, spinel $Li_4Ti_5O_{12}$ (LTO) has attracted a lot of attention as a deintercalation/intercalation anode material due to its high safety and long cycling stability, which is associated with its negligible volume variation (also known as "zero strain") during the Li^+ insertion/extraction process through the three-dimensional diffusion channels [10–12]. Meanwhile, LTO possesses a high operation voltage (1.55 V vs. Li/Li^+) that can, to some extent, avoid the formation of the solid electrolyte interphase (SEI) and Li dendrites. However, the intrinsically low electronic conductivity (10^{-13} S cm^{-1}) and limited lithium diffusion coefficient (10^{-9}–10^{-13} cm^2 s^{-1}) [13–15] of LTO, originating from the absence of electrons in the Ti 3d orbitals, leads to its large band gap (2 eV), thus preventing its more intensive applications.

To address these drawbacks, metal atom doping could be quite effective, such as Cr, Na, Gd and W, which would have a positive impact on the structure and stability of LTO during lithium intercalation and de-intercalation [16–19]. On the other hand, several works have indicated that the electrochemical properties of LTO could also be improved by the introduction of oxygen vacancies (OVs), which can narrow the band gap by creating defeats, thus enhancing the electrical conductivity of different materials [20–24] such as TiO_2, Co_3O_4 and LTO. The OVs are usually generated by treating the materials under reducing atmosphere (i.e., H_2), argon or vacuum [25], reacting with metal or hydride, hydrothermal reactions and plasma treatments [24,26]. Even though it has been demonstrated that plasma treatments under reducing atmospheres could be efficient in generating OVs in LTOs which led to the reduction of Ti^{4+} to Ti^{3+} while at the same time enhancing lithium storage performance [20,21,24], this method might not be commercially feasible at a large scale due to its high cost and complicated operation process.

Herein, we facilely treated commercialized LTO under a reduced atmosphere of Ar/H_2 (5%) for the successful generation of OVs in LTO. As demonstrated by density functional theory calculations, the increasing concentration of OVs could lead to a tuned electronic structure and a low interaction of Li^+ and LTO surface. As a result, the H_2-treated LTO demonstrated much better high-rate performance and long-term cycling stability than the untreated pristine LTO. Both the theoretical and experimental analysis confirmed that the current H_2 treatment was highly efficient and cost-effective in introducing OVs into LTO, leading to greatly enhanced lithium storage properties, thus demonstrating great potential for large-scale high-power applications.

2. Materials and Methods

2.1. Modification of LTO

0.5 g commercial LTO (Tianjiao Technology Development Co., Kuiyong Town, China) was put into a porcelain boat with a size of 60 cm in length and 30 cm in width and treated in a tube furnace (OTF-1200X) by annealing under Ar/H_2 (5%) atmosphere and the size of tube was 8 cm in diameter and 1 m in length. In particular, the rate of heat rate was 2 °C min^{-1} with a 50 mL min^{-1} of gas flow rate and the temperature of 450 °C for 1 h.

2.2. Material Characterization

The crystallographic phases of all samples were investigated by X-ray diffraction (XRD Bruker, D8 Advancer; Using Cu Kα radiation in the range of 2θ = 10–80° with 50 kV 30 mA, λ = 1.54 Å). The morphologies of HLTO and LTO were characterized via transmission electron microscope (TEM; JEM2010F; FEI Talos-s, 200 kV accelerator voltage), selected area electron diffraction (SAED) and field emission scanning electron microscope (FSEM; FEI Inspect F50, Thermo Fisher Scientific, Waltham, MA, USA). X-ray photoelectron spectroscopy (XPS, Thermo Fisher Scientific Escalab 250Xi; Al Kα hv = 1486.6 eV; working voltage 12 kV and filament current 6 mA, Thermo Fisher Scientific, Waltham, MA, USA) measurements were carried out to determine the chemical state of samples. Raman spectroscopy tests were performed on a Thermo Fisher Scientific DXR Raman spectrometer with an excitation wavelength of 532 nm. The specific surface area of LTO and HTLO was acquired by N_2 adsorption–desorption Brunauer–Emmett–Teller (BET) measurement using a Kubo X1000. The OVs were tested without pretreatment through Electron Paramagnetic Resonance (EPR) (power: 20 db, modulation amplitude: 3, center field: 3510 G, range: 100 gauss). Volume resistance of LTO and HLTO was obtained using DC resistance measurements (Malvern Mastersizer 2000, ACL Staticide, Chicago, IL, USA) at the pressure of 3 MPa.

2.3. Electrochemical Measurement

The electrochemical performance of materials were tested using CR2032 coin-type cells (Duoduo Technology Co., Guangdong, China) assembled in an Ar-filled glove box (SG1200/750TS). The electrodes were prepared by mixing active material, Super P carbon (Aiweixin Chemical Technology Co., Shenzhen, China) black and polyvinylidene difluoride (PVDF) in a weight ratio of 8:1:1 and then hand milling with N-methyl pyrrolidone (NMP) (Tianchenghe Technology Co., Beijing, China) to obtain a homogeneous slurry. Subsequently, the slurry was coated on copper foil and dried at 80 °C for 12 h under vacuum. The loading mass of active materials on the current collector is about 1~1.5 mg cm^{-2}. SEM/EDX results of the LTO and HLTO electrodes (Supplementary Materials Figure S1) confirmed that both electrodes were shown with similar morphology and porosity. A pure lithium metal disc was used as the counter electrode and Celgard 2500 (Tianchenghe

Technology Co., Beijing, China) was used as the separator. The electrolyte was obtained by dissolving 1 M LiPF6 in a mixture of ethylene carbonate (EC)/diethyl carbonate (DEC) (Duoduo Technology Co., Guangdong, China) with volume ratio of 1:1.

Galvanostatic charging–discharging profiles were tested on Neware battery tester with a voltage range of 1.0–2.5 V (vs. Li/Li^{+}) at different current densities. Both cyclic voltammetry (CV) and electrochemical impedance spectroscopy (EIS) measurements were carried out on a Bio-logic SP-150 electrochemical workstation. CV was conducted with a voltage rage of 1–2.5 V (vs. Li/Li^{+}) at different scan rates. For EIS, both LTO and HLTO were prepared as anodes following the above-mentioned protocol, and the tests were carried out after charging at the open-circuit potential of approximately 2.7 V (voltage protection is a range from −5 V to 5 V) with a superimposed 5 mV sinusoidal (root-mean-square) perturbation over the frequency range from 0.1 to 10^5 Hz. At least two cells were tested for each condition, which showed very similar performance. Besides, all the tests were performed using freshly assembled cells to rule out the aging effect. All cells were assembled with a configuration of an Li (counter electrode)/Celgard polymer separator/liquid electrolyte/LTO (or HLTO) anode.

2.4. Density Functional Theory (DFT) Calculations

First-principle calculations were performed via the Density Functional Theory (DFT) method coupled with the Vienna Ab-Initio Simulation Package (VASP, University of Vienna, Austria) [27]. The generalized gradient approximation (GGA) in the formulation of Perdew–Burke–Ernzerhof (PBE) was used to treat the exchange and correlation energy [28]. The cutoff energy of 450 eV was adopted for the wave basis sets. A k-points sampling with 0.04 and 0.08 $Å^{-1}$ separation was used in the Brillouin zone for geometry optimization and density of states, respectively. The force and energy were converged to 0.02 eV $Å^{-1}$ and 2.0×10^{-5} eV, respectively. The cutoff energy of 450 eV was set for the plane wave basis. The (111) plane of LTO was adopted to construct OVs. Moreover, a 2×2 supercell was built in this work. The vacuum layer thickness of 20 Å was applied to avoid virtual interaction. The energy barrier was calculated using the Nudged Elastic Band (NEB) method, employing eight images between two end states. The constrained optimization of the transition state was used when the NEB method was inapplicable due to a high computational expense.

3. Results and Discussion

It is clear from Figure 1 that both the pristine and H_2-treated samples contain phase-pure $Li_4Ti_5O_{12}$ (JCPDS 49-0207) [29]. The bump between 20° and 30° could probably be attributed to the amorphous carbon present in the samples, which was confirmed by Raman spectroscopy (Supplementary Materials Figure S2). Typical Raman vibration bands were observed at 1348 cm^{-1} and 1588 cm^{-1}, which correspond to the D and G band of carbon. A TGA test (Supplementary Materials Figure S3) was investigated to verify such a claim. Apparently, the initial weight loss of 1.14% below 200 °C could be the evaporation of absorbed moisture content, and the subsequent loss of 2.08% between 400 and 600 °C could be due to the combustion of amorphous carbon. Furthermore, we refined XRD patterns of both samples and calculated their grain size with the Debye–Scherer formula [24,30]:

$$\mathrm{D} = \frac{K\lambda}{Bcos0} \tag{1}$$

where the value of K is a constant; λ is the wavelength of X-ray; θ is the diffraction angle; B is the full-width-at-half-maximum. As a result, by calculating the D values based on the diffraction peak at 18.5° (111), the average grain size of HLTO (21.78 nm) is bigger than LTO (16.86 nm), which is most likely caused by the annealing process promoting the growth of crystallites. Moreover, the higher degree of crystallinity of HLTO (60.37%) than LTO (53.44%), which were obtained by the refinement results, also suggested the better crystallinity in the former.

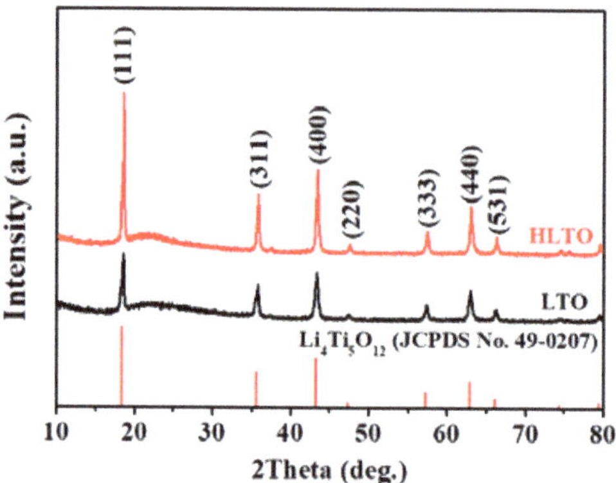

Figure 1. XRD patterns of HLTO and LTO.

On the other hand, even though both samples contained irregular submicron meter particles (Figure 2a,b), the heating process introduced subtle changes in the morphology of LTO, as HLTO seemed to have slightly larger aggregations with fewer small particles. Such a morphology was further confirmed under TEM (Figure 2c,d), and the high-resolution TEM (HRTEM) images (Figure 2e,f) showed well-defined lattice fringes with an interplanar distance of 0.48 nm, corresponding to the (111) plane of LTO in both samples. The selected-area electron diffraction (SAED) patterns display highly ordered arrangement of diffraction spots, verifying the single crystallinity of both samples. N_2 adsorption/desorption measurement was conducted to analyze the surface structure of HLTO and LTO (Supplementary Materials Figure S4). It is clear that both samples presented a typical type III isotherm [2,8] with no apparent hysteresis loop, showing a BET surface area of 10.8 $m^2 g^{-1}$ and 13.1 $m^2 g^{-1}$ for HLTO and LTO, respectively, and a pore size distribution with a well-defined peak at about 4 nm.

To confirm the presence of the OVs, X-ray photoelectron spectroscopy (XPS) was performed to investigate the surface chemical states of both samples (Figures 3a,b and S5). The O 1s spectra could be deconvoluted into three peaks which were located at 533.38 eV, 530.48 eV and 531.68 eV, corresponding to the hydroxyl species of surface-adsorbed water molecules, the Ti−O bonds and the OVs, respectively [8,11,20,24]. It can thus be quantified from the peak area that the content of OVs in HLTO is about 7.69%, which is about two times that of LTO (3.73%), confirming the higher concentration of OVs in the former. Ti 2p spectra for both samples (Figure 3b) showed two peaks at 459.09 eV and 464.78 eV, belonging to Ti^{4+}. The peaks at 458.08 eV and 460.88 eV correspond to Ti^{3+}, and HLTO possessed a higher Ti^{3+} level of 23.78%, while that of LTO is only 15.38%, also verifying the presence of more OVs in HLTO, suggesting that the H_2 treatment could not only introduce OVs in the material, but also effectively adjust the valence state of the Ti atoms in LTO for the overall charge balance [20,24]. The relative concentrations of OVs were further analyzed by Electron Paramagnetic Resonance (EPR) (Figure 3c). Judging by the *g*-values, there are two high g signals at 2.004 in HLTO and LTO, which are due to the unpaired electrons trapped by OVs, thus confirming the existence of OVs [23,29,31–33]. Meanwhile, the higher signal intensity in HLTO than LTO indicates the higher OVs concentration in the former [33–36], consistent with the above XPS analysis. Raman shift was studied to analyze the functional groups of materials and explore the influence of the OVs on Ti-O bonds (Figure 3d). Typical Raman vibration bands of LTO were observed at 227 cm^{-1}, 417 cm^{-1} and 668 cm^{-1}, which represents the bending vibration of the O–Ti–O, the stretching–bending vibrations of the Li–O bonds in LiO_4 polyhedral and the vibrations of Ti–O bonds in TiO_6 octahedra [37–39], respectively. After bringing in the OVs, the Ti-O peaks were blue-shifted, which may be caused by the asymmetric vibrations due to the replacement of Ti^{4+} by Ti^{3+} [37,38].

Figure 2. SEM images of (**a**) HLTO and (**b**) LTO; TEM images of (**c**) HLTO and (**d**) LTO; HRTEM images of (**e**) HLTO and (**f**) LTO. The insets in (**e**,**f**) show the SEAD patterns of the respective sample.

As illustrated in Figure 4a, both samples exhibited similar discharge capacity of 166 mAh g^{-1} and 161 mAh g^{-1} for LTO and HLTO at 1 C, respectively, and could be probably attributed to the slightly higher surface area of LTO than HLTO. Generally, such a difference is quite negligible at low current rates; however, HLTO demonstrated significantly higher capacities than LTO as the charge/discharge rate reached 5 C and beyond, further confirming the more efficient charge transfer process in the former [40,41]. Similar trend was also reflected in the cycling stability test at 1 C (Supplementary Materials Figure S6) and 5 C (Figure 4b), where a capacity advantage could be maintained in HLTO at 5 C for 300 cycles, while no noticeable difference could be observed at 1 C. The reason for this phenomenon could be that, at a low current rate, the Li^+ insertion/deinsertion and the dual-phase transformation are slow processes, the Li ions in both samples could have enough time to diffuse to the respective vacancy sites, thus producing comparable storage capacities. While testing at higher rates, the interaction between the Li ions and the active materials would be greatly limited, and only the sample with a higher Li diffusion coefficient could allow the efficient intercalation/deintercalation of Li ions within such a short reaction interval [20,24,26]. In the case of HLTO, the OVs could cause an unbalanced charge distribution in the local vicinity, which generated a built-in electric field [36,42], providing an extra driving force to the diffusion of Li^+, giving rise to a higher specific capacity at high charge/discharge rate. A long-term stability test at 20 C (Figure 4c) presented a gradual decrease in specific capacity for 1000 cycles with almost 99.3% Coulombic efficiency, which is much higher compared to that of LTO during the course of the test. For anodes, CE is calculated by charge capacity divided by discharge capacity, corresponding to the insertion and disinsertion of Li^+ into and from the LTO crystal framework. The initial CEs of LTO and HLTO were 99.8% and 99.6%, respectively and then stabilized at 99.1% and 99.3% after 1000 cycles at 20 C, suggesting that the insertion and deinsertion processes could take place to a similar extent. Even though both samples have similarly high CEs, HLTO delivered much higher capacities than LTO at high rates, suggesting that the diffusion of Li^+ was much more efficient in the former as more Li^+ could be inserted and deinserted during the charge/discharge process. In order for the complete storage of Li^+, an Li metal anode was used to provide an excessive amount of Li^+ in order for both samples to uptake as much Li^+ as they can store, avoiding any possible difference in the CE values originating from the intrinsic interaction between Li^+ and LTO/HLTO, but not from the depletion of Li^+ at the electrolyte/electrode interface caused by insufficient Li^+. Based on the above analysis, the performance and kinetic properties of HLTO were significantly improved with the introduction of OVs compared with the pristine LTO.

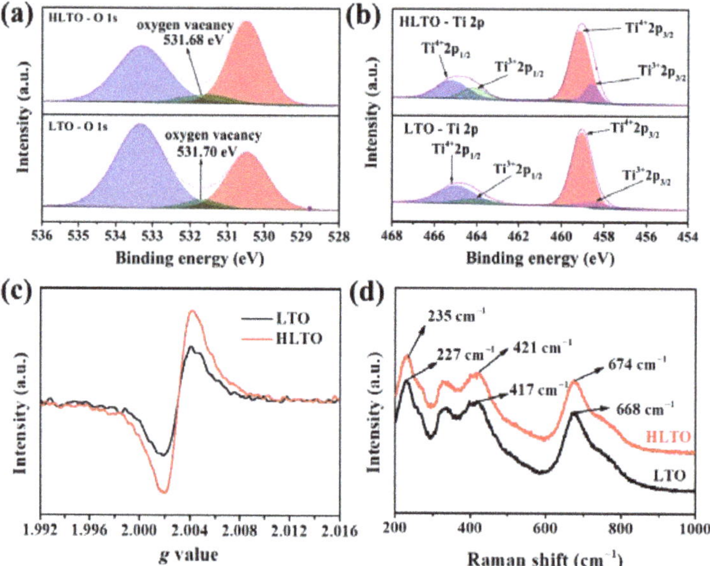

Figure 3. Characterization results of HLTO and LTO: High-resolution XPS spectra of (**a**) O 1s and (**b**) Ti 2p; (**c**) electron paramagnetic resonance (EPR) spectra; (**d**) Raman scattering spectra.

Subsequently, the galvanostatic charge/discharge curves of HLTO and LTO at various rates from 0.5 C to 30 C were investigated to inquire the capacity contributions in both samples (Figure 5a,b). It is apparent that HLTO demonstrated higher specific capacity than LTO at higher rates. Based on the analysis of the difference between the charge/discharge voltage plateaus (Supplementary Materials Figure S7), corresponding to the potential value of the distinct voltage platform from the galvanostatic charge/discharge curves, a much smaller polarization could be observed in HLTO at a high charge/discharge rate than LTO, further confirming the more efficient kinetic diffusion of Li ions. Figure 5c,d compares the discharge curves of HLTO and LTO at 1 C and 10 C, respectively. Each curve could be divided into three phases according to the potential range: the region from the open-cycle potential to ~1.55 V (noted as P1), the discharging platform at ~1.55 V (noted as P2) and a third potential region from ~1.55 V to 1 V (noted as P3) [43–45]. These three processes are related to three different reactions during discharge. P1 corresponds to the insertion of Li^+ into the LTO solid solution. The dual-phase transformation, where $Li_4Ti_5O_{12}$ transforms into $Li_7Ti_5O_{12}$ (as shown in Equation (2) below) [46,47], is related to P2. P3 corresponds to the storage of Li^+ at solid−liquid and solid−solid interfaces.

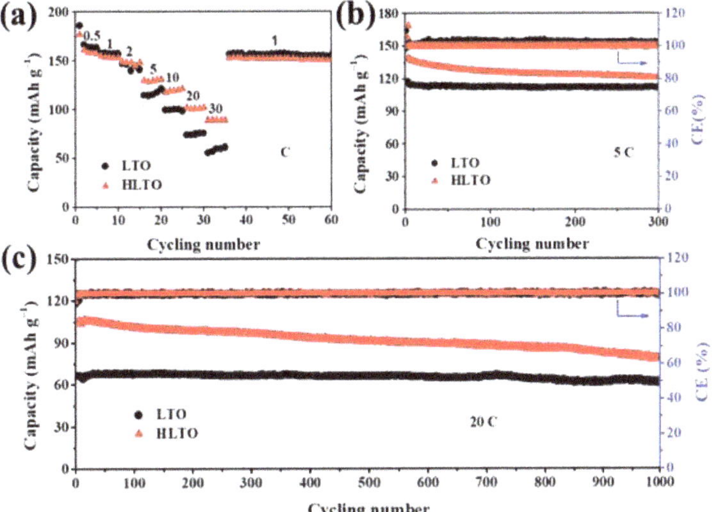

Figure 4. Electrochemical tests of LTO and HLTO: (**a**) Rate performance; (**b**) Cycling performance at 5 C; (**c**) Long-term cycling stability test at 20 C. All the capacities displayed in this figure are the discharge capacities.

$$Li_4Ti_5O_{12} + 3Li^+ + 3e^- \leftrightarrow Li_7Ti_5O_{12} \quad (2)$$

In all cases, the P1 and P3 phases contribute only a relatively small portion of the discharge capacity, and it is also very clear that P2 played the dominant role in delivering the main capacity during discharge (Figure 5e,f). Evidently, both samples demonstrated similar capacity contributions from the three phases at the low current rate of 1 C. When the charge/discharge rate increased to 10 C, the contributions from P1 and P3 remained quite comparable, and the major difference originated from P2; that is to say, the main reason for HLTO having a higher capacity than LTO is because the former had a more efficient dual-phase transformation process than the latter [48,49].

To further study the reaction kinetics of HLTO and LTO, cyclic voltammetry (CV) analysis at different scan rates from 0.2 mV s^{-1} to 5 mV s^{-1} was conducted. The first cycles of LTO and HLTO at 0.2 mV s^{-1} are displayed in Figure 6a,c, where both samples showed a pair of well-defined redox peaks at ~1.5 V/1.65 V (vs. Li/Li^+), corresponding to the Li^+ insertion/desertion of $Li_4Ti_5O_{12}$ [11–13]. It should be pointed out that HLTO demonstrated higher peak intensities compared to LTO, suggesting that the presence of OVs would enhance the electrochemical processes during charge and discharge [50,51]. When increasing the scan rate to 5 mV s^{-1}, the difference in the peak intensities becomes even more prominent, suggesting a faster kinetic process in HLTO than LTO [52]. Similar to previous measurements, HLTO displayed current peaks with higher intensities than LTO at high scan rates. The relationship between the peak current (i) and the scan rate (v) could be described by an equation of $i = av^b$ [53–55], which can be transformed into:

$$\log(i) = b \log(v) + \log(a) \quad (3)$$

where b represents the charge storage behavior, and its value is usually within a range of 0.5–1. If the b is close to 0.5, the electrochemical process is mastered by ionic diffusion. On the other hand, the process is controlled by faradaic reactions when the value of b is approaching 1 [56]. According to this theory, the b values of anodic and cathodic peaks for LTO are determined to be 0.46 and 0.36 (Figure 6b), respectively, indicating that the electrochemical process of LTO is basically controlled by ionic diffusion. In contrast, the same values of b for HLTO are 0.62 and 0.45, suggesting that there are also faradaic reactions, which could be caused by the introduced OVs in the material [57,58].

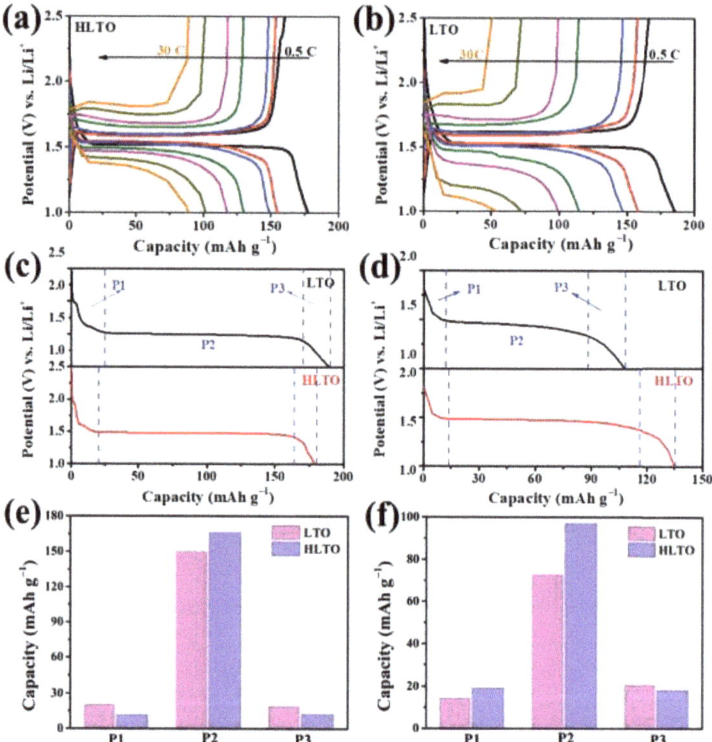

Figure 5. Charge/discharge curves of (**a**) HLTO and (**b**) LTO at various rates of 0.5 C, 1 C, 2 C, 5 C, 10 C, 20 C and 30 C; discharge curves at (**c**) 1 C and (**d**) 10 C; and the corresponding capacity contribution of different phases for HLTO and LTO at (**e**) 1 C and (**f**) 10 C.

Electrochemical impedance measurements (Figure 7) were also performed for both specimens to analyze the resistance properties. The specific frequencies at some data points have been specified (labelled in the figure). The impedance curves were fitted with the equivalent circuit model (inset of Figure 7a), where Rs represents ohmic resistance in the high frequency region and exhibits the internal resistance of electrode and electrolyte in LIBs [59], which could be obtained by the left intersection of the Nyquist plot with the Z′ axis; Rct refers to the charge transfer resistance at the electrolyte/LTO interface, presenting the resistance incurred on the Li ions when they inserted from the electrolyte into the LTO/HLTO crystal structure [11,60,61], which is illustrated by the semicircle in the middle frequency region; CPE corresponds to the double-layer capacitance, which could be probably attributed to the accumulation of charges on the surface of the electrode that were not inserted into the active material. Warburg impedance represents the resistance of the Li^+ ions transporting through the active material [50,62]. The diffusion coefficient of Li^+ (D_{Li+}) through the active material LTO or HLTO could be calculated using the following equation [63,64]:

$$D_{Li+} = \frac{R^2T^2}{2A^2n^4F^4C^2\sigma^2} \quad (4)$$

where R is the gas constant (value is 8.314 J K^{-1} mol^{-1}); T is the absolute temperature (298.15 K); A is the surface area of the electrode with a diameter of 12 mm (1.13×10^{-4} m^2); n is 3, which is the number of electrons transferred in the half-reaction for the redox reaction and can be acquired from Formula (2); F is the Faraday constant (96,500 C mol^{-1}); C is the concentration of lithium ions (8.3×10^{-3} mol cm^{-3}) [64]; σ is the Warburg factor which can be acquired from the slope of Z′ ~ $\omega^{1/2}$ curves, as shown in Figure 7b, and the related results for both samples are displayed in Table 1. As summarized in Table 1, HLTO exhibited a larger D_{Li+} than LTO, which could be attributed to the improved conductivity via the introduction of OVs. Other fitting results are also listed in Table 1, and apparently, HLTO had a smaller ohmic resistance and charge transfer resistance than LTO, which was also consistent with its better high-rate performance. Besides, DC resistance measurement also showed that HLTO had a lower resistance of 5720 Ω·m than LTO (16,900 Ω·m).

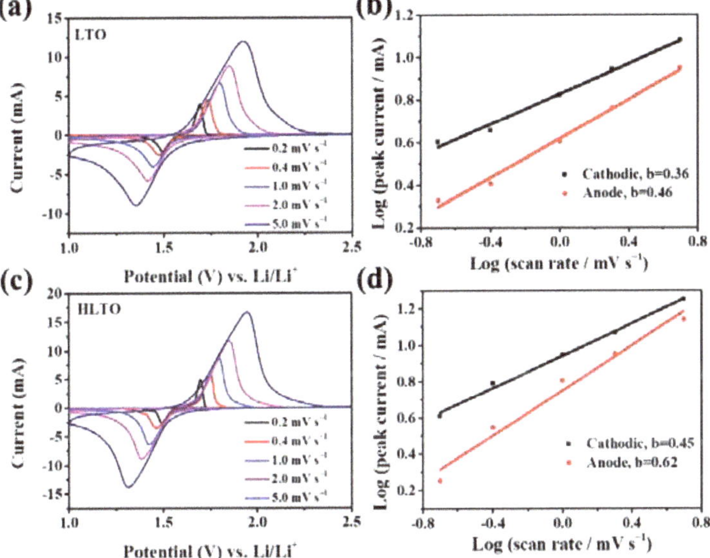

Figure 6. Kinetic analysis of the LTO and HLTO anodes: (**a**) CV curves at different scan rates for LTO and (**b**) corresponding log i versus log v plot; (**c**) CV curves at different scan rates for HLTO and (**d**) corresponding log i versus log v plot.

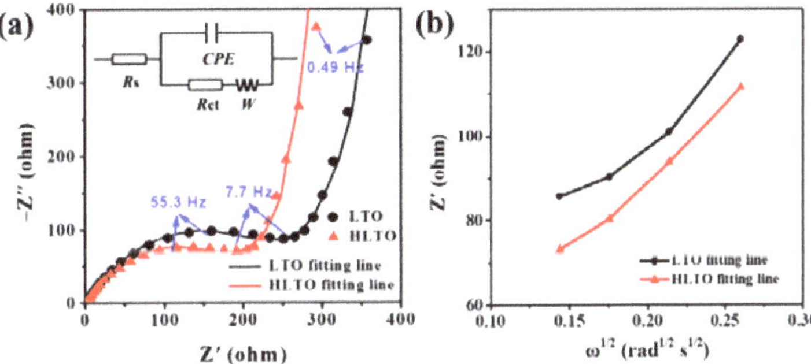

Figure 7. (**a**) EIS spectra of LTO and HLTO electrodes; (**b**) plot curves of Z′ versus $\omega^{1/2}$ of three LTO electrodes in the low-frequency region. The inset in (**a**) shows the equivalent circuit model.

Table 1. Summary of the EIS fitting results of LTO and HLTO.

Electrodes	R_s (Ω cm^{-2})	R_{ct} (Ω cm^{-2})	σ	D_{Li+} (cm^2 s^{-1})
LTO	1.48	314.9	373.73	3.29×10^{-17}
HLTO	0.87	254.3	297.58	5.18×10^{-17}

To discover the mechanism of OVs in improving the lithium storage capability, density functional theory (DFT) calculations were carried out to investigate the electronic conductivity and Li^+ diffusion behaviors in LTO lattice. Based on the XPS results, two different models of LTO with a concentration of OVs of 3.9% and 7.8% were constructed, named model I and II, respectively. Considering that the coordinate numbers of Ti atoms will affect the chemical activity of LTO [65], we thus selected the (111) surface of the two models with fivefold-coordinated and fourfold-coordinated Ti atoms, corresponding to LTO with low and high OV content, respectively. The Li^+ diffusion path on the two models are displayed in Figure 8a,b, and the corresponding diffusion barriers are shown in Figure 8c. The (111) surface of LTO with high OV content (model II) exhibits a diffusion energy barrier of 0.78 eV, which is much lower than that of the sample with low OV concentration (model I; 1.62 eV), indicating that the Li^+ diffusion is faster on the surface with high OV content. This phenomenon could be attributed to the reduced interaction between Li^+ and under-coordinated Ti atoms by the OVs, thus facilitating the migration of Li^+. The calculated density of stats (DOS) of two models are shown in Figure 8d. At high OV concentration, significant changes can be observed in the DOS of LTO, giving rise to a narrowed band gap of 1.6 eV in model II. Moreover, the corresponding Fermi level exhibited an upshift from the valence band to conduction band, suggesting an enhanced electronic conductivity. These calculation results indicated that increasing oxygen vacancy concentration could enhance the electronic conductivity and lower the diffusion energy barrier of Li^+, resulting in a fast electrochemical process and superior high-rate performance.

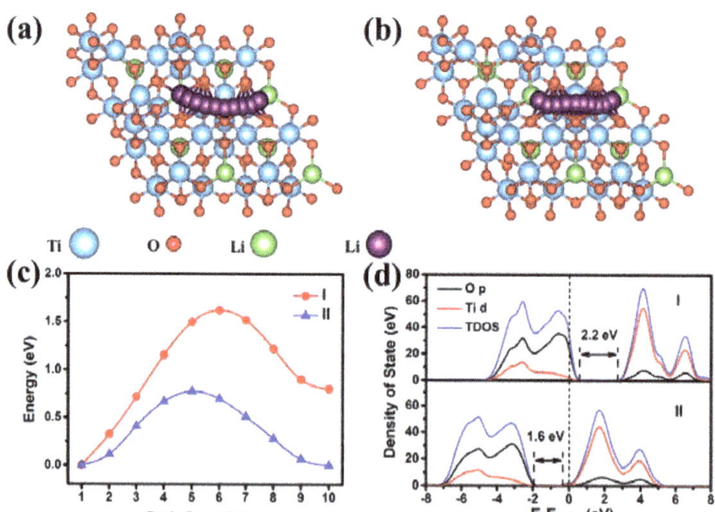

Figure 8. Li migration paths of (**a**) model I and (**b**) model II; (**c**) corresponding energy barriers and (**d**) calculated DOS of model I and model II. Model I and II correspond to the (111) surface of LTO with the OVs concentration of 3.9% and 7.8%, respectively.

4. Conclusions

In summary, by facilely treating commercial LTO under H_2 atmosphere, OVs could be efficiently introduced into the active material, giving rise to a greatly enhanced electrochemical properties for lithium storage. Specifically, as confirmed by both XPS and EPR measurements, the concentration of OVs in LTO was significantly increased after H_2 treatment, and based on density functional theory calculations, the presence of OV would enhance the electronic conductivity and lower the energy barrier of Li^+ diffusion. As expected, HLTO with more OVs demonstrated greatly improved high-rate performance than the pristine LTO, with a reversible capacity of 85.1 mAh g^{-1} at a very high charge/discharge rate of 30 C, while that of LTO was only 59.9 mAh g^{-1}. For prolonged cycling at 20 C, HLTO also exhibited higher capacities than LTO. These results confirmed that this facile method of introducing OVs into the active material would be highly effective in enhancing the high-rate performance of the electrode, shedding light on the design of high-power anode material for LIBs at an industrial scale.

Supplementary Materials: The following are available online at https://www.mdpi.com/article/10.3390/pr9091655/s1, Figure S1: SEM/EDX images of LTO (a), HLTO (b), Figure S2: Partial Raman spectra of HLTO and LTO, Figure S3: TGA results of commercial LTO, Figure S4: N_2 adsorption/desorption curves of HLTO and LTO. The inset shows the pore size distribution of both samples from adsorption branch, Figure S5: XPS spectra of HLTO and LTO, Figure S6: Cycling performance of LTO and HLTO at 1 C, Figure S7: The polarization results of both samples at different rates.

Author Contributions: Conceptualization, Y.Z.; Methodology, Y.Z. and X.L.; Software, S.X.; Formal Analysis, Y.Z. and Z.L.; Investigation, Y.Z., and Z.L.; Data Curation, Y.Z. and S.X.; Writing—Original Draft Preparation, Y.Z. and R.W.; Writing—Review and Editing, Y.Z. and J.C.; Supervision, J.L. and J.C. All authors have read and agreed to the published version of the manuscript.

Funding: This work was financially supported by Fundamental Research Funds for the Central Universities (ZYGX2019J030).

Institutional Review Board Statement: Not applicable.

Informed Consent Statement: Not applicable.

Data Availability Statement: Not applicable.

Conflicts of Interest: The authors declare no conflict of interest.

References

1. Gangaja, B.; Nair, S.; Santhanagopalan, D. Surface-Engineered $Li_4Ti_5O_{12}$ Nanostructures for High-Power Li-Ion Batteries. *Nano-Micro Lett.* **2020**, *12*, 30. [CrossRef] [PubMed]
2. He, Y.; Muhetaer, A.; Li, J.; Wang, F.; Liu, C.; Li, Q.; Xu, D. Ultrathin $Li_4Ti_5O_{12}$ Nanosheet Based Hierarchical Microspheres for High-Rate and Long-Cycle Life Li-Ion Batteries. *Adv. Energy Mater.* **2017**, *7*, 1700950. [CrossRef]
3. Lee, S.H.; Huang, C.; Grant, P.S. Multi-layered composite electrodes of high power $Li_4Ti_5O_{12}$ and high capacity SnO_2 for smart lithium ion storage. *Energy Storage Mater.* **2021**, *38*, 70–79. [CrossRef]
4. Li, Z.; Xiao, S.; Liu, J.; Niu, X.; Xiang, Y.; Li, T.; Chen, J.S. Highly Efficient Na^+ Storage in Uniform Thorn Ball-Like α-MnSe/C Nanospheres. *Acta Metall. Sin. Engl. Lett.* **2021**, *34*, 373–382. [CrossRef]
5. Sun, J.; Guo, L.; Gao, M.; Sun, X.; Zhang, J.; Liang, L.; Liu, Y.; Hou, L.; Yuan, C. Solid-state template-free fabrication of uniform Mo_2C microflowers with lithium storage towards Li-ion batteries. *Chin. Chem. Lett.* **2020**, *31*, 1670–1673. [CrossRef]
6. Wei, C.; Fei, H.; Tian, Y.; An, Y.; Tao, Y.; Li, Y.; Feng, J. Scalable construction of SiO/wrinkled MXene composite by a simple electrostatic self-assembly strategy as anode for high-energy lithium-ion batteries. *Chin. Chem. Lett.* **2020**, *31*, 980–983. [CrossRef]
7. Zheng, H.; Zhang, H.; Fan, Y.; Jin, G.; Zhao, H.; Fang, J.; Zhang, J.; Xu, J. A novel Mo-based oxide β-$SnMoO_4$ as anode for lithium ion battery. *Chin. Chem. Lett.* **2020**, *31*, 210–216. [CrossRef]
8. Liu, Y.; Liu, J.; Hou, M.; Fan, L.; Wang, Y.; Xia, Y. Carbon-coated $Li_4Ti_5O_{12}$ nanoparticles with high electrochemical performance as anode material in sodium-ion batteries. *J. Mater. Chem. A* **2017**, *5*, 10902–10908. [CrossRef]
9. Piffet, C.; Vertruyen, B.; Caes, S.; Thomassin, J.-M.; Broze, G.; Malherbe, C.; Boschini, F.; Cloots, R.; Mahmoud, A. Aqueous processing of flexible, free-standing $Li_4Ti_5O_{12}$ electrodes for Li-ion batteries. *Chem. Eng. J.* **2020**, *397*, 125508. [CrossRef]
10. Qi, S.; He, J.; Liu, J.; Wang, H.; Wu, M.; Li, F.; Wu, D.; Li, X.; Ma, J. Phosphonium Bromides Regulating Solid Electrolyte Interphase Components and Optimizing Solvation Sheath Structure for Suppressing Lithium Dendrite Growth. *Adv. Funct. Mater.* **2020**, *31*, 2009013. [CrossRef]
11. Wang, D.; Liu, H.; Shan, Z.; Xia, D.; Na, R.; Liu, H.; Wang, B.; Tian, J. Nitrogen, sulfur Co-doped porous graphene boosting $Li_4Ti_5O_{12}$ anode performance for high-rate and long-life lithium ion batteries. *Energy Storage Mater.* **2020**, *27*, 387–395. [CrossRef]
12. Wang, H.; Wang, L.; Lin, J.; Yang, J.; Wu, F.; Li, L.; Chen, R. Structural and electrochemical characteristics of hierarchical $Li_4Ti_5O_{12}$ as high-rate anode material for lithium-ion batteries. *Electrochim. Acta* **2021**, *368*, 137470. [CrossRef]

13. Wang, R.; Cao, X.; Zhao, D.; Zhu, L.; Xie, L.; Li, J.; Miao, Y. Enhancing Lithium Storage Performances of the $Li_4Ti_5O_{12}$ Anode by Introducing the CuV_2O_6 Phase. *ACS Appl. Mater. Interfaces* **2020**, *12*, 39170–39180. [CrossRef] [PubMed]
14. Wen, K.; Tan, X.; Chen, T.; Chen, S.; Zhang, S. Fast Li-ion transport and uniform Li-ion flux enabled by a double–layered polymer electrolyte for high performance Li metal battery. *Energy Storage Mater.* **2020**, *32*, 55–64. [CrossRef]
15. Zhang, L.; Zhang, X.; Tian, G.; Zhang, Q.; Knapp, M.; Ehrenberg, H.; Chen, G.; Shen, Z.; Yang, G.; Gu, L.; et al. Lithium lanthanum titanate perovskite as an anode for lithium ion batteries. *Nat. Commun.* **2020**, *11*, 3490. [CrossRef]
16. Gong, S.H.; Lee, J.H.; Chun, D.W.; Bae, J.-H.; Kim, S.-C.; Yu, S.; Nahm, S.; Kim, H.-S. Effects of Cr doping on structural and electrochemical properties of $Li_4Ti_5O_{12}$ nanostructure for sodium-ion battery anode. *J. Energy. Chem.* **2021**, *59*, 465–472. [CrossRef]
17. Zhang, Q.; Verde, M.G.; Seo, J.K.; Li, X.; Meng, Y.S. Structural and electrochemical properties of Gd-doped $Li_4Ti_5O_{12}$ as anode material with improved rate capability for lithium-ion batteries. *J. Power Sources* **2015**, *280*, 355–362. [CrossRef]
18. Zhang, Q.; Zhang, C.; Li, B.; Jiang, D.; Kang, S.; Li, X.; Wang, Y. Preparation and characterization of W-doped $Li_4Ti_5O_{12}$ anode material for enhancing the high rate performance. *Electrochim. Acta* **2013**, *107*, 139–146. [CrossRef]
19. Zhao, F.; Xue, P.; Ge, H.; Li, L.; Wang, B. Na-Doped $Li_4Ti_5O_{12}$ as an Anode Material for Sodium-Ion Battery with Superior Rate and Cycling Performance. *J. Electrochem. Soc.* **2016**, *163*, A690–A695. [CrossRef]
20. Liang, K.; He, H.; Ren, Y.; Luan, J.; Wang, H.; Ren, Y.; Huang, X. Ti^{3+} self-doped $Li_4Ti_5O_{12}$ with rich oxygen vacancies for advanced lithium-ion batteries. *Ionics* **2020**, *26*, 1739–1747. [CrossRef]
21. Lyu, P.; Zhu, J.; Han, C.; Qiang, L.; Zhang, L.; Mei, B.; He, J.; Liu, X.; Bian, Z.; Li, H. Self-Driven Reactive Oxygen Species Generation via Interfacial Oxygen Vacancies on Carbon-Coated TiO_{2-x} with Versatile Applications. *ACS Appl. Mater. Interfaces* **2021**, *13*, 2033–2043. [CrossRef]
22. Shin, J.-Y.; Joo, J.H.; Samuelis, D.; Maier, J. Oxygen-Deficient $TiO_{2-\delta}$ Nanoparticles via Hydrogen Reduction for High Rate Capability Lithium Batteries. *Chem. Mater.* **2012**, *24*, 543–551. [CrossRef]
23. Xiong, T.; Yu, Z.G.; Wu, H.; Du, Y.; Xie, Q.; Chen, J.; Zhang, Y.W.; Pennycook, S.J.; Lee, W.S.V.; Xue, J. Defect Engineering of Oxygen-Deficient Manganese Oxide to Achieve High-Performing Aqueous Zinc Ion Battery. *Adv. Energy Mater.* **2019**, *9*, 1803815. [CrossRef]
24. Zhu, J.; Chen, J.; Xu, H.; Sun, S.; Xu, Y.; Zhou, M.; Gao, X.; Sun, Z. Plasma-Introduced Oxygen Defects Confined in $Li_4Ti_5O_{12}$ Nanosheets for Boosting Lithium-Ion Diffusion. *ACS Appl. Mater. Interfaces* **2019**, *11*, 17384–17392. [CrossRef]
25. Dong, C.; Dong, W.; Lin, X.; Zhao, Y.; Li, R.; Huang, F. Recent progress and perspectives of defective oxide anode materials for advanced lithium ion battery. *EnergyChem* **2020**, *2*, 100045. [CrossRef]
26. Liu, Y.; Xiao, R.; Fang, Y.; Zhang, P. Three-Dimensional Oxygen-Deficient $Li_4Ti_5O_{12}$ Nanospheres as High-Performance Anode for Lithium Ion Batteries. *Electrochim. Acta* **2016**, *211*, 1041–1047. [CrossRef]
27. Blöchl, P.E. Projector augmented-wave method. *Phys. Rev. B* **1994**, *50*, 17953–17979. [CrossRef]
28. Perdew, J.P.; Burke, K.; Ernzerhof, M. Generalized Gradient Approximation Made Simple. *Phys. Rev. Lett.* **1996**, *77*, 3865–3868. [CrossRef] [PubMed]
29. Wang, H.; Zhang, J.; Hang, X.; Zhang, X.; Xie, J.; Pan, B.; Xie, Y. Half-metallicity in single-layered manganese dioxide nanosheets by defect engineering. *Angew. Chem. Int. Edit.* **2015**, *54*, 1195–1199. [CrossRef]
30. Wang, Q.; Chen, S.; Jiang, J.; Liu, J.; Deng, J.; Ping, X.; Wei, Z. Manipulating the surface composition of Pt-Ru bimetallic nanoparticles to control the methanol oxidation reaction pathway. *Chem. Commun.* **2020**, *56*, 2419–2422. [CrossRef]
31. Wang, A.; Cao, Z.; Wang, J.; Wang, S.; Li, C.; Li, N.; Xie, L.; Xiang, Y.; Li, T.; Niu, X.; et al. Vacancy defect modulation in hot-casted NiO film for efficient inverted planar perovskite solar cells. *J. Energy. Chem.* **2020**, *48*, 426–434. [CrossRef]
32. Ye, J.; Zhai, X.; Chen, L.; Guo, W.; Gu, T.; Shi, Y.; Hou, J.; Han, F.; Liu, Y.; Fan, C.; et al. Oxygen vacancies enriched nickel cobalt based nanoflower cathodes: Mechanism and application of the enhanced energy storage. *J. Energy. Chem.* **2021**, *62*, 252–261. [CrossRef]
33. Zhang, J.; Yin, R.; Shao, Q.; Zhu, T.; Huang, X. Oxygen Vacancies in Amorphous InOx Nanoribbons Enhance CO_2 Adsorption and Activation for CO_2 Electroreduction. *Angew. Chem. Int. Edit.* **2019**, *58*, 5609–5613. [CrossRef]
34. Li, J.; Shu, C.; Liu, C.; Chen, X.; Hu, A.; Long, J. Rationalizing the Effect of Oxygen Vacancy on Oxygen Electrocatalysis in Li-O_2 Battery. *Small* **2020**, *16*, 2001812. [CrossRef] [PubMed]
35. Ni, W.; Liu, Z.; Zhang, Y.; Ma, C.; Deng, H.; Zhang, S.; Wang, S. Electroreduction of Carbon Dioxide Driven by the Intrinsic Defects in the Carbon Plane of a Single Fe-N_4 Site. *Adv. Mater.* **2021**, *33*, 2003238. [CrossRef] [PubMed]
36. Sadighi, Z.; Huang, J.; Qin, L.; Yao, S.; Cui, J.; Kim, J.-K. Positive role of oxygen vacancy in electrochemical performance of $CoMn_2O_4$ cathodes for Li-O_2 batteries. *J. Power Sources* **2017**, *365*, 134–147. [CrossRef]
37. Guo, M.; Chen, H.; Wang, S.; Dai, S.; Ding, L.-X.; Wang, H. TiN-coated micron-sized tantalum-doped $Li_4Ti_5O_{12}$ with enhanced anodic performance for lithium-ion batteries. *J. Alloy. Compd.* **2016**, *687*, 746–753. [CrossRef]
38. Liao, J.-Y.; Chabot, V.; Gu, M.; Wang, C.; Xiao, X.; Chen, Z. Dual phase $Li_4Ti_5O_{12}$–TiO_2 nanowire arrays as integrated anodes for high-rate lithium-ion batteries. *Nano Energy* **2014**, *9*, 383–391. [CrossRef]
39. Tang, Y.; Huang, F.; Zhao, W.; Liu, Z.; Wan, D. Synthesis of graphene-supported $Li_4Ti_5O_{12}$ nanosheets for high rate battery application. *J. Mater. Chem.* **2012**, *22*, 11257. [CrossRef]
40. Luo, S.; Zhang, P.; Yuan, T.; Ruan, J.; Peng, C.; Pang, Y.; Sun, H.; Yang, J.; Zheng, S. Molecular self-assembly of a nanorod N-$Li_4Ti_5O_{12}/TiO_2/C$ anode for superior lithium ion storage. *J. Mater. Chem. A* **2018**, *6*, 15755–15761. [CrossRef]

41. Qin, T.; Zhang, X.; Wang, D.; Deng, T.; Wang, H.; Liu, X.; Shi, X.; Li, Z.; Chen, H.; Meng, X.; et al. Oxygen Vacancies Boost delta-Bi_2O_3 as a High-Performance Electrode for Rechargeable Aqueous Batteries. *ACS Appl. Mater. Interfaces* **2019**, *11*, 2103–2111. [CrossRef] [PubMed]
42. Hou, C.; Hou, Y.; Fan, Y.; Zhai, Y.; Wang, Y.; Sun, Z.; Fan, R.; Dang, F.; Wang, J. Oxygen vacancy derived local build-in electric field in mesoporous hollow Co_3O_4microspheres promotes high-performance Li-ion batteries. *J. Mater. Chem. A* **2018**, *6*, 6967–6976. [CrossRef]
43. Wang, S.; Yang, Y.; Quan, W.; Hong, Y.; Zhang, Z.; Tang, Z.; Li, J. Ti^{3+}-free three-phase $Li_4Ti_5O_{12}/TiO_2$ for high-rate lithium ion batteries: Capacity and conductivity enhancement by phase boundaries. *Nano Energy* **2017**, *32*, 294–301. [CrossRef]
44. Xu, G.; Tian, Y.; Wei, X.; Yang, L.; Chu, P.K. Free-standing electrodes composed of carbon-coated $Li_4Ti_5O_{12}$ nanosheets and reduced graphene oxide for advanced sodium ion batteries. *J. Power Sources* **2017**, *337*, 180–188. [CrossRef]
45. Yang, Z.; Huang, Q.; Li, S.; Mao, J. High-temperature effect on electrochemical performance of $Li_4Ti_5O_{12}$ based anode material for Li-ion batteries. *J. Alloy. Compd.* **2018**, *753*, 192–202. [CrossRef]
46. Ma, J.; Wei, Y.; Gan, L.; Wang, C.; Xia, H.; Lv, W.; Li, J.; Li, B.; Yang, Q.-H.; Kang, F.; et al. Abundant grain boundaries activate highly efficient lithium ion transportation in high rate $Li_4Ti_5O_{12}$ compact microspheres. *J. Mater. Chem. A* **2019**, *7*, 1168–1176. [CrossRef]
47. Wu, Q.; Xu, J.; Yang, X.; Lu, F.; He, S.; Yang, J.; Fan, H.J.; Wu, M. Ultrathin Anatase TiO_2 Nanosheets Embedded with TiO_2-B Nanodomains for Lithium-Ion Storage: Capacity Enhancement by Phase Boundaries. *Adv. Energy Mater.* **2015**, *5*. [CrossRef]
48. Chen, C.; Xu, H.; Zhou, T.; Guo, Z.; Chen, L.; Yan, M.; Mai, L.; Hu, P.; Cheng, S.; Huang, Y.; et al. Integrated Intercalation-Based and Interfacial Sodium Storage in Graphene-Wrapped Porous $Li_4Ti_5O_{12}$Nanofibers Composite Aerogel. *Adv. Energy Mater.* **2016**, *6*, 1600322. [CrossRef]
49. Feng, X.-Y.; Li, X.; Tang, M.; Gan, A.; Hu, Y.-Y. Enhanced rate performance of $Li_4Ti_5O_{12}$ anodes with bridged grain boundaries. *J. Power Sources* **2017**, *354*, 172–178. [CrossRef]
50. Huang, C.; Zhao, S.-X.; Peng, H.; Lin, Y.-H.; Nan, C.-W.; Cao, G.-Z. Hierarchical porous $Li_4Ti_5O_{12}$–TiO_2 composite anode materials with pseudocapacitive effect for high-rate and low-temperature applications. *J. Mater. Chem. A* **2018**, *6*, 14339–14351. [CrossRef]
51. Li, G.; Blake, G.R.; Palstra, T.T. Vacancies in functional materials for clean energy storage and harvesting: The perfect imperfection. *Chem. Soc. Rev.* **2017**, *46*, 1693–1706. [CrossRef]
52. Xu, H.; Chen, J.; Li, Y.; Guo, X.; Shen, Y.; Wang, D.; Zhang, Y.; Wang, Z. Fabrication of $Li_4Ti_5O_{12}$-TiO_2 Nanosheets with Structural Defects as High-Rate and Long-Life Anodes for Lithium-Ion Batteries. *Sci. Rep.* **2017**, *7*, 2960. [CrossRef] [PubMed]
53. Hong, Z.; Zhou, K.; Huang, Z.; Wei, M. Iso-Oriented Anatase TiO_2 Mesocages as a High Performance Anode Material for Sodium-Ion Storage. *Sci. Rep.* **2015**, *5*, 11960. [CrossRef] [PubMed]
54. Xiao, S.; Li, Z.; Liu, J.; Song, Y.; Li, T.; Xiang, Y.; Chen, J.S.; Yan, Q. SeC Bonding Promoting Fast and Durable Na^+ Storage in Yolk-Shell $SnSe_2$ @SeC. *Small* **2020**, *16*, 2002486. [CrossRef] [PubMed]
55. Xu, Y.; Zhou, M.; Zhang, C.; Wang, C.; Liang, L.; Fang, Y.; Wu, M.; Cheng, L.; Lei, Y. Oxygen vacancies: Effective strategy to boost sodium storage of amorphous electrode materials. *Nano Energy* **2017**, *38*, 304–312. [CrossRef]
56. Jiang, Y.; Song, D.; Wu, J.; Wang, Z.; Huang, S.; Xu, Y.; Chen, Z.; Zhao, B.; Zhang, J. Sandwich-like SnS_2/Graphene/SnS_2 with Expanded Interlayer Distance as High-Rate Lithium/Sodium-Ion Battery Anode Materials. *ACS Nano* **2019**, *13*, 9100–9111. [CrossRef]
57. Deng, X.; Wei, Z.; Cui, C.; Liu, Q.; Wang, C.; Ma, J. Oxygen-deficient anatase TiO_2@C nanospindles with pseudocapacitive contribution for enhancing lithium storage. *J. Mater. Chem. A* **2018**, *6*, 4013–4022. [CrossRef]
58. Kim, H.S.; Cook, J.B.; Lin, H.; Ko, J.S.; Tolbert, S.H.; Ozolins, V.; Dunn, B. Oxygen vacancies enhance pseudocapacitive charge storage properties of MoO_{3-x}. *Nat. Mater.* **2017**, *16*, 454–460. [CrossRef]
59. Yi, T.-F.; Fang, Z.-K.; Deng, L.; Wang, L.; Xie, Y.; Zhu, Y.-R.; Yao, J.-H.; Dai, C. Enhanced electrochemical performance of a novel $Li_4Ti_5O_{12}$ composite as anode material for lithium-ion battery in a broad voltage window. *Ceram. Int.* **2015**, *41*, 2336–2341. [CrossRef]
60. Wang, C.; Wang, S.; Tang, L.; He, Y.-B.; Gan, L.; Li, J.; Du, H.; Li, B.; Lin, Z.; Kang, F. A robust strategy for crafting monodisperse $Li_4Ti_5O_{12}$ nanospheres as superior rate anode for lithium ion batteries. *Nano Energy* **2016**, *21*, 133–144. [CrossRef]
61. Ge, H.; Cui, L.; Zhang, B.; Ma, T.-Y.; Song, X.-M. Ag quantum dots promoted $Li_4Ti_5O_{12}/TiO_2$ nanosheets with ultrahigh reversible capacity and super rate performance for power lithium-ion batteries. *J. Mater. Chem. A* **2016**, *4*, 16886–16895. [CrossRef]
62. Yuan, T.; Yu, X.; Cai, R.; Zhou, Y.; Shao, Z. Synthesis of pristine and carbon-coated $Li_4Ti_5O_{12}$ and their low-temperature electrochemical performance. *J. Power Sources* **2010**, *195*, 4997–5004. [CrossRef]
63. Wang, D.; Shan, Z.; Na, R.; Huang, W.; Tian, J. Solvothermal synthesis of hedgehog-like mesoporous rutile TiO_2 with improved lithium storage properties. *J. Power Sources* **2017**, *337*, 11–17. [CrossRef]
64. Wang, X.; Hao, H.; Liu, J.; Huang, T.; Yu, A. A novel method for preparation of macroposous lithium nickel manganese oxygen as cathode material for lithium ion batteries. *Electrochim. Acta* **2011**, *56*, 4065–4069. [CrossRef]
65. Wang, Y.; Sun, H.; Tan, S.; Feng, H.; Cheng, Z.; Zhao, J.; Zhao, A.; Wang, B.; Luo, Y.; Yang, J.; et al. Role of point defects on the reactivity of reconstructed anatase titanium dioxide (001) surface. *Nat. Commun.* **2013**, *4*, 2214. [CrossRef] [PubMed]

 *processes*

Article

Understanding Slovakian Gas Well Performance and Capability through ArcGIS System Mapping

Gabriel Wittenberger, Jozef Cambal, Erika Skvarekova, Andrea Senova * and Ingrid Kanuchova

Department of Montaneous Sciences, Institute of Earth's Resources, Faculty of Mining, Ecology, Process Control and Geotechnologies Technical University of Kosice, Park Komenskeho 19, 040 01 Kosice, Slovakia; gabriel.wittenberger@tuke.sk (G.W.); jozef.cambal@tuke.sk (J.C.); erika.skvarekova@tuke.sk (E.S.); ingrid.kanuchova@student.tuke.sk (I.K.)
* Correspondence: andrea.senova@tuke.sk; Tel.: +421-55-6022985

Abstract: There are two important territories in Slovakia with functioning gas well operations: the Eastern Slovak Lowland and the Vienna Basin. This article focuses on the creation of electronic monitoring and graphical mapping of the current technical conditions of gas wells in the Eastern Slovak Lowland. An analysis of the gas wells' current state in the terrain is available. The aim of the article is to draw attention to the current state of gas wells, such as the insufficient processing of gas wells, the lack of summary and uniform records concerning them, and the lack of an electronic system for monitoring the technical security of the wells. The scientific contribution of this article lies in its ability to interpret and address operational problems related to gas wells. Through analogy, the step algorithm expresses the possibility of also using gas wells for oil, geothermal and hydrogeological wells. The intention was to highlight the importance of the need to create a database for the security and strategic needs of the state regarding the storage of natural gas. As of yet, no computer or graphic system has been used in Slovakia to monitor, unify and clarify the actual technical condition of gas wells. Using the ArcGIS electronic and graphical software tool, the mapping and recording of gas wells was carried out in the area under investigation. The mapping was completed with the mentioned technical patterns. These patterns have the information found on individual gas wells. After the information is added to the database, this mapping can also be carried out in another important area with functioning gas wells, such as in the Vienna Basin, which could be another theme for further research in this area.

Keywords: natural gas; gas well; gas extraction; monitoring; mapping; ArcGIS software

Citation: Wittenberger, G.; Cambal, J.; Skvarekova, E.; Senova, A.; Kanuchova, I. Understanding Slovakian Gas Well Performance and Capability through ArcGIS System Mapping. *Processes* **2021**, *9*, 1850. https://doi.org/10.3390/pr9101850

Academic Editors: Albert Ratner, Adam Smoliński and Sergey Zhironkin

Received: 2 August 2021
Accepted: 14 October 2021
Published: 18 October 2021

Publisher's Note: MDPI stays neutral with regard to jurisdictional claims in published maps and institutional affiliations.

1. Introduction

Today, Slovakia is almost entirely dependent on the import of oil and natural gas. Despite the current situation, the history of oil production in Slovakia, which was previously part of Austria-Hungary and later Czechoslovakia, is one of the oldest in the world, and its origins date back roughly to the middle of the 19th century.

At the beginning, extraction from a geological point of view was situated in the area of the western and eastern part of the flysch belt of Slovakia. Over time, extraction in these territories declined, and in the period between the First and Second World Wars, almost completely disappeared. During the First World War, however, intensive exploration and then extraction began in the Vienna Basin area, and then in the 1960s in the East Slovak Neogene Basin area as well.

These extraction locations reached their maximum production in the 1960s and 1980s. The Vienna and East Slovak Basin were then gradually joined by the Danube Basin. These are rated among the traditional areas of oil exploration and continue to be some of the most promising exploration areas in Slovakia. These basins have a degree of oil exploration that is relatively uneven in terms of quality and volume [1,2].

The wells must be situated in suitable rock-bearing environments, e.g., in sandstones. Because of its structure, sandstone allows the accumulation of hydrocarbons such as oil and natural gas. For drilling applications in the evaluation of cases such as borehole instability and many other drilling problems, it is necessary to analyze the effects of bound tension and pore pressure in the bearing rocks [3,4].

It is useful to analyze the history of measurements and perform continuous optimization in order to reduce the number of measurements on wells where problems have not been registered for a long time and focus more on potential problematic wells [5].

In the territory of Slovakia, the largest Neogene basin is the Danube Basin. When compared to the Vienna Basin and East Slovak Basin, verified deep borehole source rocks contain less organic carbon and have a lower hydrocarbon potential. In Slovakia, the Vienna Basin is definitely one of the very important gas and oil regions, and important deposits with predominantly gasoline types of oil can be found in the East Slovak Basin (Figure 1).

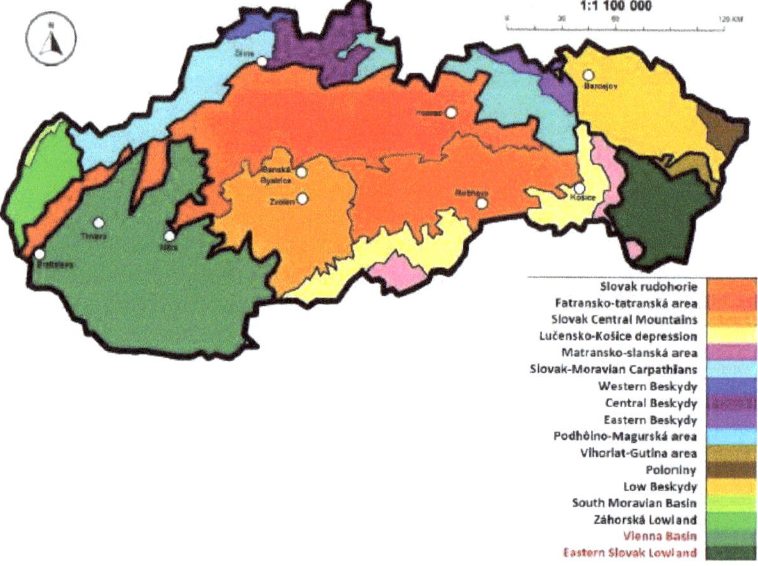

Figure 1. The Eastern Slovak Lowland and Vienna Basin [6].

Energy security, as one of the essential prerequisites for sustainable development, is an objective pursued by all countries in the world [7].

The dynamics of the world oil and gas market have changed drastically over the last decade. The discovery and extraction of large, new high-quality oil and gas reservoirs, combined with the involvement of states in the development of low-carbon alternative energies, lead to the fact that the global market is currently shaped not only by demand but also by supply. The volatility of oil prices creates uncertainty, which negatively affects decision-making processes in all energy sectors. The price of oil is the reference unit for the investment and development in the alternative energy sector. In the context of the Paris Climate Agreement, it is necessary to mitigate the impact of oil prices on the development of alternative energies [8].

The current global economy is also making progress by using oil and natural gas in several sectors of the economy. Their use is visible in the cosmetic, pharmaceutical, alimentary, chemical, and metallurgical industries, but especially in the energy industry.

Oil is not recyclable and is now more frequently substituted with alternative fuel in the energy industry. Currently, there is a trend in the energy industry towards the use of low-carbon energy.

An alternative solution to the use of low-carbon fuel sources is also natural gas. One of the possibilities of the efficient use of this energy source is its storage in excavated natural gas (NG) deposits [9]. Slovakia has the largest storage capacity of underground storage tanks in Central Europe.

Another possibility for obtaining natural gas from unconventional sources, especially from shale, is hydraulic fracturing, currently mainly used in Poland. Hydraulic fission (HF) techniques are particularly effective in stimulating hydrocarbon production from shale gas or oil formations [10,11].

Hydraulic fracturing and induced seismicity monitoring are operating procedures for the safe and effective production of oil and gas [12].

The aim of the article is to draw attention to the current state of wells, in which there is insufficient processing of gas wells, a lack of summary and uniform records concerning them, as well as a lack of an electronic system for monitoring the technical security of the wells. In the area of fuel production and use, oil derivatives are increasingly being replaced by fuels of plant origin such as biofuels, while many aggregates and mechanisms have also started to use electric drive. At present, the trend in energy is to force out aggregates burning hydrocarbon fuels because they produce a carbon footprint. Fossil fuels are exchanged in favor of alternative fuels such as biofuels. Nevertheless, the traditional hydrocarbon fuels will still be used in many other industry sectors.

In the territory of Slovakia, the largest extraction company is NAFTA a.s., which is also our largest natural gas storage company with a maximum annual storage capacity of approximately 2.5 billion m^3. It operates in two exploration areas (Figure 2) with a total area of approximately 3050 km^2. NAFTA is currently the most important player in Slovakia's oil and gas exploration in the production sector. Table 1 has a list of valid exploration areas and permits for oil and gas extraction for 2020 used by the international energy company NAFTA Ltd., (London, UK). The original collection of gas centers where natural gas had been stored provisionally were replaced by new modern technology. Underground gas storage in the Slovak Republic represents the conversion of natural gas deposits to the conditions of the underground storage of natural gas. The technological equipment for gas processing, measurements as well as the control system have a very high technical level. The other companies such as Vermilion Slovakia Exploration s.r.o., Bratislava, Alpine Oil & Gas Slovakia s.r.o., Bratislava and ROMGAZ s.r.o., Bratislava are small shareholders in Slovakia (Table 1).

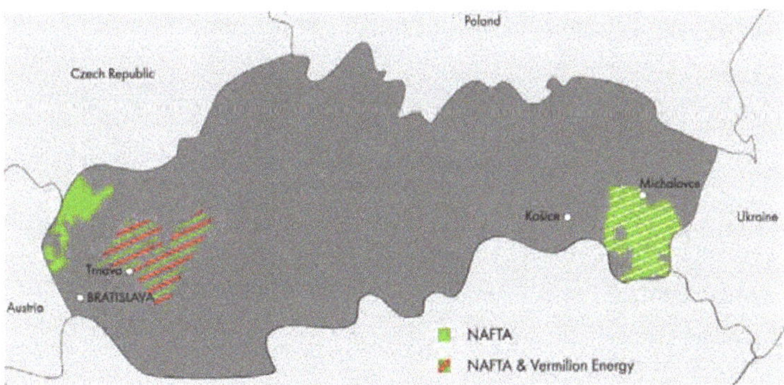

Figure 2. Exploration territories in the Slovak Republic [13].

Out of concern for the social interest, the national strategy is to store natural gas in underground storage facilities, as part of the sustainability of the country's energy potential for the future [2].

In recent years, it has been discovered that there are several operational problems in the fields around underground reservoirs. During mining, changes in the geological environment may occur that affect the mining process, e.g., compaction and sandblasting.

With increases in the understanding of mining procedures and methods, the work on underground gas storage facilities will become more efficient [14].

Well mapping allows the creation of efficient and safe gas reservoirs based on their properties, which is a way of locating wells during the development of oil and gas reservoirs and during extraction. Well modeling using an optimization algorithm automatically finds the most suitable well locations, can help increase efficiency, provide objective decision-making and reduce risk [15]. The exact identification of localized wells by mapping via ArcGIS software leads to the accurate localization of wells. The excavated boreholes with suitable geological conditions (e.g., porosity and the boundary of the deposit with impermeable layers) are potentially suitable for use as underground storage tanks.

The stochastic drilling approach with the usage of well property algorithms is used to find the optimal drilling depth measured in vertical wells in three-dimensional space. The algorithms created in such a way are then used to demonstrate the determination of the actual measured depth of the vertical wells [16].

Based on the above-mentioned information, we can demonstrate that it is possible to quickly identify deposits with good productivity and to identify the most promising areas with the presence of natural gas (according to Table 1). In the article, all available data and information obtained from the East Slovak Basin from the available gas wells are mapped and electronically processed into the summary registration and mapping system. The authors used the existing software, ArcGis, and showed the possibility of filling its database with the verified data and information from the existing wells. This information was not completely processed in the electronic form. This software more easily provides research into the technical characteristics of wells. The main aim is to show the newly realized approaches to the electronic registration and mapping of wells, which, in Slovakia, have not been mapped so far. This method of electronic evidence is possible to use to process data and information, as well as for geothermal wells.

We can use various automated systems and programming in other software such as DELPHI and MATLAB [17,18] to monitor, map and digitize information on the conditions of wells. For example, a computational fluid dynamics (CFD) simulation tool is used to build computational models, which together, creates modelling support for employees at the Centre for Renewable Resources at the Technical University of Košice [19].

Table 1. List of all valid exploration territories for oil and gas in Slovakia (2020), (source: elaborated by authors based on [13,20]).

Name of the Exploration Area	Keeper of the Exploration Area	Area [km^2]	Validity Up to
Bažantnica oil and natural gas	NAFTA a. s., Bratislava	451.4	14.05.2020
Beša natural gas	NAFTA a. s., Bratislava	770.45	26.09.2029
Gbelyoil and natural gas	NAFTA a. s., Bratislava	419.00	21.05.2020
Topoľčany natural gas	NAFTA a. s., Bratislava	1190.93	31.12.2024
Trnava natural gas	NAFTA a. s., Bratislava (50%), Vermilion Slovakia Exploration s.r.o., Bratislava (50%)	91.3	31.03.2028
Vienna Basin—north oil and natural gas	NAFTA a. s., Bratislava	146.51	28.06.2024
Svidník oil and natural gas	Alpine Oil & Gas Slovakia, s.r.o., Bratislava (66.67%), ROMGAZ s.r.o., Bratislava (33.33%)	34.22	01.08.2021

As we have demonstrated in the introduction, up until recently, information on the location and condition of historical oil and gas wells in Slovakia has been poorly recorded and archived, including important geological, fluid and hydrocarbon production information collected for each well.

As a result, this has hampered the efficient new exploration of oil and gas, has prevented the repurposing of historical wells for carbon or hydrogen storage or the assessment of geothermal resources, and has impaired the ability to effectively identify and remediate wells in poor condition that pose an environmental risk.

In this article we demonstrate how we have developed a new database and interrogation workflow utilizing ArcGIS technology that can address these concerns. We provide case examples of how the new system can be used to easily identify wells that require remediation, depleted gas reservoirs that can be repurposed for carbon storage, and areas of future hydrocarbon perspectivity.

2. Materials and Methods

For our research we have chosen gas wells in the territory of the East Slovak Lowland.

One model example of information inserted into the ArcGis database is from the Stretava 45 well. The individual steps show the process of creating the database and entering information into it. The valuation of the current status of all available wells was carried out. After a detailed analysis, it was found that in 2020, there were six active extractions of hydrocarbon deposits in the territory of the East Slovak Lowland. We considered all gas and oil wells in Slovakia, which are registered by NAFTA Ltd. (270 wells). There are 270 wells in these mining areas, of which 61 are drilled and equipped, 11 planned and 198 that have been plugged and abandoned. These wells were used to create a program for their records, electronic mapping and the collection of their basic information.

We entered the data of production wells, testing wells and exploration wells into the software database. The abandoned wells were no longer important when considering data entry into the database as they are excluded from the study.

After obtaining the measured data (the location, quantity, and state of oil and gas wells), the data were evaluated using the following analyses:

(a) We analyzed up-to-date information on the status and number of gas wells in the selected territory. In the territory of the Slovak Republic, the production of oil, gasoline, and natural gas as well as the underground storage of natural gas in natural rock structures is localized in the districts of competence of the OBU (District Mining Office) in Bratislava and the OBU (Mining Office) in Košice. A list of all valid exploration territories in Slovakia for 2020 is given in Table 1.

Due to its suitable geological structure, the territory of Slovakia has several oil and gas deposits (Tables 2 and 3). This favorable geological structure is necessary for their formation, accumulation and subsequent extraction, or more precisely, for underground storage.

Table 2. List of currently valid extracted natural gas deposits, (source: elaborated by authors).

Natural Gas—14 Deposits	
From which 5 are mined deposits	
No.	Name of the deposit
1	Bánovce nad Ondavou
2	Trhovište—Pozdišovce
3	Stretava
4	Senné
5	Ptrukša

Table 3. Listing of the actual valid mined oil deposits, (source: elaborated by authors).

Oil (Gasoline)—6 Deposits	
From which 4 are mined deposits	
No.	Name of the deposit
1	Bánovce nad Ondavou
2	Trhovište—Pozdišovce
3	Stretava
4	Senné

However, not all the discovered deposits were extracted because they were not financially efficient. Extracted oil or gas deposits can be used for other extraction activities in Slovakia as well as in other countries. Depleted oil and gas reservoirs can be subsequently used for other activities such as gas storage, carbon storage, and the disposal of mine waste water, sulphide, CO_2 etc. [21,22].

Most of these suitable underground geological structures are used for the storage of natural gas, which are located mainly near the villages of Jakubov, Gajary, Suchohrad, Plavecký Thursday, Láb and Križovany nad Váhom.

An analysis of the current state of deposits was carried out from the total current number of natural gas deposits (Table 2) and oil (Table 3) located in the east of Slovakia in the East Slovak Basin.

The above mined oil deposits (Table 3) are located in Eastern Slovakia. Currently, the total production of oil in our territory is symbolic and accounts for only 1% of domestic consumption. The amount of mined oil and gasoline, which is around 58,000 tons and 10,500 tons per year, respectively, is not sufficient for the needs of the country, so oil has to be purchased abroad.

More important are natural gas deposits, which produce one million m^3 per year, while geologically suitable mined deposits are mainly used in Slovakia for the underground storage of natural gas, and are among the most important in Central Europe.

Table 4 gives an overview of the hydrocarbon deposits and the oil and gas wells in the extraction territory of the Eastern Slovak Lowland.

Table 4. Actual state and number of wells in the territory of the Eastern Slovak Lowland, (source: elaborated by authors).

No.	Name of Hydrocarbon Deposits	Realized Wells	Abandoned Wells	Planned Wells
1	Bánovce nad Ondavou	10	28	0
2	Pavlovce nad Uhom	10	46	2
3	Pavlovce nad Uhom I.	13	36	2
4	Pozdišovce I.	12	39	1
5	Kapušianske Kl'ačany	14	42	6
6	Trebišov	2	7	0
	Total	61	198	11

Table 4 shows that abandoned wells are largest in number and are also in the last, irreversible phase in the life of the gas and oil well. Liquidation shall be taken if there is no use of the well or if its further use would be inefficient from an economic point of view. Abandoned wells can be used for waste storage, the pumping of mining water, heating with heat pumps, or heating through the usage of low-level heat. The wells are controlled, plugged and abandoned if they are filled with special cement mixtures.

These wells have to then be deepened and cleaned, and then undergo the installation of underground equipment [23,24].

(b) We gathered and collected the necessary technical data and relevant information from 2020 available in the annual reports published by NAFTA a.s.

(c) We obtained and processed the collected data from the Information System of Geodesy, Cartography and Cadaster in the Slovak Republic (ISGKK).
(d) We verified the collected data using a functional algorithm, where 270 wells were processed, which will provide basic information about these wells simply by searching for them. ArcGIS software was used for recording, monitoring and mapping, from which the electronic and graphical records of gas wells were obtained. The selection of well parameters (depth, diameter of casings, casing wall thickness, the type and composition of the lower part of the well, material quality, the type of joint, etc.) were decided based on the amount of available information about the wells and locations of the wells in the East Slovak Basin.

ArcGIS is a geographic information system (GIS) for working with maps and geographic information maintained by the Environmental Systems Research Institute (ESRI). It is used for creating and using maps, compiling geographic data, analyzing mapped information, sharing and discovering geographic information, using maps and geographic information in a range of applications, and managing geographic information in a database [25–27].

3. Results

In the contents of the methodology, we analyzed all the indicators necessary to create an algorithm for solving the status of wells in the electronic system. A clear database of technical overground and underground information about wells was created. After the creation of the database, an image base was visualized in ArcGIS software and used to form the map based on the Slovak Republic.

The background of the map made it possible to filter different information and make a selection according to the selected parameters.

As a support for the creation of the algorithm, we used the monitoring of the well and a database created by us that contained all the basic data obtained about individual wells. These data are listed as follows:

- Hydrocarbon deposits;
- Location;
- Type of well;
- Completion of the well;
- Pumping tests;
- Underground correction of the well;
- Sample of further usage;
- Liquidation well;
- Photo documentation;
- Further obtained information.

The process of the creation of the algorithm is visualized by the map base of the area in which we were interested, as shown in Figure 3.

The methodology of the sequence of the creation of geographical maps consists of the following main steps:

(a) Opening the ArcMap tool and creating a digital geographical map base of the area of interest (Figure 3);
(b) Creating the geographical names in the format ".shp ", from the source geoportal.sk (Figure 4);
(c) Opening the ArcCatalog system module, which is necessary for the organization of the geographic database and the creation of polygon mining areas with a system file (Figure 5);
(d) Marking the mining areas and wells on a map base, where the wells were placed in the individual mining areas (Figure 6);
(e) Creating a file and inserting information about individual wells (Figure 7);
(f) Conducting the final search based on the selected well (Figure 8).

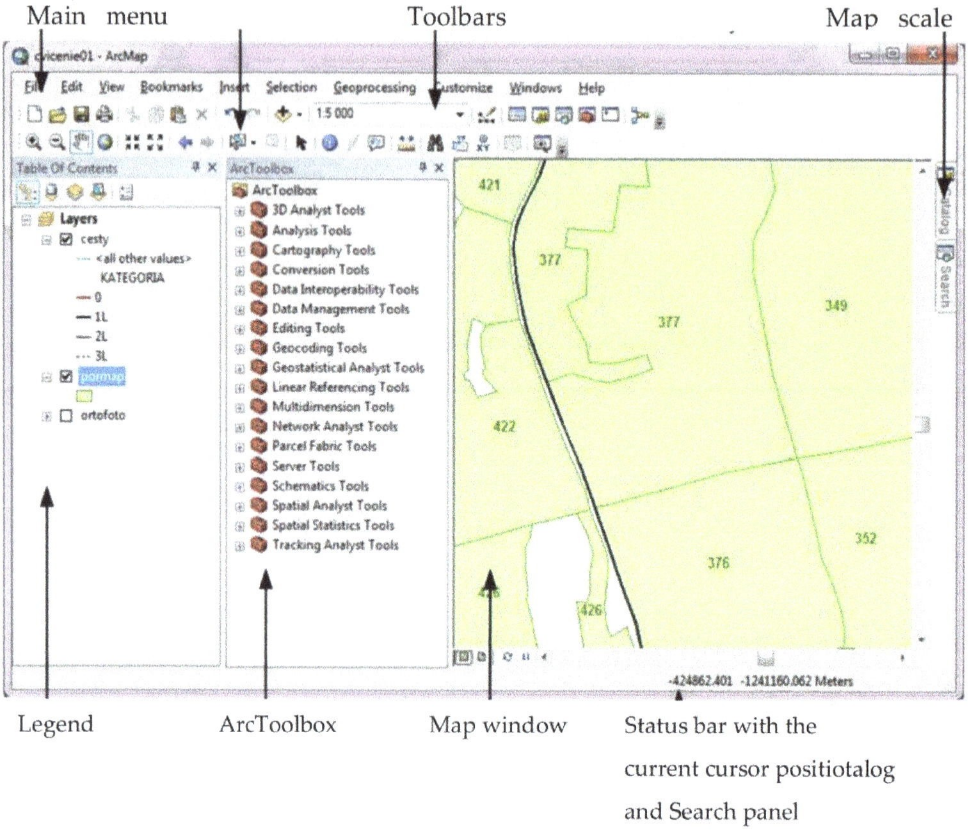

Main menu

Toolbars

Map scale

Legend

ArcToolbox

Map window

Status bar with the current cursor positiotalog and Search panel

Figure 3. The process of the creation of the algorithm maps, (source: elaborated by authors).

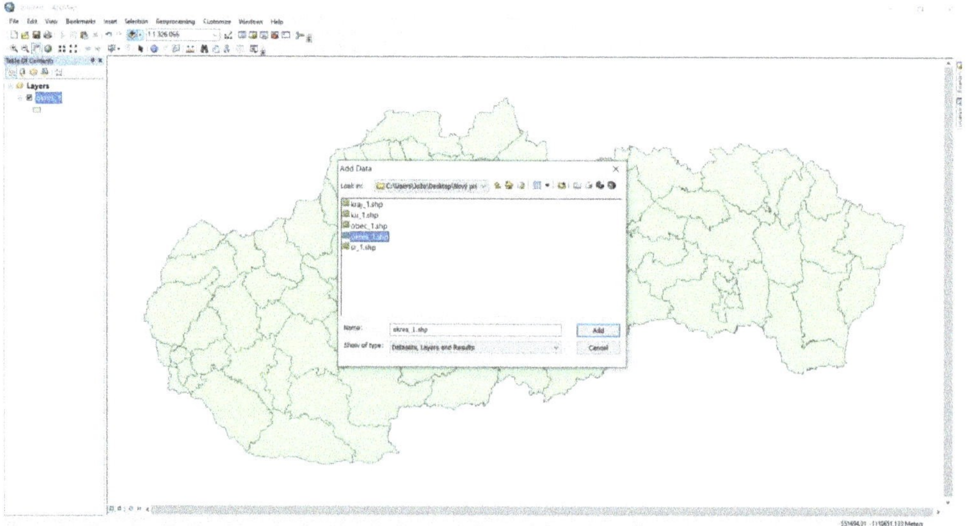

Figure 4. Inserting geographical districts and names, (source: elaborated by authors).

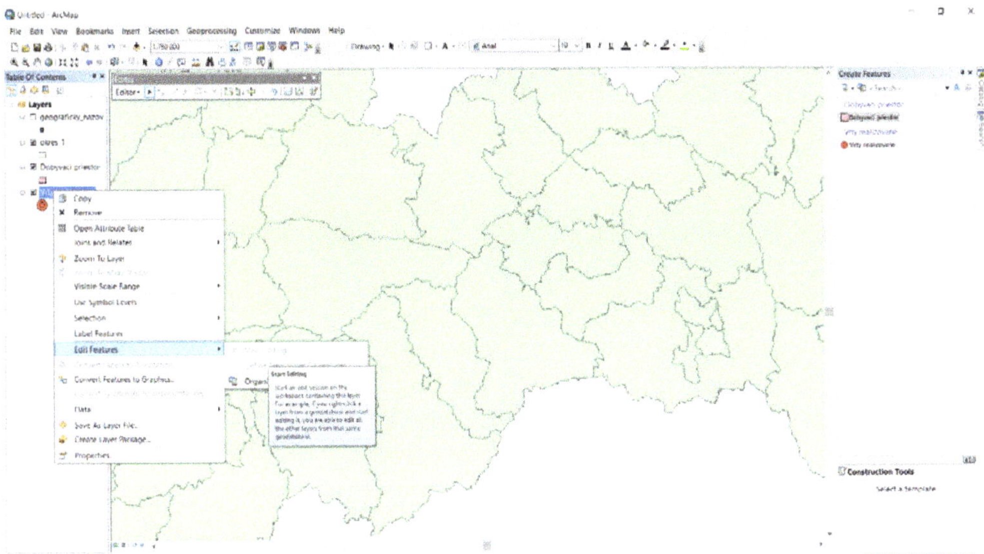

Figure 5. Creating layers for mining areas, (source: elaborated by authors).

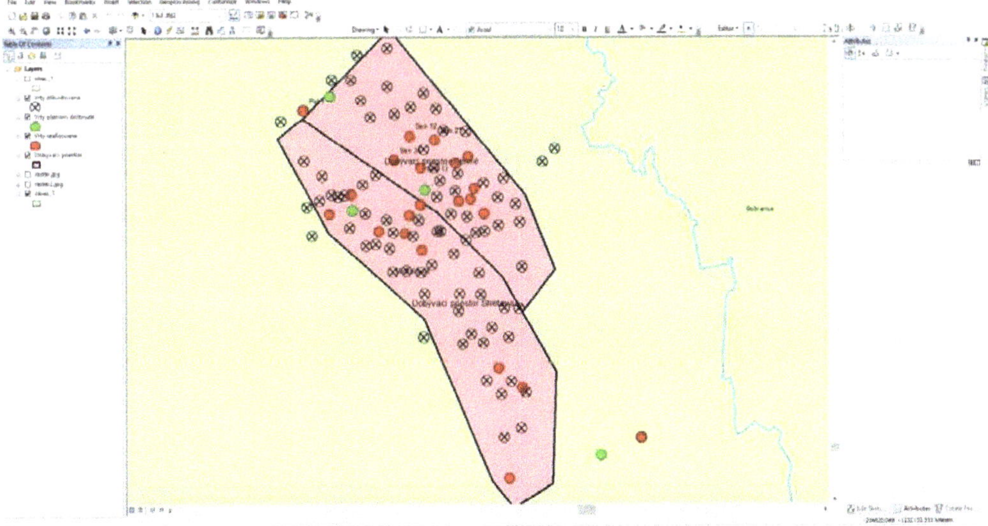

Figure 6. Placement of the wells in mining areas, (source: elaborated by authors).

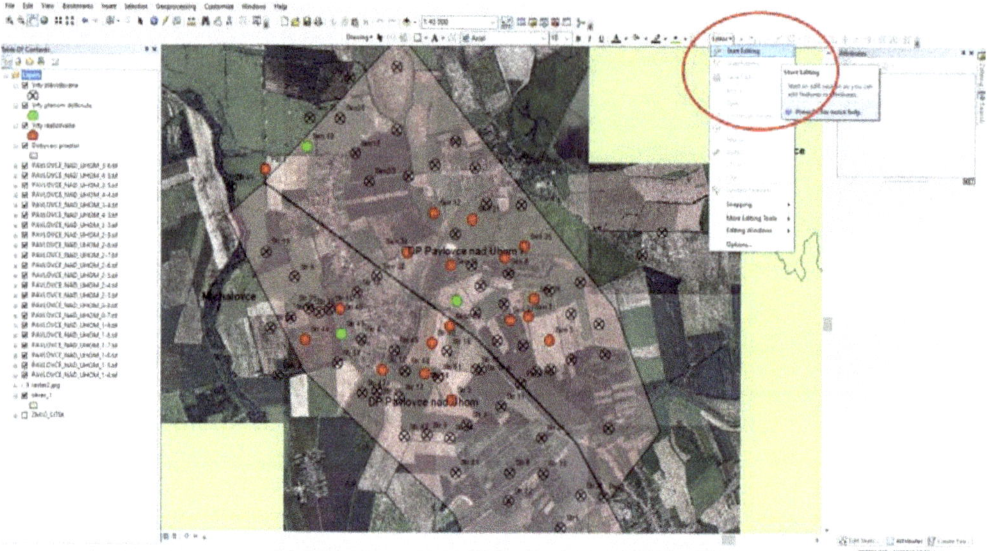

Figure 7. Entering the detected technical specifications and data, (source: elaborated by authors).

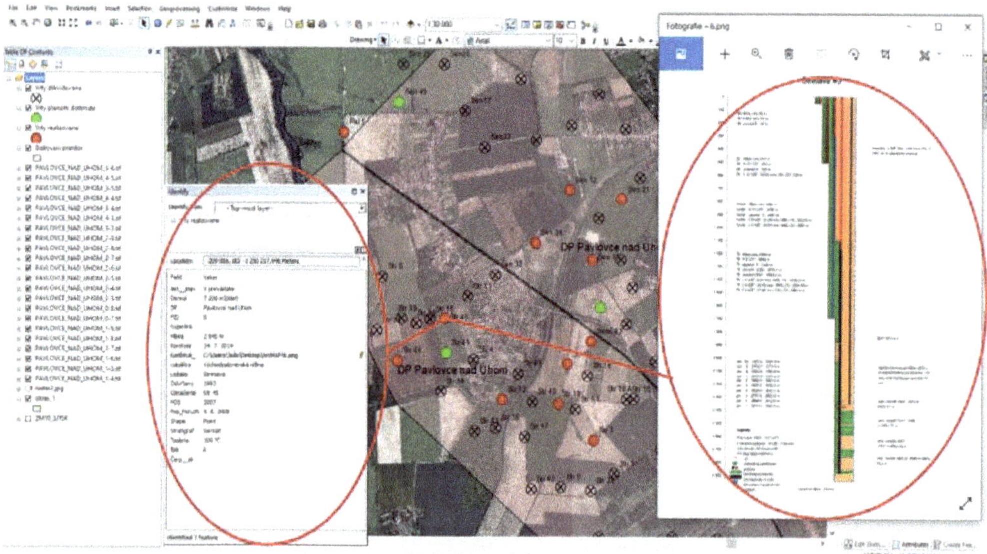

Figure 8. Preview of how to view basic information about the Stretava 45 well, (source: elaborated by authors).

The first step was to open ArcMap, one of the basic ArcGIS tools for creating digital geographic maps. This was followed by downloading and inserting the background map of the Slovak Republic in .shp format from the ZBGIS Map Client application, available on the geoportal.sk website.

The second step was to download and insert the geographical names in the ".shp" format (Figure 4 from the source geoportal.sk).

In the third step, the ArcCatalog system module was opened to organize and manage the geographic database, as well as used to connect the system file. In order to form polygon-type mining areas, a new layer (also using the ".shp" format) was created and added to the same folder (Figure 5). In a similar way, individual wells were created, which were of the point type.

After creating all the necessary layers, in step four we marked the mining areas and wells on the map base. The wells were inserted exactly according to the individual valid mining areas, and their items/categories can be seen in the table (Figure 6).

The last step was to fill the individual wells with all the basic technical information and well parameters. This was done using the top bar via the Start Editing function as shown in Figure 7.

The result of processing the data on wells was the creation of clear gas well recordings. The advantage for the user is the speed of obtaining available data according to the name of the searched well.

Alternatively, through the legend on the right side of the screen, the user can choose which hydrocarbon deposits he or she is interested in. The user then clicks on the type of well he or she wants to be displayed in the given hydrocarbon deposits. Then, the user clicks on the selected well, as shown in Figure 8 (Stretava 45 well), where he or she will see all the basic information available on the searched well.

This will show the user a clear table with windows and then he or she will be able to open photo documentation of the well.

Similarly, the user will be able to view information about pumping tests and underground repairs of the wells, as well as information on the disposal of wells and the possibilities for their further use via internet links [28].

4. Discussion

In the operation of wells by individual companies, it is a priority to have clear and accessible information about the current state of individual wells. The electronic monitoring and recording of wells allows wells to be kept under constant control. This helps companies respond in an appropriate manner to changes in wells, such as those caused by weather conditions, mechanical damage, and especially operational changes. It contributes to their smooth operation and prevents abrasion or well damage.

Another shortcoming is inaccurate legislation with a lack of solutions to the issue of well remediation after mining.

A number of obligations are unclear, such as who has the obligation to rehabilitate the territory after mining, how often the territory is to be controlled, or in what state the well is deemed to have been properly disposed of.

Each well requires regular monitoring on the surface, but mainly below the surface. This will prevent unplanned downtime due to poor technical conditions. Figure 9 shows an example of the unsatisfactory condition of a gas well at its mouth.

Figure 10 shows the correct surface and subsoil technical condition of the gas well, which is equipped with all safety devices (reinforced panel area, fencing, camera system, GPS system, information boards with basic data) and complete technical equipment (well completion) necessary for its safe operation.

Each gas well shall be subject to a so-called underground repair of the well after a certain period of operation or on the basis of its actual technical condition and production in order to continue to operate safely. Electronic records of its technical condition will also contribute to this.

The benefit of our algorithm and the work progress is to create fast and transparent monitoring of wells (i.e., well records and information about wells).

Thus, the result of the formed program is the recording of wells, where the user can find his or her sought-after well by name. Alternatively, through the legend on the right side of the screen, the user can choose what hydrocarbon deposits he or she is interested in and click on the type of well he or she wants to be displayed in those hydrocarbon deposits. The user will then click on the chosen well and then will be shown all available information about that well.

First, the user will see a table with basic information about the well. Second, the user will be able to open the photo documentation of the well. Next, it will be possible to consult the pumping tests, the underground repair of the well and the possibility of further use or disposal via internet links. All information and images entered can be edited, amended or supplemented according to the current technical condition with the possibility of comparing their status in the history of the program.

Figure 9. Example of the unsatisfactory condition of a gas well at its mouth, (source: elaborated by authors).

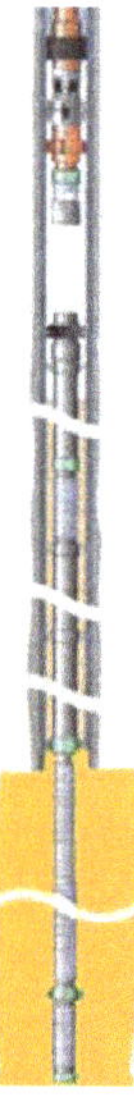

Figure 10. Correct overhead technical equipment of a gas well at its mouth, (source: elaborated by authors).

5. Conclusions

There is currently no systematic and consistent approach in Slovakia when deciding on the necessity and priorities of remediation measures of the wells implemented. There are no clearly defined criteria for assessing the importance of addressing them. There is a lack of public programs for the disposal of old wells after oil and gas extraction and for other environmental burdens. This should be done based on the objective evaluation according to the state of pollution and according to the evaluation of environmental risks.

The disposal of underground wells and underground spaces represents not only a technical but also an economically demanding task. This requires a combination of technical knowledge from the use of materials and the creation of open spaces with an emphasis on high environmental protection [29]. The modification of legislation would eliminate the potential threat of burdens to the environment.

In the case of the overall technical and electronic equipment of gas wells, the well can be considered safe for operation and control as shown in Figure 10. The benefit of this processing is to check the records in the actual operation of the well. Therefore, the purpose of this proposed electronic register is to simplify the search and control of the current technical status of wells.

Another benefit is to detect possible defects in well damage or accidents, which has a positive impact on the safety and technical security of the wells and, thus, on its service life and environmental quality.

The mapping created by us using ArcGIS software is unique in Slovakia in such a way that it allows an overview of individual wells and allows users to find out their current technical condition [30]. The purpose was to inform the general public about the latest developments in the subject under examination.

The results presented and the differences between our findings and the literature data obtained in some other studies indicate the need for further research.

The scientific use of the article resides in the ability to interpret technical problems and methods of processing and using information on gas wells. On this basis, gas companies can quickly and safely assess and evaluate the suitability of selected wells. Furthermore, it is possible to use them as storage devices, such as for the underground storage of natural gas. The database is processed based on the step algorithm offered by ArcGIS software. The database leads to a clear overview of the technical conditions of the wells, on the basis of which they can evaluate the overall safety of well operation. The interpretation of these data can provide gas companies with the strategic needs of the state and create additional capacity in the storage of natural gas in the Slovak Republic. The algorithm interpreted in the article is applicable not only to gas wells but also to oil, geothermal and hydrogeological wells.

Author Contributions: Conceptualization, G.W. and J.C.; methodology, G.W., A.S. and E.S.; software, J.C. and G.W.; validation, J.C. and G.W.; formal analysis, A.S., I.K. and E.S.; investigation and resources, J.C. and G.W.; data curation, A.S. and E.S.; writing—original draft preparation, J.C. and G.W.; writing—review and editing, A.S., I.K. and E.S.; visualization, G.W., A.S. and E.S.; supervision, G.W., A.S. and E.S.; project administration and funding acquisition, J.C. and G.W. All authors have read and agreed to the published version of the manuscript.

Funding: This contribution is a partial output of the implementation of the project Historical Mining–tracing and learning from ancient materials and mining technologies, No. 18111, EIT/RAW MATERIALS/SGA2019/1 supported by the European Institute of Innovation and Technology (EIT), a body of the European Union under Horizon 2020, the EU Framework Programme for Research and Innovation.

Institutional Review Board Statement: Not applicable.

Informed Consent Statement: Not applicable.

Data Availability Statement: The data presented in this study are available on request from thecorresponding author.

Conflicts of Interest: The authors declare no conflict of interest.

References

1. Rudinec, R. *Zdroje ropy, zemného plynu a geotermálnej energie na východnom Slovensku.*, 1st ed.; Bratislava: Alfa, Slovak Republic, 1989; p. 162.
2. Nafta a. s. Available online: https://www.nafta.sk/en/ (accessed on 11 March 2021).
3. Knez, D.; Rajaoalison, H. Discrepancy between measured dynamic poroelastic parameters and predicted values from Wyllie's equation for water-saturated Istebna sandstone. *ACTA Geophys.* **2021**, *69*, 673–680. [CrossRef]
4. Malvić, T.; Barudžija, U.; Pašić, B.; Ivšinović, J. Small Unconventional Hydrocarbon Gas Reservoirs as Challenging Energy Sources, Case Study from Northern, Croatia. *Energies* **2021**, *14*, 3503. [CrossRef]
5. Rubesova, M.; Bujok, P.; Klempa, M. Application of production of production well logging for the monitoring of tightness underground gas storage. *Int. Multidiscip. Sci. GeoConference SGEM* **2017**, *17*, 27–32. [CrossRef]
6. Prehľadné geologické mapy. Available online: http://apl.geology.sk/pgm/ (accessed on 4 April 2021).
7. Cherp, A.; Jewell, J. The concept of energy security: Beyond the four As. *Energy Policy* **2014**, *75*, 415–421. [CrossRef]
8. Failler, P. Special Issue on Global Market for Crude Oil. *Energies* **2021**, *14*, 1199. [CrossRef]
9. Zadsarb, M.; Zareipourc, H.; Kazemid, M. Resilient operation planning of integrated electrical and natural gas systems in the presence of natural gas storages. *Int. J. Electr. Power Energy Syst.* **2021**, *130*, 106936. [CrossRef]
10. Li, L.; Tan, J.; Wood, A.D.; Zhao, Z.; Becker, D.; Lyu, Q.; Shu, B.; Chen, H. A review of the current status of induced seismicity monitoring for hydraulicfracturing in unconventional tight oil and gas reservoirs. *Fuel* **2019**, *242*, 195–210. [CrossRef]
11. Quosay, A.A.; Knez, D.; Ziaja, J. Hydraulicfracturing: New uncertainty based modeling approach for process design using Monte Carlo simulation technique. *PLoS ONE* **2020**, *15*, e0236726. [CrossRef] [PubMed]
12. Mavko, G.; Mukerji, T.; Dvorkin, J. Tools for seismic analysis of porous media. In *The Rock Physics Handbook;* Cambridge University Press: Cambridge, UK, 2010.
13. Zoznam prieskumných území k 1.7.2019. Available online: https://www.minzp.sk/files/sekcia-geologie-prirodnych-zdrojov/zoznam-platnych-prieskumnych-uzemi-k-01.07.2019.pdf (accessed on 10 April 2021).
14. Settari, A. Reservoir Compaction. *J. Pet. Technol.* **2002**, *54*, 62–69. [CrossRef]
15. Jung, J.; Han, D.; Kwon, S. Well placement optimisation using a productivity potential area map. *Int. J. Oil Gas Coal Technol.* **2021**, *27*, 41–53. [CrossRef]
16. Atashnezhad, A.; Wood, A.D.; Fereidounpour, A.; Khosravanian, R. Designing and optimizing deviated wellbore trajectories using novel particle swarm algorithms. *J. Nat. Gas Sci. Eng.* **2014**, *21*, 1184–1204. [CrossRef]
17. Laciak, M.; Raškayová, D.; Flegner, P.; Kačur, J.; Durdan, M. Automated system for optimizing input parameters of the UCG process. In Proceedings of the 20th International Carpathian Control Conference (ICCC), Krakow-Wieliczka, Poland, 26–29 May 2019. [CrossRef]
18. Saderova, J.; Rosova, A.; Kacmary, P.; Sofranko, M.; Bindzar, P.; Malkus, T. Modelling as a Tool for the Planning of the Transport System Performance in the Conditions of a Raw Material Mining. *Sustainibility* **2020**, *12*, 8051. [CrossRef]
19. Rybár, R.; Kudelas, D.; Beer, M.; Horodníková, J. Elimination of Thermal Bridges in the Construction of a Flat Low-Pressure Solar Collector by Means of a Vacuum Thermal Insulation Bushing. *J. Sol. Eng.—Trans. ASME* **2015**, *137*, 054501. [CrossRef]
20. Správa o činnosti Hlavného banského úradu a obvodných banských úradov Slovenskej republiky za rok 2018. Available online: https://www.hbu.sk/files/documents/spravy/2018/spr%C3%A1va%20o%20%C4%8Dinnosti%20-%20text.pdf (accessed on 1 April 2021).
21. Franklin, M.; Orr, J. Storage of Carbon Dioxide in Geologic Formations. *J. Pet. Technol.* **2004**, *56*, 90–97. [CrossRef]
22. Geertsma, J. Land subsidence above compacting oil and gas reservoirs. *J. Pet. Technol.* **1973**, *25*, 734–744. [CrossRef]
23. Pinka, J. *Vyhľadávanie a ťažba nekonvenčných zdrojov ropy a zemného plynu. Monografia.*; VŠB TU Ostrava: Ostrava, Czech Republic, 2013; pp. 1–136.
24. Pinka, J.; Wittenberger, G.; Engel, J. *Borehole Mining*, 1st ed.; FBERG TU: Košice, Slovakia, 2007; p. 233.
25. Koreň, M. *Databázové Systémy;* Zvolen: Lesnícka fakulta, Slovakia, 2009; p. 90.
26. Worboys, M.F.; Duckham, M. *GIS. A Computing Perspective*, 2nd ed.; CRC Press: Boca Raton, FL, USA, 2004; p. 448.
27. Map and Interact with Your Location Data. Available online: https://www.esri.com/en-us/arcgis/products/arcgis-online/overview (accessed on 4 April 2021).
28. Alexander, L.G. Real Time, Low Cost, Diagnostic Tool for Understanding Oil and Gas Well Performance. *J. Can. Pet. Technol.* **2001**, *40*. [CrossRef]
29. Klempa, M.; Malis, J.; Sancer, J.; Slivka, V. Research of Stowing Material for Filling of Free Underground Spaces of Old Mine Works. *Inz. Miner.-J. Pol. Miner. Eng. Soc.* **2018**, *2*, 321–325. [CrossRef]
30. Girija, R.R.; Mayappan, S. Mapping of mineral resources and lithological units: A review of remote sensing techniques. *Int. J. Image Data Fusion* **2019**, *10*, 79–106. [CrossRef]

Article

Molten-Salt-Assisted Synthesis of Nitrogen-Doped Carbon Nanosheets Derived from Biomass Waste of Gingko Shells as Efficient Catalyst for Oxygen Reduction Reaction

Wei Hong [1,†], Xia Wang [1,†], Hongying Zheng [1], Rong Li [1,*], Rui Wu [2,*] and Jun Song Chen [2,*]

1 Chemical Synthesis and Pollution Control Key Laboratory of Sichuan Province, College of Chemistry and Chemical Engineering, China West Normal University, Nanchong 637000, China; hongweicqu@163.com (W.H.); luwckxy@163.com (X.W.); 18380585773@163.com (H.Z.)

2 School of Materials and Energy, University of Electronic Science and Technology of China, Chengdu 611731, China

* Correspondence: lirong406b@126.com (R.L.); ruiwu0904@uestc.edu.cn (R.W.); jschen@uestc.edu.cn (J.S.C.)

† These authors contributed equally to this work.

Abstract: Developing superior efficient and durable oxygen reduction reaction (ORR) catalysts is critical for high-performance fuel cells and metal–air batteries. Herein, we successfully prepared a 3D, high-level nitrogen-doped, metal-free (N–pC) electrocatalyst employing urea as a single nitrogen source, NaCl as a fully sealed nanoreactor and gingko shells, a biomass waste, as carbon precursor. Due to the high content of active nitrogen groups, large surface area (1133.8 m^2 g^{-1}), and 3D hierarchical porous network structure, the as-prepared N–pC has better ORR electrocatalytic performance than the commercial Pt/C and most metal-free carbon materials in alkaline media. Additionally, when N–pC was used as a catalyst for an air electrode, the Zn–air battery (ZAB) had higher peak power density (223 mW cm^{-2}), larger specific-capacity (755 mAh g^{-1}) and better rate-capability than the commercial Pt/C-based one, displaying a good application prospect in metal-air batteries.

Keywords: N-doped carbon nanosheet; gingko shells; oxygen reduction reaction; Zn–air battery

Citation: Hong, W.; Wang, X.; Zheng, H.; Li, R.; Wu, R.; Chen, J.S. Molten-Salt-Assisted Synthesis of Nitrogen-Doped Carbon Nanosheets Derived from Biomass Waste of Gingko Shells as Efficient Catalyst for Oxygen Reduction Reaction. *Processes* **2021**, *9*, 2124. https://doi.org/10.3390/pr9122124

Academic Editors: Sergey Zhironkin and Radim Rybar

Received: 14 October 2021
Accepted: 22 November 2021
Published: 25 November 2021

Publisher's Note: MDPI stays neutral with regard to jurisdictional claims in published maps and institutional affiliations.

1. Introduction

The oxygen reduction reaction (ORR) plays a key role in fuel cells and metal–air batteries to implement electrochemical energy conversion [1,2]. However, the sluggishness of ORR kinetics at the cathode limits their commercial application [3,4]. Pt and its alloys materials, recognized as the best ORR electrocatalysts, suffer from the resource deficiency, prohibitive cost and inferior durability [5,6]. Consequently, development of earth-abundance, cost-effective and performance-stable catalysts to substitute the precious-based catalysts for ORR is crucial yet challenging [7–9].

Heteroatom-doped carbon materials, such as P-doped carbon xerogel [10], N-doped nanotube [11,12], S-doped carbon [13,14] and B-doped graphene [15], are among the most promising electrocatalysts for ORR due to their advantages of relatively low cost, abundant reserves, good electronic conductivity and environmental friendliness [16,17]. Among them, N-doping has been considered to be one of the most effectual ways to enhance the electrocatalytic activities of carbon materials for ORR [18,19] because the N-doping can improve the electrical conductivity of carbon and change the charge distribution of the carbon atoms close to the dopants and then make them positively charged, which is beneficial to the adsorption and reduction of oxygen at the corresponding sites and improving the ORR activity [7,20,21].

However, the morphology of N-doped carbon materials prepared by traditional methods is hardly controlled, and the content of the N-dopant is relatively low because of the volatilization of the N element at high temperature, which has an undesirable effect on the electrocatalytic performance of the prepared materials. The employment of nanoreactor during the synthesis process can help relieve these challenges. The assistance of inorganic salts, such as NaCl [22,23], MgO [24], SiO_2 [25,26] and Fe_3O_4 [27], has been considered to be an effective strategy to prepare porous carbon materials. Among them, NaCl is one of the most commonly used because NaCl can not only act as a confining agent to reduce the weight loss of precursors and promote the N-doping during the high-temperature pyrolysis process but also as pore-forming agents to promote the production of macro-pores and a 3D network structure. In addition, NaCl can be conveniently removed by water-washing with no use of any corrosive and harmful chemicals and can be facilely recycled by recrystallization. However, a cost-effective and large-scale synthetic production method is still highly desirable but challenging. During the past decade, various biomass-derived carbon ORR catalysts have aroused ever-growing attention [28–31]. Gingko is mainly used for food and medicine, but the shells are usually thrown away as garbage, which not only causes a huge waste of resources but also leads to environmental pollution.

Herein, we report a molten-salt-assisted strategy to fabricate a 3D high-level nitrogen-doped porous carbon nanosheets network (N–pC) by a facile high-temperature pyrolysis method, employing the carbon-enriched biomass waste of gingko shells and urea as a carbon precursor and NaCl and urea as a fully sealed nanoreactor and a nitrogen source, respectively. The merits, such as large specific surface area, hierarchical pore structure and high active nitrogen (pyridinic and graphitic-N) content, endow the obtained N–pC excellent electrocatalytic ORR performance in alkaline solution, much superior to commercial Pt/C and most of the reported N-doped carbon materials. Inspired by the excellent ORR catalytic performance, the prepared N–pC was further used to assemble Zn–air batteries, which showed high peak power density, large specific-capacity and good rate-capability, making it a competitive alternative to the application in metal–air batteries.

2. Materials and Methods

2.1. Synthesis of the Materials

N–pC was prepared by pyrolyzing the mixture of gingko shells, NaCl and urea. Typically, 0.5 g gingko shells rinsed by deionized water, 1.4 g NaCl and 4 g urea were mixed to refined powder by planetary ball-milling for 4 h. Then, the mixture was initially carbonized under Ar atmosphere at 750 °C for 2 h with a ramp rate of 3 °C min^{-1}. Subsequently, the obtained product was refluxed in 3 M HCl (50 mL) for 6 h and washed with deionized water 3 times. After being dried at 80 °C for 24 h, the obtained black powder was further pyrolyzed at 900 °C for 2 h, and then the final product (N–pC) was harvested. For comparison, N–C catalyst was fabricated through the same procedure in the absence of NaCl. C catalyst was also synthesized by direct pyrolysis of the gingko shell powder.

2.2. Characterizations

Rigaku Ultima IV diffractometer (Cu Kα radiation) was employed to obtain the X-ray diffraction (XRD) spectra. Microstructural morphologies were observed employing field emission scanning electron microscopy (SEM, ZEISS SUPRA 55) and transmission electron microscope (TEM, JEM-2100F). To analyze the specific surface areas and pore-size distribution, nitrogen adsorption–desorption isotherms were tested on ASAP 2000 analyzer. X-ray photoelectron spectroscopy (XPS) measurement was implemented employing a Thermo Scientific Escalab 250 Xi.

2.3. Electrochemical Measurements

All the electrochemical tests were implemented on bipotentiostat (CHI760E, Shanghai, China) assembled with a rotation system (Pine Research Instrumentation, Durham, NC, USA) in a three-electrode system. A carbon rod, an Ag/AgCl elctrode (in saturated KCl solution) and a glassy carbon electrode (GCE) loaded with various catalyst (N–pC, N–C, C or benchmark 20 wt.% Pt/C) were used as the counter electrode, reference electrode, and working electrode, respectively. To prepare the working electrode, 14 μL the catalyst ink obtained by dispersing 1 mg catalyst powder into 0.5 mL Nafion solution (0.05 wt.%) via sonication was dropped onto the GCE, keeping the catalyst loading of 141.6 μg cm^{-2}. All potentials versus Ag/AgCl electrode ($E_{Ag/AgCl}$) were converted to the reversible hydrogen electrode (E_{RHE}) based on the formula of $E_{RHE} = E_{Ag/AgCl} + E^{\theta}{}_{Ag/AgCl} + 0.059pH$.

Cyclic voltammetry (CV) and linear sweep voltammetry (LSV) measurements were carried out in 0.1 M KOH electrolyte at a scanning rate of 100 mV s^{-1} and 10 mV s^{-1}, respectively. The electron transfer number (n) was calculated using the Koutecky–Levich (K-L) equation:

$$\frac{1}{J} = \frac{1}{J_k} + \frac{1}{B\omega^{1/2}} \quad (1)$$

$$B = 0.62nFC_0D_0^{2/3}v^{1/6} \quad (2)$$

where J and J_k are the measured disk's current density and the kinetic current density, respectively, ω is the angular velocity of rotation. F is the Faraday constant (96,485 C mol^{-1}), C_0 is the O_2 concentration in electrolyte (1.2 × 10^{-6} mol cm^{-3}), D_0 is the O_2 diffusion coefficient (1.9 × 10^{-5} cm^2 s^{-1}), and v represents the kinematic viscosity of the electrolyte (1.13 × 10^{-2} cm^2 s^{-1}).

The yield of HO_2^- and the n values for ORR via the RRDE measurements were determined by the equations as below:

$$HO_2^-\% = 200 \times \frac{I_r/N}{I_d + I_r/N} \quad (3)$$

$$n = 4 \times \frac{I_d}{I_d + I_r/N} \quad (4)$$

where I_d and I_r are the current density on disk and ring electrode, respectively, and N is the current collection efficiency of Pt ring, which is equal to 0.37.

The durability and the possible methanol crossover effect of N–pC and benchmark Pt/C catalyst were evaluated by chronoamperometric measurement at a rotation of 1600 rpm and a fixed potential of 0.7 V (versus RHE) in the O_2-saturated 0.1 M KOH solution.

The homemade Zn–air battery (ZAB) was equipped with a polished pure Zn plate (0.5 mm in thickness) as the anode, carbon paper (2 cm × 2 cm) uniformly coated with the catalyst (N–pC or benchmark 20 wt.% Pt/C, mass loading: 1 mg cm^{-2}) as the air cathode and 6 M KOH solution as the electrolyte.

3. Results and Discussion

The synthetic procedures of N–pC are illustrated in Scheme 1. Gingko shells, NaCl and urea were first mixed to a refined powder by ball-milling. Then, the powder was initially carbonized at 750 °C (slightly lower than the melting point of NaCl to avoid the structural collapse of newly formed carbon materials). After the removal of NaCl by water washing and further pyrolysis at 900 °C to increase the graphitization degree and conductivity, the final product (N–pC) was obtained. The compositions of the C, N–C and N–pC were characterized by XRD. As shown in Figure 1, the diffraction peaks of C, N–C and N–pC at 26° and 44° correspond to the (002) and (101) planes of graphite carbon, respectively, indicating that gingko shells have been successfully transformed into carbon materials under high-temperature pyrolysis. The characteristic peaks of NaCl cannot be observed in N–pC, indicating that the NaCl template can be removed by simple washing. In addition, we notice that the peak intensity weakens in the order of C, N–C and N–pC, suggesting that N-doping can increase the number of defect sites and the addition of NaCl can further induce the formation of more defect sites.

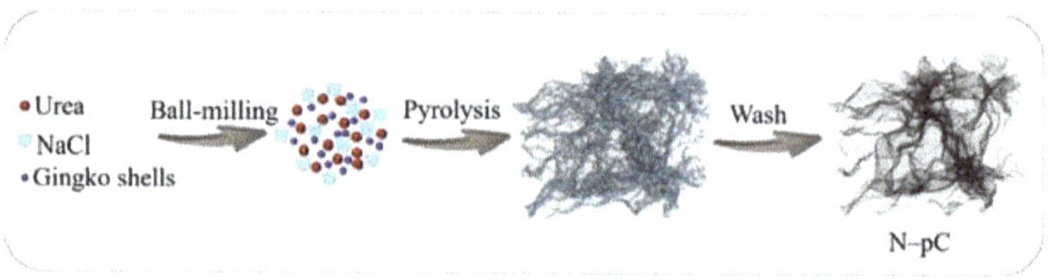

Scheme 1. Schematic illustration of the preparation of the N–pC catalyst.

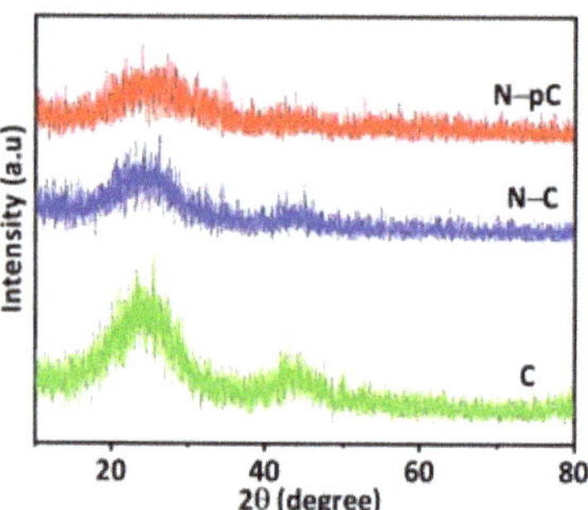

Figure 1. XRD patterns of C, N–C and N–pC.

The morphologies of the synthesized materials were observed by SEM and TEM. As shown in Figure 2a, the C catalyst obtained from the direct pyrolysis of gingko shells without the addition of urea and NaCl presents a compact and irregular block, whereas N–C has a significantly much looser and more porous morphology (Figure 2b), which mostly results from the large amount of pyrolysis gas of urea. With the further addition of NaCl, acted as sealed nanoreactor, the N–pC shows a 3D porous networks structure composed of graphene-like nanosheets (Figure 2c), which is beneficial to the increase of the specific surface area, the exposure of more active sites, and the improvement of transport rate of various substances involved in the reaction. TEM results are consistent with SEM. As shown in Figure 2d,e, the N–pC consists of a large number of graphene-like thin nanosheets, which interleave with each other, resulting in the formation of hierarchical pore structure and somewhat distorted of the carbon crystal faces. According to the elemental mapping (Figure 2f), the elements C, N and O are homogeneously distributed in the nanosheet, indicating the successful doping of N in N–pC catalyst.

The effect of urea and NaCl was further investigated by nitrogen adsorption/desorption measurements. As shown in Figure 3a, a distinct nitrogen uptake occurs at relative low pressure and a hysteresis loop appears at medium pressure in the isotherm curves of C, N–C and N–pC, indicating the existence of micro- and meso-pores in the three samples, which may be ascribed to the gas released from the decomposition of urea and gingko shells in the carbonization process. In addition, the isothermal adsorption curve of N–pC shows a sharp adsorption in the relative high-pressure region, demonstrating the existence of macropores on the material as the result of the occupying space of NaCl. The corresponding pore size distribution of the three materials (Figure 3b) further confirms their porous structures. Due to the different pore structure, the specific surface area and pore volume of C (641.9 m^2 g^{-1}, 0.365 cm^3 g^{-1}), N–C (1102.6 m^2 g^{-1}, 1.026 cm^3 g^{-1}) and N–pC (1133.8 m^2 g^{-1}, 1.22 cm^3 g^{-1}) are different. The large specific surface area and pore volume of N–pC means it can fully expose the active sites and most effectively accelerate mass transport, thus improving the electrochemical catalytic performance.

Figure 2. SEM images of (**a**) C, (**b**) N–C and (**c**) N–pC; (**d**) TEM; (**e**) HRTEM; and (**f**) element mapping images of N–pC.

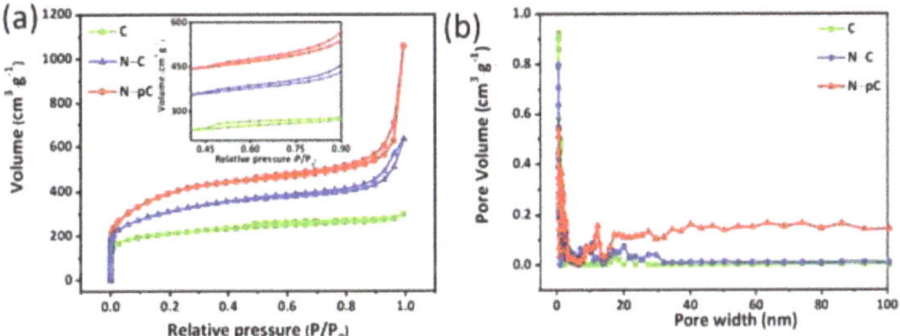

Figure 3. (**a**) N_2 adsorption/desorption isotherms (inset shows the enlarged plots) and (**b**) the related pore size distribution curves of C, N–C and NpC.

The surface composition of the C, N–C and N–pC samples were investigated by XPS measurements. Figure 4a shows the full spectrum of the three materials. Compared with C, both N–C and N–pC contain N, indicating that the addition of urea can successfully realize N-doping. However, the N content of N–pC (5.18%) is higher than that of N–C (4.65%) due to the existence of NaCl's confining effect acting as a fully sealed nanoreactor, which makes the volatilization and escape of N-containing gas generated by urea pyrolysis reduced. As can be observed from the high-resolution C 1s spectrum of N–pC (Figure 4b), there are four peaks located at 284.67, 285.72, 286.98 and 289.50 eV assigned to the C = C, C = N/C–O, C–N/C = O and O–C = O, respectively [32]. The presence of C = N and C–N bonds suggests that N atoms have been successfully incorporated into the carbon skeleton. Additionally, the N 1s spectra (Figure 4c) can be well separated into four species, corresponding to pyridine N at 398.07 eV, pyrrole N at 399.1 eV, graphitic N at 400.72 eV and oxidized N at 403.4 eV [33]. It has been reported that pyridine N and graphitic N can improve the electrocatalytic ORR performance of the materials, in which pyridine N can promote O_2 adsorption on the neighboring carbon and graphitic N can enhance the electrical conductivity [34–36]. Combining Figure 4a,d, it is easy to find that N–pC possesses a much higher content of pyridine N and graphitic N than N–C. Therefore, the excellent ORR electrocatalytic performance of N–pC will be expected.

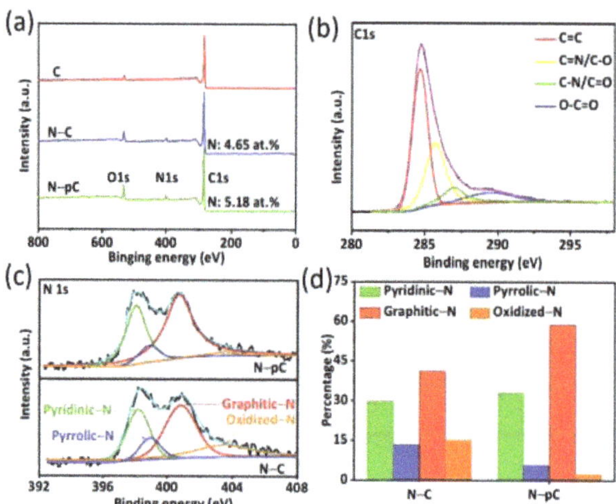

Figure 4. (**a**) XPS spectra of C, N–C and N–pC; (**b**) C1s high-resolution spectra of N–pC; (**c**) N1s high-resolution spectra of N–C and N–pC; (**d**) The percent of different forms of doping nitrogen in N–C and N–pC.

Cyclic voltammetry (CV) was conducted to evaluate the electrocatalytic ORR performance of C, N–C and N–pC in O_2-saturated 0.1 M KOH electrolyte. As it can be observed from Figure 5a, the CV curves of C, N–C and N–pC show an obvious cathode reduction peak, among which N–pC gives the most positive oxygen reduction peak and the highest peak current density, indicating that N–pC exhibits the highest electrocatalytic activity among the three catalysts. The linear sweep voltammetry (LSV) curves were also measured to further evaluate the ORR activity. As shown in Figure 5b, the C, N–C and commercial Pt/C catalysts have the onset potential (E_{onset}) of 0.756 V, 0.913 V and 1.012 V and the half-wave potential ($E_{1/2}$) of 0.585 V, 0.773 V and 0.836 V, respectively. Compared with those catalysts, the N–pC has the more positive onset potential (1.020 V) and half-wave potential (0.863 V) and higher limiting current density. In addition, as presented in Figure 5c, the Tafel slop of N–pC (59 mV dec^{-1}) is slightly smaller than that of commercial Pt/C (62 mV dec^{-1}), suggesting that N–pC has a comparable ORR kinetic process to that of commercial Pt/C. All these results suggest that N–pC has better electrocatalytic ORR activity than C, N–C, Pt/C and most of the reported metal-free catalysts (Table S1) in alkaline media. The superior catalytic performance of N–pC can be attributed to the incorporation of highly active N and the 3D hierarchical porous network structure.

To further evaluate the reaction kinetics of the ORR, the polarization curves at different rotation rates of N–pC were acquired (Figure 5d). Based on the corresponding Koutecky–Levich (K–L) plots, the transferred electron number (n) of the C catalyst are calculated to be about 2~2.5, suggesting that the ORR on the C electrode follows a two-electron and two-step mechanism. For the N–C catalyst, the n values are about 3.5, meaning that the ORR current efficiency is widely increased after the N-doping. On the contrary, the values of n for N–pC are close to 4 in a wide potential range, implying that a $4e^{-1}$ process occurs on the N–pC electrode (Figure 5e). The RRDE experiments of the C, N–C and N–pC catalysts were carried out to further quantify the n and HO_{2-} yield. As shown in Figure 5f, the n values for the C, N–C and N–pC are about 2.4, 3.5 and 4, respectively, entirely consistent with the RDE measurement results. Meanwhile, the HO_2^- yield at the N–pC is approximately equal to 0 in the investigated potential range, which is much lower than that at N–C and C.

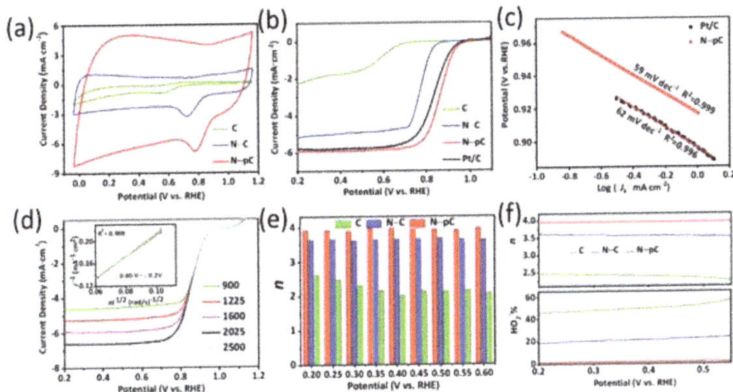

Figure 5. (**a**) CV curves of different catalysts in 0.1 M KOH solutions at a scan rate of 100 mV s^{-1}; (**b**) LSV curves of different catalysts in 0.1 M KOH solutions at a sweep rate of 10 mV s^{-1} and a rotation speed of 1600 rpm; (**c**) Tafel plots for N–pC and Pt/C; (**d**) LSV curves of N–pC in 0.1 M KOH recorded at various rotation speeds (900–2500 rpm) at scan rate of 10 mV s^{-1} (inset: corresponding K–L plots of N–pC at various potentials); (**e**) Electron transfer number (*n*) of ORR on N–pC electrode based on RDE test; (**f**) Electron transfer number (*n*) and corresponding peroxide yield of ORR on N–pC electrode at different potentials based on RRDE test.

In addition, the tolerance for methanol and durability are also important factors in fuel cells. The chronoamperometric response of N–pC and Pt/C catalysts to methanol crossover was tested in the O_2-saturated 0.1 M KOH (Figure 6a). After the injection of 1 M methanol, an obvious current attenuation is observed on Pt/C due to methanol oxidation, while there was a negligible change in N–pC, indicating that the N–pC catalyst shows much better tolerance for methanol than commercial Pt/C catalyst. Furthermore, the durability was also evaluated (Figure 6b). After a continuous operation for 16,000 s, the current density on N–pC remains at about 90%, whereas it only maintains 79% on Pt/C under the same condition, confirming that the N–pC catalyst has higher durability than the Pt/C catalyst.

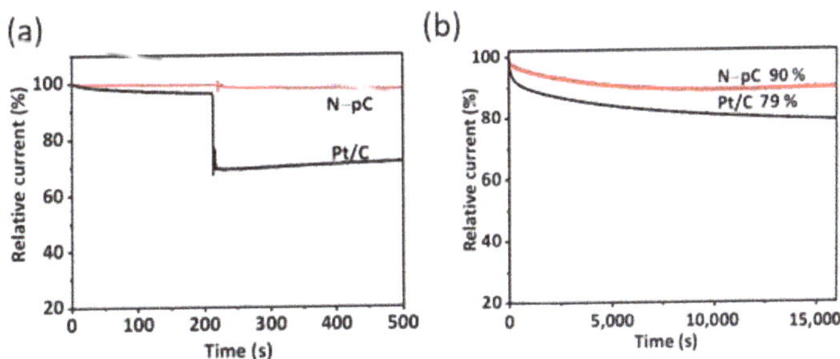

Figure 6. Comparison of (**a**) methanol tolerance and (**b**) stability for N–pC and 20% Pt/C catalysts measured at 0.7 V versus RHE in O_2-saturated 0.1 M KOH at a rotation speed of 1600 rpm.

To investigate the potential use of the N–pC catalyst in real batteries, the home made Zn–air battery (ZAB) was constructed using 6 M KOH solution as the electrolyte, polished pure Zn plate as the anode and N–pC or Pt/C as the air cathode, respectively. The open circuit voltage provided by ZAB using the N–pC catalyst is 1.50 V (Figure 7a), higher than that of 1.43 V for the commercial Pt/C-based one. The polarization and power density curves of the ZAB were given in Figure 7b. The peak power density of the battery equipped with N–pC catalyst achieves 223 mW cm^{-2} at a high current density of 347.8 mA cm^{-2}, higher than the Pt/C catalyst (182.6 mW cm^{-2} at 304.9 mA cm^{-2}), suggesting the N–pC can be employed as a potential cathode electrocatalyst for ZAB. The discharge curves of N–pC-based ZAB at different current densities were measured by typical constant current discharge test. As shown in Figure 7c, the voltage can quickly reach equilibrium when the current changes, confirming that the structure of N–pC is beneficial to the mass-transfer. When the discharge current density recovers to the initial value (10 mA cm^{-2}) after the high current discharge (50 mA cm^{-2}), the discharge voltage can be well restored to the initial value, suggesting the good stability of the N–pC. Finally, the ZABs were galvanostatically discharged at 50 mA cm^{-2} to investigate their stability and specific capacity. As shown in Figure 7d, the N–pC-based ZAB has higher and more stable voltage than that of the Pt/C-based one during the discharge process. When normalized to the mass of consumed Zn, the specific capacity of the ZAB prepared with N–pC exhibits 755 mAh g^{-1}, higher than that of 709 mAh g^{-1} for Pt/C and similar or even better than recently reported catalysts (Table S2). All these results suggest that the N–pC catalyst has a good application prospect in ZAB.

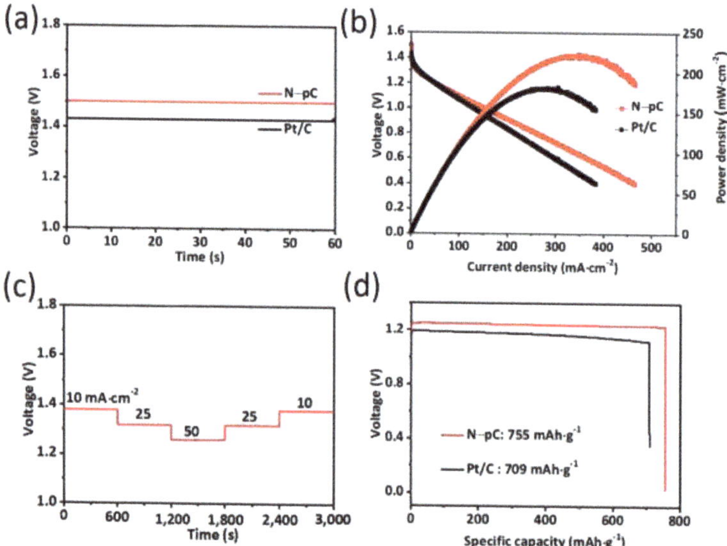

Figure 7. (**a**) Open circuit voltages and (**b**) discharge polarization curves of N–pC and Pt/C-based batteries; (**c**) The rate capability behavior of N–pC-based battery; (**d**) The long-term galvanostatic curves of the Zn–air batteries at 50 mA cm^{-2} using N–pC and Pt/C as ORR catalysts.

4. Conclusions

In summary, we successfully prepared a 3D, high-level nitrogen-doped, metal-free (N–pC) electrocatalyst employing urea as a single nitrogen source, NaCl as a fully sealed nanoreactor and gingko shells, constantly renewable and naturally available, as a carbon precursor. Due to the high content of active nitrogen, large surface area (1133.8 m^2 g^{-1}) and 3D hierarchical porous network structure, the N–pC showed a high onset potential of 1.020 V and a half-wave potential of 0.863 V, and superior tolerance for methanol and stability in alkaline media. Additionally, in ZAB, N–pC showed higher peak power density (223 mW cm^{-2}), larger specific-capacity (755 mAh g^{-1}) and better rate-capability than the commercial Pt/C-based ZAB, displaying a good application prospect in metal–air batteries.

Supplementary Materials: The following are available online at https://www.mdpi.com/article/10.3390/pr9122124/s1 Table S1: Comparison of electrocatalytic ORR performance between N-pC and state-of-the-art metal-free catalysts reported in the literatures in alkaline electrolyte; Table

S2: Comparison of the performance of primary Zn-air batteries assembled with various cathodic electrocatalysts.

Author Contributions: X.W. and H.Z. designed and carried out the experiment; W.H. and R.L. contributed to the formal analysis and data curation; writing—original draft preparation was mainly accomplished by W.H.; writing—review and editing was implemented by R.L., R.W. and J.S.C. All authors have read and agreed to the published version of the manuscript.

Funding: This research was funded by the Education Department of Sichuan Province (Grant 18ZB0494) and Sichuan Science and Technology Program (Grant 2020YJ0299).

Institutional Review Board Statement: Not applicable.

Informed Consent Statement: Not applicable.

Conflicts of Interest: The authors declare no conflict of interest.

References

1. Wang, W.-T.; Batool, N.; Zhang, T.-H.; Liu, J.; Han, X.-F.; Tian, J.-H.; Yang, R. When MOFs meet MXenes: Superior ORR performance in both alkaline and acidic solutions. *J. Mater. Chem. A* **2021**, *9*, 3952–3960. [CrossRef]
2. Cao, F.; Yang, X.; Shen, C.; Li, X.; Wang, J.; Qin, G.W.; Li, S.; Pang, X.; Li, G. Electrospinning synthesis of transition metal alloy nanoparticles encapsulated in nitrogen-doped carbon layers as an advanced bifunctional oxygen electrode. *J. Mater. Chem. A* **2020**, *8*, 7245–7252. [CrossRef]
3. Pan, K.-J.; Hussain, A.M.; Huang, Y.-L.; Gong, Y.; Cohn, G.; Ding, D.; Wachsman, E.D. evolution of solid oxide fuel cells via fast interfacial oxygen crossover. *ACS Appl. Energy Mater.* **2019**, *2*, 4069–4074. [CrossRef]
4. Hong, W.; Feng, X.; Tan, L.; Guo, A.; Lu, B.; Li, J.; Wei, Z. Preparation of monodisperse ferrous nanoparticles embedded in carbon aerogelsvia in situsolid phase polymerization for electrocatalytic oxygen reduction. *Nanoscale* **2020**, *12*, 15318–15324. [CrossRef]
5. Zhao, L.; Wu, R.; Wang, J.; Li, Z.; Wei, X.; Chen, J.S.; Chen, Y. Synthesis of noble metal-based intermetallic electrocatalysts by space-confined pyrolysis: Recent progress and future perspective. *J. Energy Chem.* **2021**, *60*, 61–74. [CrossRef]
6. Nie, Y.; Li, L.; Wei, Z. Achievements in Pt nanoalloy oxygen reduction reaction catalysts: Strain engineering, stability and atom utilization efficiency. *Chem. Commun.* **2021**. [CrossRef] [PubMed]
7. Yanhua, L.; Tan, N.; Zhu, Y.; Huo, D.; Zhang, Y.; Sun, S.; Gao, G. Synthesis of porous N-rich carbon/MXene from MXene@Polypyrrole hybrid nanosheets as oxygen reduction reaction electrocatalysts. *J. Electrochem. Soc.* **2020**, *167*, 116503. [CrossRef]
8. Ji, Y.; Dong, H.; Liu, C.; Li, Y. The progress of metal-free catalysts for the oxygen reduction reaction based on theoretical sim-ulations. *J. Mater. Chem. A* **2018**, *6*, 13489–13508. [CrossRef]
9. Zhang, X.; Wang, L. Research progress of carbon nanofiber-based precious-metal-free oxygen reaction catalysts synthesized by electrospinning for Zn-Air batteries. *J. Power Sources* **2021**, *507*, 230280. [CrossRef]
10. Wu, J.; Yang, Z.; Sun, Q.; Li, X.; Strasser, P.; Yang, R. Synthesis and electrocatalytic activity of phosphorus-doped carbon xerogel for oxygen reduction. *Electrochim. Acta* **2014**, *127*, 53–60. [CrossRef]
11. Yokoyama, K.; Sato, Y.; Yamamoto, M.; Nishida, T.; Motomiya, K.; Tohji, K.; Sato, Y. Work function, carrier type, and con-ductivity of nitrogen-doped single-walled carbon nanotube catalysts prepared by annealing via defluorination and efficient oxygen reduction reaction. *Carbon* **2019**, *142*, 518–527. [CrossRef]
12. Xu, Z.; Zhou, Z.; Li, B.; Wang, G.; Leu, P. Identification of efficient active sites in nitrogen-doped carbon nanotubes for oxygen reduction reaction. *J. Phys. Chem. C* **2020**, *124*, 8689–8696. [CrossRef]
13. Sun, Y.; Wu, J.; Tian, J.; Jin, C.; Yang, R. Sulfur-doped carbon spheres as efficient metal-free electrocatalysts for oxygen reduction reaction. *Electrochim. Acta* **2015**, *178*, 806–812. [CrossRef]
14. Wong, C.H.A.; Sofer, Z.; Klímová, K.; Pumera, M. Microwave exfoliation of graphite oxides in H_2S plasma for the synthesis of sulfur-doped graphenes as oxygen reduction catalysts. *ACS Appl. Mater. Interfaces* **2016**, *8*, 31849–31855. [CrossRef] [PubMed]
15. Van, T.T.; Kang, S.G.; Babu, K.F.; Oh, E.-S.; Lee, S.G.; Choi, W.M. Synthesis of B-doped graphene quantum dots as a metal-free electrocatalyst for the oxygen reduction reaction. *J. Mater. Chem. A* **2017**, *5*, 10537–10543.
16. Gong, K.; Du, F.; Xia, Z.; Durstock, M.; Dai, L. Nitrogen-doped carbon nanotube arrays with high electrocatalytic activity for oxygen reduction. *Science* **2009**, *323*, 760–764. [CrossRef]
17. Liu, R.; Wu, D.; Feng, X.; Müllen, K. Nitrogen-doped ordered mesoporous graphitic arrays with high electrocatalytic activity for oxygen reduction. *Angew. Chem. Int. Ed.* **2010**, *49*, 2565–2569. [CrossRef]
18. Dai, L.; Xue, Y.; Qu, L.; Choi, H.-J.; Baek, J.-B. Metal-free catalysts for oxygen reduction reaction. *Chem. Rev.* **2015**, *115*, 4823–4892. [CrossRef]
19. Dong, Y.; Yu, M.; Wang, Z.; Zhou, T.; Liu, Y.; Wang, X.; Zhao, Z.; Qiu, J. General synthesis of zeolitic imidazolate frame-work-derived planar-N-doped porous carbon nanosheets for efficient oxygen reduction. *Energy Storage Mater.* **2017**, *7*, 181–188. [CrossRef]

20. Wang, T.; Chen, Z.-X.; Chen, Y.-G.; Yang, L.-J.; Yang, X.-D.; Ye, J.-Y.; Xia, H.-P.; Zhou, Z.-Y.; Sun, S.-G. Identifying the active site of n-doped graphene for oxygen reduction by selective chemical modification. *ACS Energy Lett.* **2018**, *3*, 986–991. [CrossRef]
21. Wang, M.; Wu, Z.; Dai, L. Graphitic carbon nitrides supported by nitrogen-doped graphene as efficient metal-free electrocatalysts for oxygen reduction. *J. Electroanal. Chem.* **2015**, *753*, 16–20. [CrossRef]
22. Ding, W.; Li, L.; Xiong, K.; Wang, Y.; Li, W.; Nie, Y.; Chen, S.; Qi, X.; Wei, Z. Shape fixing via salt recrystallization: A morphology-controlled approach to convert nanostructured polymer to carbon nanomaterial as a highly active catalyst for oxygen reduction reaction. *J. Am. Chem. Soc.* **2015**, *137*, 5414–5420. [CrossRef]
23. Wang, W.; Luo, J.; Chen, W.; Li, J.; Xing, W.; Chen, S. Synthesis of mesoporous Fe/N/C oxygen reduction catalysts through NaCl crystallite-confined pyrolysis of polyvinylpyrrolidone. *J. Mater. Chem. A* **2016**, *4*, 12768–12773. [CrossRef]
24. Wang, D.; Xu, H.; Yang, P.; Xiao, L.; Du, L.; Lu, X.; Li, R.; Zhang, J.; An, M. A dual-template strategy to engineer hierarchically porous Fe-N-C electrocatalysts for the high-performance cathodes of Zn-air batteries dagger. *J. Mater. Chem. A* **2021**, *9*, 9761–9770. [CrossRef]
25. Sa, Y.J.; Seo, D.-J.; Woo, J.; Lim, J.T.; Cheon, J.Y.; Yang, S.Y.; Lee, J.M.; Kang, D.; Shin, T.J.; Shin, H.S.; et al. A general approach to preferential formation of active Fe-N-x sites in Fe-N/C electrocatalysts for efficient oxygen reduction reaction. *J. Am. Chemical Soc.* **2016**, *138*, 15046–15056. [CrossRef] [PubMed]
26. Hu, B.-C.; Wu, Z.-Y.; Chu, S.-Q.; Zhu, H.-W.; Liang, H.-W.; Zhang, J.; Yu, S.-H. SiO_2-protected shell mediated templating synthesis of Fe–N-doped carbon nanofibers and their enhanced oxygen reduction reaction performance. *Energy Environ. Sci.* **2018**, *11*, 2208–2215. [CrossRef]
27. Meng, F.-L.; Wang, Z.-L.; Zhong, H.-X.; Wang, J.; Yan, J.-M.; Zhang, X.-B. Reactive multifunctional template-induced prepa-ration of Fe-N-doped mesoporous carbon microspheres towards highly efficient electrocatalysts for oxygen reduction. *Adv. Mater.* **2016**, *28*, 7948–7955. [CrossRef] [PubMed]
28. Hou, J.; Wen, S.; Chen, J.; Zhao, Q.; Wang, L. Large-scale fabrication of biomass-derived N, S co-doped porous carbon with ultrahigh surface area for oxygen reduction. *Mater. Chem. Phys.* **2021**, *267*, 124601. [CrossRef]
29. Zhang, J.J.; Sun, Y.; Guo, L.K.; Sun, X.N.; Huang, N.B. Ball-Milling Effect on biomass-derived nanocarbon catalysts for the oxygen reduction reaction. *Chemistry* **2021**, *6*, 6019–6028. [CrossRef]
30. Borghei, M.; Lehtonen, J.; Liu, L.; Rojas, O.J. Advanced biomass-derived electrocatalysts for the oxygen reduction reaction. *Adv. Mater.* **2018**, *30*, 1703691. [CrossRef] [PubMed]
31. Gao, S.; Wei, X.; Fan, H.; Li, L.; Geng, K.; Wang, J. Nitrogen-doped carbon shell structure derived from natural leaves as a potential catalyst for oxygen reduction reaction. *Nano Energy* **2015**, *13*, 518–526. [CrossRef]
32. Arrigo, R.; Hävecker, M.; Wrabetz, S.; Blume, R.; Lerch, M.; McGregor, J.; Parrott, E.P.J.; Zeitler, J.A.; Gladden, L.F.; Knop-Gericke, A.; et al. tuning the acid/base properties of nanocarbons by functionalization via amination. *J. Am. Chem. Soc.* **2010**, *132*, 9616–9630. [CrossRef] [PubMed]
33. Tao, G.; Zhang, L.; Chen, L.; Cui, X.; Hua, Z.; Wang, M.; Wang, J.; Chen, Y.; Shi, J. N-doped hierarchically macro/mesoporous carbon with excellent electrocatalytic activity and durability for oxygen reduction reaction. *Carbon* **2015**, *86*, 108–117. [CrossRef]
34. Ferre-Vilaplana, A.; Herrero, E. Understanding the chemisorption-based activation mechanism of the oxygen reduction reac-tion on nitrogen-doped graphitic materials. *Electrochim. Acta* **2016**, *204*, 245–254. [CrossRef]
35. Guo, D.; Shibuya, R.; Akiba, C.; Saji, S.; Kondo, T.; Nakamura, J. Active sites of nitrogen-doped carbon materials for oxygen reduction reaction clarified using model catalysts. *Science* **2016**, *351*, 361–365. [CrossRef] [PubMed]
36. Lai, L.; Potts, J.R.; Zhan, D.; Wang, L.; Poh, C.K.; Tang, C.; Gong, H.; Shen, Z.; Lin, J.; Ruoff, R.S. Exploration of the active center structure of nitrogen-doped graphene-based catalysts for oxygen reduction reaction. *Energy Environ. Sci.* **2012**, *5*, 7936–7942. [CrossRef]

MDPI

Article

Designing a Tool for an Innovative, Interdisciplinary Learning Process Based on a Comprehensive Understanding of Sourcing: A Case Study

Radim Rybár, Ľubomíra Gabániová *, Jana Rybárová, Martin Beer and Lucia Bednárová

Institute of Earth Resources, Faculty of Mining, Ecology, Process Control and Geotechnology, Technical University of Kosice, Letna 1/9, 040 01 Kosice, Slovakia; radim.rybar@tuke.sk (R.R.); jana.rybarova@tuke.sk (J.R.); martin.beer@tuke.sk (M.B.); lucia.bednarova@tuke.sk (L.B.)

* Correspondence: lubomira.gabaniova@tuke.sk or lubka.gabaniova@gmail.com

Abstract: The paper presents a case study describing the process of creating and validating the benefits of two Innovative Learning Tools (ILTs) aimed at more effective knowledge acquisition in the interdisciplinary field of earth resource extraction with links to the status of renewable energy. The philosophy behind the original designs and the design of the two tools, the way they are used, and the results of their application in the educational process are presented in a framework. The opinions and attitudes of both students and educators towards the tools were surveyed, and some research questions related to this form of knowledge acquisition were validated. The presented results show the students' interest in the educational form as well as the attractive content that goes beyond conventional educational subjects, with its connections.

Keywords: education; RES; Innovative Learning Tools

Citation: Rybár, R.; Gabániová, Ľ.; Rybárová, J.; Beer, M.; Bednárová, L. Designing a Tool for an Innovative, Interdisciplinary Learning Process Based on a Comprehensive Understanding of Sourcing: A Case Study. *Processes* **2022**, *10*, 9. https://doi.org/10.3390/pr10010009

Academic Editor: Jorge Cunha

Received: 24 November 2021
Accepted: 20 December 2021
Published: 22 December 2021

1. Introduction

The issues of the impacts of anthropogenic climate change on human society appear to be one of the most pressing issues of our time, reflected in political, economic, social, and personal terms [1–6]. Addressing the issues is the subject of major events such as summits (Climate Adaptation Summit) [7], meetings of international organizations (Glasgow Climate Change Conference) [8], scientific conferences (2nd International Conference on Environmental Science and Applications (ICESA'21)) [9], and so on.

In addressing, or rather finding solutions to, the identified problems, the massive take-up of Renewable Energy Sources (RES) is often seen as a solution to the problem. While in 2010 the world's energy production from RES was at 761.2 TWh, in 2020 it was already 3147 TWh [10]. In 2021, 70% of the 530 billion USD spent on energy production should be invested in renewable energy sources [11]. There are also various youth demonstrations taking place around the world for sustainable development and climate change mitigation measures, as well as for the reduction of emissions and decarbonizations of human activities [12,13].

However, aspects of the problem remain in the background, which are related to the overall shape and size of the current human society with a population of 7.9 billion [14] built on the intensive use of natural resources (raw materials, water, land, forests), the global economy, regional differentiation, transport, etc. [15] Understanding the material base and the energy background of these aspects is essential for the creation of a feasible framework of measures that can somehow lead to at least partial elimination of the problems or to the creation of more effective tools for solving some of the current problems [16]. For this reason, it is important to understand the needs and extent of the application of renewable energy sources in practice and this is not possible without a consistent and comprehensive understanding of the source of raw materials and energy in their historical and current form in the context of the global economy and society [17–19]. This concerns the people working in the industries concerned (mining, energy, transport, processing industry, waste management, environmental protection, etc.), politicians, people creating a legislative framework for the implementation of new solutions, access to and technologies using portfolio RES, but also the common public, which is the bearer of certain ideas due to public opinions or awareness, which create the basis for potentially global applicable and effective changes in socio-economic activities. An important role in this is played by the building of awareness and knowledge among the next generation, which takes place in the school environment through the educational process [20]. Additionally, on the basis of many years of teaching experience in the field of higher technical education, we can state that young people often take very courageous and ambitious attitudes to problems, which, however, are often not based on sufficient knowledge of the issues or are not based on an adequate knowledge base. It is in this area that attitudes to the issue of securing the energy and raw material base of human society today and in the future also fall. Education on RES as well as on earth resource extraction in general is currently carried out conventionally and is tied to isolated topics within individual scientific disciplines. Conventional educational forms often do not lead to the desired results, if only because of the low degree of attractiveness, which is illustrative in relation to the empirically acquired subjective perception of reality. For this reason, various other educational forms are emerging, in the creation of which their authors try to include elements with added attractive and motivational value, as well as integrating inter- and multi-disciplinary approaches [21–23].

In the FMEPCG (Faculty of Mining, Ecology, Process Control and Geotechnologies) environment, we create many interesting activities in this direction, which have been subject to a certain natural evolution, conditioned both by the feedback and achievements of previous work, triggered by external factors. The key factor for faster development in the last years was the pandemic situation related to COVID-19, and that was an imaginary wheel, which led us into faster changes in the teaching process.

The present paper presents a view of the process of creating an innovative educational form and its content and the results achieved, as a kind of case study that can be an important source of data, information, and experience for other analogous activities.

The process of developing those tools for this type of education includes didactic, creative/innovative, and professional knowledge in the disciplines and much information, which are included in a broader context point of view.

Looking at the learning process itself, the theoretical aspects of the mechanism have been comprehensively elaborated and published in many publications [24]. According to Vester [25], the types of learning are based on five types of perception, according to which channel is dominant in the reception of information, and, thus, we distinguish: auditory type, visual type, learning by reading, tactile type, and combined type. In this context, for example, the source [26] cites Roy's (1995) [27] data on what and how we remember, which accounts for 10% of what we hear, 15% of what we see, 20% of what we simultaneously see and hear, 40% of what we discuss, 80% of what we directly experience or do, and 90% of what we try to teach others. The motivational aspect of curiosity-play may play a very important role here, as several studies suggest. According to the source [28], maintaining the "curious spirit of the young child" may be one of the keys to successful learning. Here, there is a possible connection between all three domains of learning according to Bloom [29], namely, the cognitive domain (learning to recall certain facts, to think, make decisions, and to get better at it, to acquire a certain skill), the psychomotor domain (learning practical skills), and the affective domain (to develop values, feelings, and beliefs).

There are several studies similar to the one which is presented in this article that have investigated the incorporation of innovative elements into the teaching process and their impact on students' ability to retain information for longer periods of time. Under innovations or innovative educational toolkits, we understand the implementation of materials, games, group projects, and toolkits that provide information and knowledge to students in a form other than convectional. There are more interesting ways to show very essential information to people and, as was mentioned before, this area is more attractive for Z generation. This generation is more educated in the use of new technology and the app that was created, giving them opportunity to be in familiar fields and giving them more perspective and an interesting education. Kvam investigated the effect of Active Learning Methods on student retention in Engineering statistics, and his findings suggest that active learning can help increase retention rates for average or below average performing students [30]. Zakaria found that cooperative learning improved math scores by nearly 11%, and 13% of students had a positive attitude towards math after this style of learning [31]. Risnita and Bashori compared conventional and contextual learning, finding that the contextual learning group outperformed the conventional learning group by 20.9% [32]. According to Ponomarenko, innovative teaching at selected universities is more prevalent in economics majors, less so in engineering majors, which causes students to be underprepared for an interdisciplinary perception of knowledge [33]. The positive effect of innovation in the educational process was also confirmed by Setiawan [34]. The results of Sadeghi's study were that the combined method of education is more satisfying for students, but it is more time consuming and more demanding for the teacher to prepare; moreover, the results of both methods (classical lectures and e-learning) were very similar [35].

In order to develop creative abilities and to link the different cognitive domains in education in the field of earth resource extraction and the use of renewable energies, the Laboratory of Earth Resource Extraction (LERE) was established at FMEPCG in 2012 (Figure 1) [36], which soon brought results not only in stimulating a deeper interest in the unconventional way of rendering reality but also in new application possibilities of this creative environment, producing a number of outputs in the forms of original bachelor and diploma theses, several of which have won leading positions in faculty and international rounds of the Student Scientific Competition and Student Scientific Conference as well as prizes of the Association of Slovak Scientific and Technical Societies. Students quickly understood the potential of such "playful" teaching and showed a real interest in applying their creative abilities during their studies and in their development under appropriate pedagogical guidance.

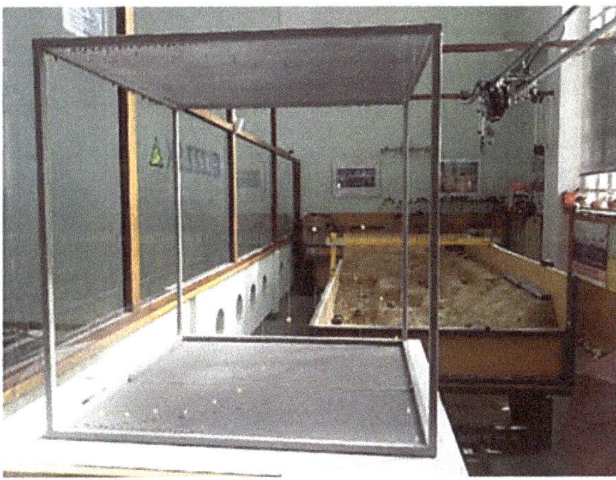

Figure 1. Laboratory of Earth Resource Extraction, Cube [37].

In 2020, the LERE was expanded with the ECL Engineering Creativity Laboratory, which provided a space for the development of creative abilities of gifted students and managed to integrate creative workers and students into effective working teams capable of solving complex tasks and creating new solutions. One such task was the implementation of game or competition elements in the process of knowledge acquisition or mastering tasks. It was possible to design the basic framework of a platform that combined the real-artifact and virtual worlds in an ICT environment and to create a hierarchically structured model, which was piloted in the field of earth resource acquisition and applied in the teaching of students in several high schools (under the guidance of university teachers). The proposed educational product won the first prize of the V4 Raw Materials Ambassadors Schools in Budapest in September 2019.

We have proposed two IETs, one with a tangible platform (preCOVID), the other without a tangible platform. Both had a common and different concept compared to other tools in the way of game implementation and competition elements (level conquest) and variability of individual tasks and items. The innovativeness of the presented tools is on two levels. For the first IET, besides the content itself, there is the rendering of the IET itself in the form of a structured box with progressively unlockable cartridges with individual tasks and the rendering of some of the tasks (the educational set), the designs of which are also subject to intellectual property protection. In the case of the second IET, it is primarily an informal link between a timeline and a quiz built on a broad base of disciplines and areas of human activity. The aim of the research was to verify the functionality of the proposed innovative IET educational tools focused on interdisciplinary issues of raw materials and energy, understanding their role and complexity and the diversity of relationships with emphasis on the attractiveness of education and building stable attitudes to students.

The contribution of the paper is the emphasis on the importance of the positive role of non-conventional educational tools, which use knowledge from different scientific disciplines and allow covering more complex issues of RES. Overall, it will be given more information about effective extraction of the earth's resources in a form that can be more durable and helps to build students' personal attitudes towards the issue, which can then be reflected in the level of practice as well as in the political level.

2. Materials and Methods

The materials were created by the IET and the methodology was oriented towards testing them.

2.1. Content: Professional Background in the Field of Earth Resources' Extraction

When creating the content of the new educational tool, i.e., its content, the complexity of the content was taken into account in terms of the interconnectedness of the individual phases of the formation, acquisition, and use or practical application of individual raw materials, chemical elements, or chemical compounds of which they are the source. For the purpose of the primary design of the innovative educational tool, 12 items (raw materials/elements/compounds) were selected and divided into groups: metals (titanium, chromium, copper, lead), energy raw materials (coal, oil, natural gas, uranium), non-metals (gypsum, talc, salt, sulphur), and mix (quartz/SiO_2, limestone, iron, clay). In addition to the criterion of overall importance (position in the industry), aspects of coverage of variability of the different aspects were taken into account when selecting the individual items.

Energy raw materials (fossil fuels) and their importance are key to understanding the position of renewables in the role of alternative energy sources (from the beginning of the industrial revolution to current climate impacts). The volume of global daily oil consumption, which is around 92 million barrels [38], can serve as a representative indicator, which, if adequately presented, is a good indicator for understanding the quantitative aspect of humanity's use of fossil fuels. The mineral energy source, uranium (a radioactive raw material), is, on the other hand, exceptional in its nature (origin of the element itself; understanding of the meaning of isotopes; accumulation of the useful component in the Earth's crust; and the process of mining, extraction, enrichment, fuel production, fission reaction, reprocessing, production of the artificial element/fuel Plutonium, radioactivity of the fission products, waste management, etc.). The category "energy feedstocks", thus, provides a good basis for understanding the current shape and structure of the sector in terms of providing for the energy needs of society.

The metals' category largely represents the material and technical base of contemporary technology. Copper as a key metal for electrical engineering reflects the state of the industry. Chromium represents a metal that is important for the production of noble steel types of corrosion-resistant and refractory alloys. It is also an element whose useful minerals come from ultrabasic rocks [39], which is important for understanding the evolution of raw material deposits. Titanium represents a metal with a high added value in its chemical and physical properties, on the one hand, and a paradoxical disparity between its abundant percentage in the Earth's crust [40] and the rarity of its deposits [41]. Lead represents a metal that has been of great importance in technical practice in the past but, which, as an element, is an excellent example to understand the evolution of the elements (the final product of many decay series [42]) and their stability (it is a landmark between radioactive and stable elements in the periodic table of elements [43]). It also serves to document the process of ionizing radiation shielding.

The non-metals' category adds to the range and variability of materials in terms of their origin and importance in their application. Gypsum is a paradox in that not all gypsum is made from gypsum (energy gypsum from the flue gas desulphurization process in coal-fired power stations) [44]. In addition to pointing out the fact that talc is a bulk constituent of drugs, it is useful to use the example of talc to highlight the issue of ensuring the stability of underground works in relation to cohesiveness and other geotechnical properties of the geological environment. Salt is one of the raw materials that has been sought and obtained by humans since time immemorial and whose deposits show the geological evolution of the environment related to the evaporation of seawater (evaporites) [45]. Sulfur, in addition to its occurrence in elemental form, is the basis for the production of sulfuric acid, an important input in the processing of ores, petroleum refining, and a number of manufacturing industrial processes.

The mix category covers raw materials that are used in large volumes as the basis for construction and building materials. Quartz/SiO_2 represents, on the one hand, the basic input for semiconductor production [46] and, on the other hand, quarried aggregates and glass raw materials [47]. Limestone represents the basic raw material for the production of cement [48], lime, crushed aggregate, building stone, and input for production and treatment processes in industry (metallurgy, paper, sugar production) and energy (desulphurization) [44]. Although iron is undoubtedly a metal, on the other hand, in terms of the genesis of deposits, it has a special position because, according to the geologist H.S. Washington [49], iron figures in the group of both petrogenic and metallogenic elements. Iron, as an element located on the boundary between the two groups, has a dual role in nature, participating in the formation of rock-forming minerals as petrogenic elements but also forming typical heavy metal minerals as metallogenic elements. The importance of iron to human society has been increasing since the Iron Age, with the construction industry currently the largest consumer of iron/steel [50]. Clay represents an important raw material with a wide range of uses (production of bricks and building and sanitary ware, ceramics, etc.) and an important link to the development of man and human society [51,52].

2.2. The Process of Creating an Innovative Educational Tool Oriented to the Field of Earth Resource Extraction

The creation of the innovative learning tool was linked to the EIT RawMaterials: RM@Schools project, for which reason standard items such as name/acronym, benefits to be achieved, definition, and structuring of objectives were defined. At the same time, a credo was formulated that captures the main idea of the proposal and reads as follows: "The adventure of knowledge is in the context". There are several other toolkits created under this project to enrich the educational process and to highlight the broader connections between the current economic situation in the world, the circular economy, renewable energy sources, and raw materials, such as FIND the elements (Treasure Hunt 2.0, RawMatCards, RawMatCards, RAWsiko), Materials Around Us (Board Game or Digital version), RockGame, MineralCheck, Recycling of silicon based PV modules, etc. [53].

The proposed innovative educational tool was given the form of a game. The primary task for the players was not to learn or understand something, but to successfully complete all the tasks of the game in the shortest time. By transferring the tasks to a competitive level and not emphasizing the process of knowledge acquisition or by presenting knowledge of an interdisciplinary nature that is not the primary focus of their curricula in each differentiated learning subject students' prejudices and overall reticence towards the educational process were eliminated.

The game design pursued three sets of objectives:

1. Cognitive goals:
    - To understand the complexity of the issue of resource extraction and earth resources in general.
    - To learn to perceive and identify the relationships between reality and its material background and to understand the key importance of earth resources in this relationship.
    - Gain the ability to understand analogies and model problem descriptions.
    - Gain the ability to extract a problem from its complex verbal formulation or broader description and develop critical thinking skills.
2. Value Goals:
    - Acquire the ability to take an attitude towards things, phenomena, and work activities in their environment primarily in terms of their social, material, and energetic value.
3. Operational Objectives:
    - To acquire the ability to work cooperatively in a team and to deal operationally with tasks of a diverse nature.
    - To learn how to actively use information technology in the educational process and in play at the same time.

As the basic concept of an innovative educational tool, the game was designed into two platforms.

1. Material platform, which is a box containing boxes with objects (artifacts necessary for solving tasks).
2. A virtual platform, which is a tablet containing an application necessary for the implementation of the individual tasks and stages of the game.

2.2.1. Description of the Material Platform

The wooden box itself contains eight separate, removable wooden boxes, each divided by dividers into four compartments. The boxes are stacked in the box to create two horizontal levels, each containing a quartet of boxes. Each level of boxes represents one level of tasks. Each compartment of the box contains a tangible (3D) artifact needed to solve the problem or to better understand the problem under analysis (e.g., a sample of a mineral, a product or material containing the chemical in question, and element or compound, a model or mock-up of an object, etc.).

The boxes placed on top of each other are physically connected to each other by a mechanism containing a lock with a numerical code accessible from the top. The numerical code is four digits. Knowledge of each digit in the code is the result of correctly answering a problem/question from one compartment of the box. In this way, after solving the four problems from one box, the code to unlock the lock can be obtained, so that the box in question can be removed from the box and, thus, the box located underneath it can be accessed. The bottom level box is connected to the bottom of the box in an analogous way. Upon removal of the bottom level box, the bottom of the box is revealed to contain a third level of play consisting of 2D artifacts, panels that are pictures or other two-dimensional aids (such as the periodic table of elements). The sequence and continuity of steps in the game itself is analogous to the sequence and continuity of steps in natural processes associated with the formation of deposits and anthropogenic activities associated with exploitation, work operations associated with extraction of a useful component, the processing of a raw material into a usable form or its subsequent use, or steps associated with other activities such as waste disposal, recycling, environmental impacts, and the like.

The technical design of the material platform itself has become the subject of intellectual property protection and is protected by a utility model.

2.2.2. Description of the Virtual Platform

The virtual platform consists of a tablet with a web application. The web application is designed and configured to guide the users of the game in a hierarchical, logical, and intuitive way from the beginning of the game until its end. Each question arising from a material artifact contained in the 'material platform' of the game will be accompanied by an application, whether at the level of the formulation of the task, support in mastering an activity arising from the task, entering the answer to the question or the results of the task, the use of additional information (aids), the generation of numerical codes, the accompaniment of the game itself, the timekeeper, etc.

Each of the three levels of play, based on a material platform, is ideologically differently oriented. For each problem, items such as the accompanying text (providing relevant information and indicating relationships and connections in addition to an introduction to the problem itself), the wording of the question or problem (formulated in such a way as to clearly define the subject of the solution), the answer choices (multiple (four) answer formulations, one of which is correct), and the aids in the case of an incorrect answer (helping to draw attention to the point of the problem) are formulated.

The first level of the game is focused on the identification of the utility component of an element, a chemical compound, and its source mineral. Either mineral samples (utility mineral, rock) or objects containing or unambiguously representing the main utility component serve as artifacts. Examples of first-level artifacts include:

Coal, a sample of lignite with a distinct structure;

Uranium, a sample of uranium glass and a UV lamp (to induce a luminescent effect);

Copper, chalcopyrite sample; and

Chromium, a sample of electroplated chrome-plated plastic (in this case the product appears to be more illustrative compared to the mineral itself).

The individual boxes of the second tier will represent the same groups of raw materials as for the first tier, and the individual compartments will follow each other vertically.

The second level of the game focuses on the extraction, processing, treatment, transportation, trading, and technological aspects in relation to the raw material or its utility component. Various objects will serve as artifacts, either demonstrating an important characteristic or method, most often using analogies or helping to understand the specifics of the area of raw material extraction and processing.

Examples of artifacts with second- level task descriptions include:

Coal, a wooden block with holes arranged in a regular grid. Marked "pins" are inserted into the holes so that they protrude above the plane of the surface and can be pulled out of the block. This object represents a block of rock containing one or more coal seams. The pins represent the boreholes and will be colored so that the course of the seam in the object under examination can be clearly seen. The thickness of the seam will vary in different places, including its onset or dislocation. The present solution is the educational file and has become the subject of intellectual property protection and is protected by a utility model.

The task is to find out the volume of coal in the examined block, in such a way that the players successively pull out individual pins, which they place on the tablet display at a place where there will be a pair of runners (cursors), which they set to point to the lower and upper boundaries of the strata (Figure 2). They repeat this process with all the pins, at which point the app calculates the volume of coal in the bed. If they have followed the procedure correctly, they will get a result from a certain interval, which will be the correct answer. Among other things, this task is aimed at understanding that we can only "see" into the rock environment indirectly, and always our idea of the deposit is only a better or worse approximation to reality conveyed indirectly.

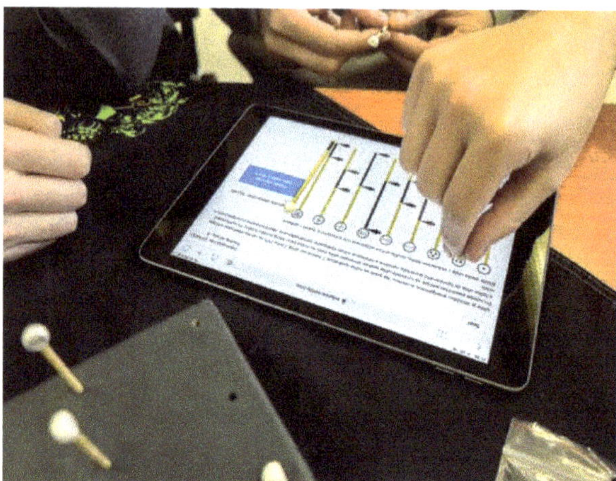

Figure 2. Students using educational toolkit [authors].

Copper, a similar block to coal, but the spikes will have cross lines on them, the density (spacing) of which will represent the pay grade of the deposit. By placing a pin on the scale displayed on the tablet and setting the runner to the appropriate part of the scale corresponding to the line on the pin, the player (competitor) will get an indication of the pay grade ranging from 0.1–1.0%.

The challenge is to match the mining in different parts of the deposit to achieve the desired pay grade in the mined material. This task is aimed at understanding that, in addition to knowing where the useful mineral is located, we also need to know its quality or the content of the useful or, conversely, the harmful component.

Natural Gas, a plastic sphere containing a liquid, which is an analogy for the spherical tanks on LNG carriers, due to the understanding of the behavior of the liquid when moving in a tank of this shape and the importance of the global transport of fossil fuels.

The third level of the game focuses mainly on the area of the use of the raw materials itself or its utility component and the products made from it.

Examples of artifacts include:

Copper, a Picture of a copper brick (used for investment purposes), and

Lead, the Periodic table of elements with radioactive elements clearly marked.

In the final phase of the game, after all tasks have been solved, students/players are presented with a representation of a complex object (in the presented case study it was a hospital) in a virtual platform (via a created application) in a visual, simplified form in cross section, while the task of the contestants is to insert into the picture the symbols of the individual items (chemical elements/materials) that are significantly represented in each object or are behind the supply of energy. Assigning all the items to all the required boxes completes the process of the game itself and the timer stops.

2.3. *Evolution of an Innovative Educational Tool (IET) in the Light of the Need for Distance Learning*

Due to the limitations of the measures related to the pandemic of COVID-19, it was decided to modify the originally designed educational tool into a form that does not include a material platform and can, therefore, be implemented remotely.

The first part is the creation of a timeline of the use of an item (chemical element/compound/compound). The second part is a quiz covering the different disciplines or areas of human activity in which the item in question figures. Both parts have been incorporated into the application.

The timeline covers the range from the Paleolithic to the present, and each time milestone is represented by symbols such as hieroglyphics on artifacts from the ancient Egyptian Empire, the Mona Lisa, or the Woodstock Festival.

In the first phase, the competition was set up for two chemical elements or two substances, gold and silicon/SiO_2 (two competing teams).

Substances were deliberately chosen that have an exceptionally long history of use by humans, in order to maximize the range of the timeline and also to highlight the changes in the function and use of these substances throughout history, which were directly related to the development, progress, and changes in society. In the first part, students/competitors were tasked with placing particular events on the timeline (e.g., gold, the first documented gold hoard, the first gold money, the introduction of the gold standard; Si/SiO_2, dating the first flint tools, the first glass, the first silicon semiconductor components).

For the quiz part, questions were worked out in 10 areas (Gold: physics, chemistry, circular economy, sports, astronomy, education, geography, electrical engineering, metallurgy, art; Si/SiO_2: physics, chemistry, agriculture, astronomy, watches, personalities, computing, engineering, geology, geography), the composition of which aimed to cover as broad a base as possible for the use of each of the substances, from the key to the unexpected applications. For each of the 20 quiz questions, a video was produced explaining the answer to the question in an illustrative manner, ranging from 0:27 min to 2:20 min. Examples and forms of demonstration are used, using original perspectives, techniques, and punchlines. The time scale and content are conditioned by the desire to gain maximum attention while fixing the presented information to content with sold value (visual, humorous).

The proposed educational tool has been used in the context of a secondary school curriculum oriented towards interdisciplinary issues of earth resources, environmental protection, climate change awareness, and renewable energy sources.

2.4. Collecting Data from Visits to Secondary Schools and Using Games in the Education Process

Following the completion of the first IET, which took place in 2020, three grammar schools were visited (one day for one school, with the visit lasting 2 h) where 86 students (15–19 years old) and five teachers were involved in the IET trial. Their opinion about the toolkit was collected by interviews, and citations of their answers can be seen in Table 1.

Table 1. Citation of answers obtained in the framework of the implementation of the feedback to the first IET.

Group of Goals	The Text of the Target	Text of the Answers
Cognitive goals	Understand the complexity of the issue of obtaining raw materials and land resources in general.	"We were surprised by the connection between the common things around us and the energy sources and raw materials behind them."
		"It looks like everything around us is completely dependent on raw materials and energies that we don't even know where they come from or what is happening to them."
	To learn to perceive and identify the relationships between reality and its material background and to understand the key importance of earth resources in this relationship.	"Without people knowing to use resources" efficiently, "our society probably wouldn't have developed much and today we wouldn't be where we are in both positive and negative ways"
		"It is already clear that RES are only part of the energy mix and the role of individual sources depends on many factors and is different for each country and economy, and therefore RES cannot be considered as a flat rate solution."
		"Many fundraising activities take place completely outside of us, so we don't think they exist, the more shocked they are of great importance to us."

Table 1. *Cont.*

Group of Goals	The Text of the Target	Text of the Answers
Cognitive goals	Understand the complexity of the issue of obtaining raw materials and land resources in general.	"We were surprised by the connection between the common things around us and the energy sources and raw materials behind them."
		"It looks like everything around us is completely dependent on raw materials and energies that we don't even know where they come from or what is happening to them."
	To learn to perceive and identify the relationships between reality and its material background and to understand the key importance of earth resources in this relationship.	"Without people knowing to use resources" efficiently, "our society probably wouldn't have developed much and today we wouldn't be where we are in both positive and negative ways."
		"It is already clear that RES are only part of the energy mix and the role of individual sources depends on many factors and is different for each country and economy, and therefore RES cannot be considered as a flat rate solution."
		"Many fundraising activities take place completely outside of us, so we don't think they exist, the more shocked they are of great importance to us."
	Gain the ability to understand analogies and model problem description.	"It's great as you can find in common things chemistry, mathematics, geometry, physics, biology and the like, that's how it makes sense."
		"When we learn such things at school, the connections are not there or no one will tell us."
		"I really liked that we went through the individual levels and it was necessary to complete things in order to find the code and go on, we wanted to win."
	Gain the ability to extract the problem from its complex wording or broader description, development of critical thinking.	"It was probably the hardest thing to understand what the role involved in the coal and copper ore models was, but when you try to imagine it and realize that we have only indirect information about things underground, for example, it started to make sense. "
		"Those tasks with 'sticks' looked pretty scientific, we had to read it a few times to understand what we were supposed to do in the end, but then it went well."
		"It was interesting to choose from a variety of answers that contained logical reasoning, but only one was correct."
Value goals	Gain the ability to take a stand on things, phenomena, and work activities in your environment, especially in terms of their social, material, and energy value.	"Now we understand that each person's actions and attitudes contribute to the pressure on energy and raw materials."
		"We already understand why the recyclability of some things, such as glass, copper, or iron, does not take into account the energy dimension of the problem, which is important to take into account."
		"Replacing all those 100,000,000 barrels of oil a day with another energy source is probably quite unfeasible and at some point soon unrealistic."
		"The link between the size of the human population and its material level and the degree of global environmental, economic growth, and climate impacts are still fully interlinked."
		"The behavior of an individual in the system seems to be important, because of his direct energy consumption but also because of the demands on the raw materials and energy that are in the products he buys and so on."
Operational goals	Gain the ability to cooperate in the team and operationally solve tasks of diverse nature.	"It took us a while to share who would do what, but then it worked well when the tasks were divided; it saved time."
	To learn to actively use information technology in both the educational process and the game.	"The fact that we had those things in our hands was great. The application in the tablet was well done, as such a guide, when we could not cope, so we read it again. The tools were also great after the wrong answer; you clarified enough."

The testing of the second IET, a quiz application, involved two grammar schools (one day for one school, visiting lasted 2 h) during summer 2021 with 42 students. In addition, the application was tested at the SlovakTech event where another 42 students from different high schools were involved. In total, 84 students (15–19 years old) tried the toolkit. The selection of the schools as well as the participating students was random. The procedure was the following:

1. Research questions (A, B, C) were defined for which null and alternative hypotheses (H0, H1) were formulated.
   - The Research question was A: Can education using the tools of interdisciplinary competitions or games lead to a better understanding of the context of earth resource acquisition, compared to conventional education on aspects of these issues covered in individual subjects?
     - A-H0: Education through an interdisciplinary competition or game does not provide a better understanding of the relationships in the field of earth resource extraction than conventional education within individual subjects.
     - A-H1: Education through interdisciplinary competition or games provides a better understanding of the relationships in the acquisition of earth resources than conventional education within individual subjects.
2. In order to test the hypotheses for research question A, a questionnaire was formulated with the following wording:

   Is it effective to educate on the subject in the form of a competition or a game?
3. Research question B: Does information in the subject area acquired informally, through tasks fixed to experience, have a more lasting character than that acquired through standard formal channels?
   - B-H0: Information given informally is not of a more permanent nature than the information obtained through conventional sources.
   - B-H1: Information given in an informal way is more durable than information obtained in a conventional way.
4. In order to test the hypotheses for research question B, a questionnaire was formulated with the following wording:

Can information given in an informal way be more lasting than that given in a formal way?

5. Research Question C: Does the fact that the educational competition/game contains any 3D artifacts and tangible objects have an impact on the engaging nature of the educational competition/game?
   - C-H0: The attractiveness of an educational competition/game is not enhanced by enriching it by adding 3D and tangible objects.
   - C-H1: The attractiveness of the educational competition/game will be enhanced by adding 3D artifacts and tangible objects.
6. In order to test the hypotheses for research question C, a questionnaire was formulated with the following wording:

   Would the contest be more engaging if it included some tangible artifacts?
7. The subject questions for each research question were included in a questionnaire that also included questions aimed at verifying knowledge.
8. All participants first filled in a questionnaire where there were eight questions focused on students' knowledge about gold and silicon and three questions focused on their opinion about learning through the game.
9. The participants were divided into two teams. Each team chose a captain who was to interact with the toolkit leader and chose an element.
10. A timeline was arranged.

11. There was a quiz (Figure 3), in which students could choose the number of points per question (100, 300, 500) and the area from which they wanted to answer the question. This also allowed them to practice their strategic thinking.
12. The team that had the higher number of points won.
13. Then, after the game, the students completed the same questionnaire again with the same questions.

Figure 3. Innovative educational toolkit application [authors].

Students' responses to the survey questions focusing on their opinion of the toolkit, which were the subject of hypothesis testing, were evaluated using Non-Parametric Related-Samples Wilcoxon Signed Rank Test using SPSS software. The results of the questionnaire as well as the evaluation of the hypotheses are presented in Section 3 Results.

3. Results and Discussion

3.1. Evaluation of the First IET

After the game, qualitative feedback was obtained through dialogue with the student contestants and with the teachers providing the teaching in the schools from which the students came.

The questions asked by the IET designers were largely aimed at verifying that the set cognitive, value, and operational objectives had been met. They were directed to the content area, the formal area, aspects of the novelty of the information, its relevance for further perception of the issue, and the construction of attitudes towards it.

The following findings emerged from the interviews:

- Students perceived positively that they were learning in the form of a game, focusing on problem solving and they did not see the actual taking in of information as learning but as overcoming an obstacle to progress through the game.
- They rated individual tasks from moderately to very challenging. The tasks with the highest level of abstraction (tasks with educational sets) posed the greatest challenge.
- Overall, students expressed interest in this kind of learning.
- The issues presented were largely surprising to the students and many of the facts presented were described as fundamentally new.
- The overall impression of the presented content of the game was the clarification that the human society is directly intertwined with the raw material energy resources, their availability, and rate of use, with an emphasis on energy resources.
- The students largely lost the vision of simple solutions consisting of the quantitative substitution of non-renewable resources (mainly fossil fuels) by renewable resources, which did not reflect other aspects and their limits.

- Students presented concerns about the future direction of human society and the possibility of effectively addressing the problems related to the impact of human activity on the environment and climate change.
- Teachers unanimously expressed a positive attitude towards the implementation of IET in the teaching process as a means of gaining attention and engaging students in the process of acquiring interdisciplinary knowledge through team creative work in solving competitive tasks in the game itself. Educators also expressed surprise at the presented connections between different phenomena related to the use of earth resources. Educators expressed the view that it is very difficult to translate the complex issue of earth resource extraction into individual, differentiated teaching subjects pursuing primarily different objectives.

Specific types of responses are listed in Table 1 according to the group of objectives pursued.

3.2. Evaluation of the Second IET

A total of 84 students took part in the second IET, completing the same questionnaire before and after game, with eight questions focusing on their knowledge and ability to relate information from different areas and three questions focusing on their opinion on supplementing the learning process with such playful learning.

Evaluation of the questionnaire responses revealed the following findings:

- Before playing the toolkit, the mean score for the knowledge questions was 5.4 points, while after playing it was 5.8 points.
- Before the game, none of the participants achieved full points, while after the game, eight students achieved full points.
- Before using the toolkit, 23 students had less than 5 points, while after using the toolkit only 15 students had less than 5 points.
- These findings are also presented in Figure 4.

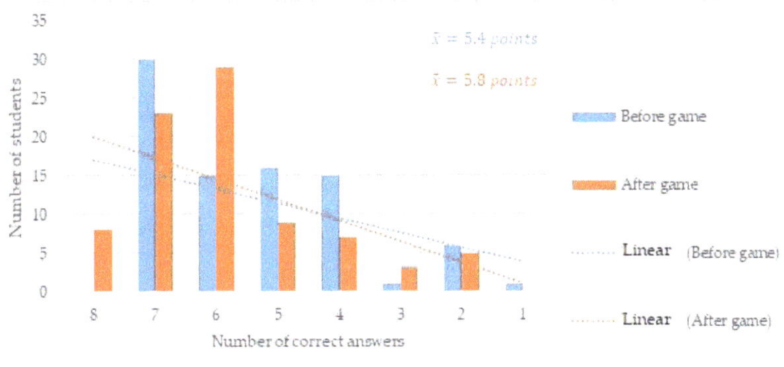

Figure 4. Results of questionnaire for knowledge questions.

In addition to the knowledge questions, hypotheses A-H0, A-H1, B-H0, B-H1, C-H0, and C-H1 for research questions mentioned in 2.4. were verified. The students' responses for the questionnaire are evaluated in the form of graphs in Figures 5–10.

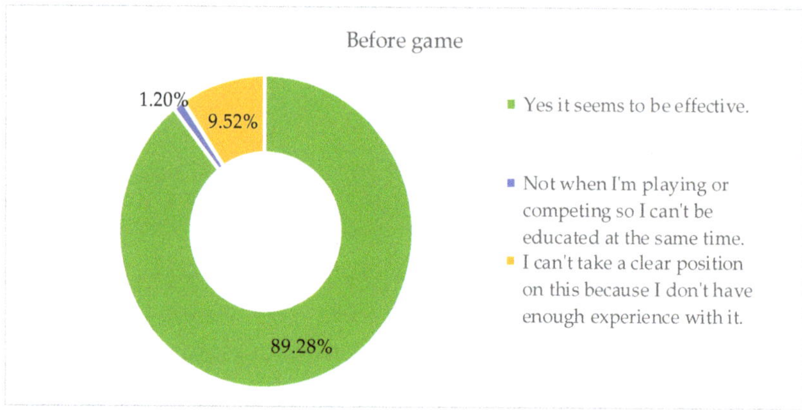

Figure 5. Results of questionnaire for question: Is it effective to educate in the form of a competition or a game? (before using toolkit).

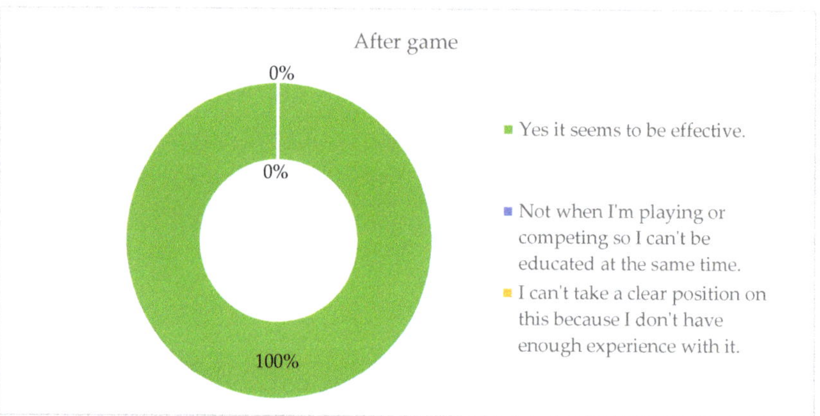

Figure 6. Results of questionnaire for question: Is it effective to educate in the form of a competition or a game? (after using toolkit).

Figures 5 and 6 show that before playing the toolkit, 1.2% of students thought that learning through a competition or game was not effective and almost 10% could evaluate. After the competition, all 84 respondents (100%) agreed that such learning is effective.

Based on the results of the Non-Parametric Related Samples Wilcoxon Signed Rank Test, the A-H1 hypothesis was accepted, i.e., education through interdisciplinary competition or play provides a better understanding of the relationships in earth resource acquisition than conventional education within individual classroom subjects. Hypothesis A-H0 was rejected.

When we asked whether information delivered in an entertaining way can be more enduring than formal education, almost 10% said no before using IET, almost 36% could not say, and over 54% said yes. After the use of IET, 90.47% agreed that information given in this way can be more permanent and almost 10% could not express their opinion.

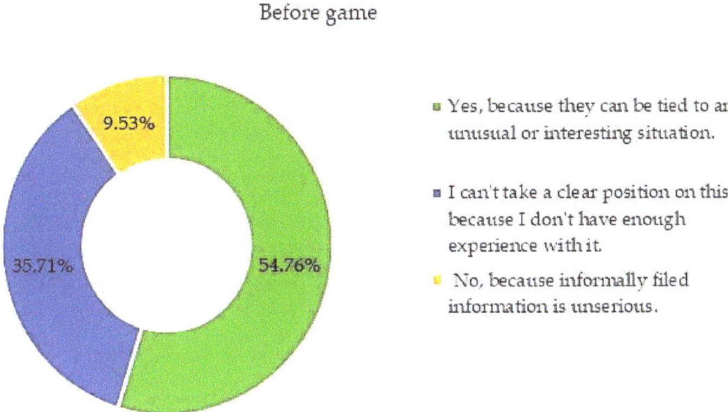

Figure 7. Results of questionnaire for question: Can information given in an informal way be more lasting than formal ones? (before using toolkit).

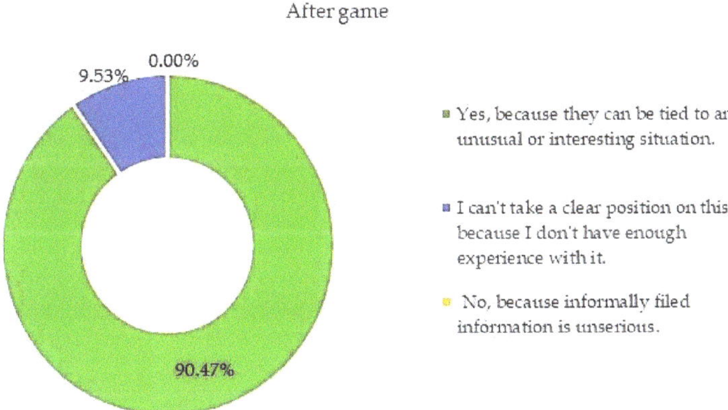

Figure 8. Results of questionnaire for question: Can information given in an informal way be more lasting than formal ones? (after using toolkit).

Based on the results of the Non-Parametric Related Samples Wilcoxon Signed Rank Test, the B-H1 hypothesis was accepted, i.e., the information given informally is of a more permanent nature than the information obtained conventionally. Hypothesis B-H0 was rejected.

Finally, students were asked to answer the question whether tangible artifacts would have made the competition more engaging. Prior to the quiz, the majority of students (54.76%) could not answer, 27.38% said yes, and almost 18% disagreed. After the quiz, all students could answer, with almost 73% of students agreeing that the game would be more interesting with real artifacts and over 27% of students disagreeing.

Based on the results of the Non-Parametric Related Samples Wilcoxon Signed Rank Test, the C-H1 hypothesis was accepted, i.e., the engaging nature of the educational competition/game will be enhanced by enriching it with 3D artifacts and tangible objects. Hypothesis C-H0 was rejected.

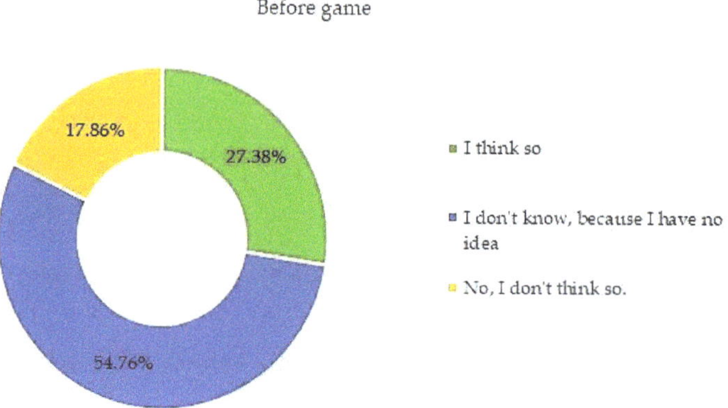

Figure 9. Results of questionnaire for question: Would the competition be more engaging if it included some tangible artifacts? (before using toolkit).

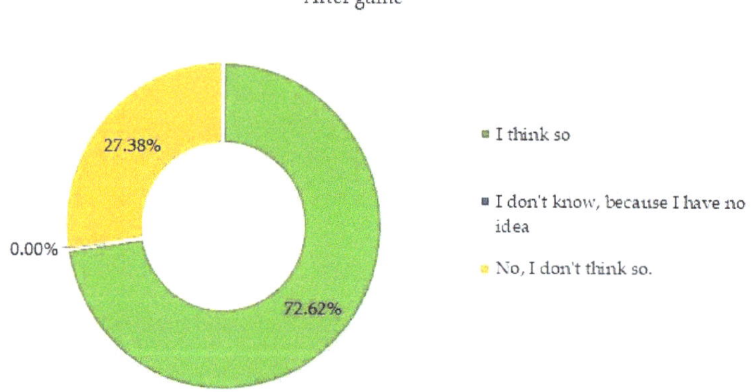

Figure 10. Results of questionnaire for question (Would the competition be more engaging if it included some tangible artifacts?) after using toolkit.

Although significant differences between the number of correct answers before and after the game were expected, we concluded that the use of the toolkit was successful. Indeed, several factors may have influenced the results:

- Before playing the game, students were relaxed and, after the game, which lasted almost 2 h, their attention and concentration were already reduced.
- Students may have felt less motivated to complete the questionnaire after the game, especially the team that lost the game.
- The students who did not improve may belong to a group of students who need a different learning approach other than listening and reading.
- The students were in a hurry to complete the questionnaire and, therefore, they did not pay the same attention to the answers as they did before the game.
- Students did not want to fill out the same questionnaire a second time and, therefore, answered "just so it is".

However, based on the students' responses to the opinion questions concerning the testing of hypotheses before and after playing the game, it was evident that the students had a positive attitude towards the toolkit and were aware of its benefits for the learning process. Additionally, most students agreed that such information is more permanent and that supplementing the quiz with tangible artifacts could further enhance the engaging nature of the competition and bring about more significant differences in the results of correct answers before and after the quiz.

The results indicated that conventional education is not sufficient for a proper understanding of complex issues such as RES and the extraction of raw materials–energy or land resources. It is important to present the information in an interdisciplinary level but at the same time in an engaging form, so that it is sufficient to build attitudes for students and so that there is a prerequisite for better implementation of knowledge in practical life. The proposed IETs can be applicable with different content at different levels of education and also for complex interdisciplinary issues, while, at the same time, they can improve students' attitudes and motivations to receive knowledge. The contribution of the solution lies in the fact that by using one tool, besides the actual knowledge acquisition, aspects of teamwork, working under pressure (time, points), aspects of using ICT tools at work, and motivational and value aspects can be addressed.

4. Research Limitations

This study contains many limitations that also offer opportunities for further research. First, the sampling and data collection were limited to the region, which means that this research has limited empirical applicability. This study focused only on middle and high school students. The age limitation of the students (12–19 years old) was determined by the EIT project itself. Sample size, number of days visited, and time spent in school were affected by the COVID-19 pandemic and schools being closed during the lockdown. If the situation improves during the next year, more schools are expected to be visited, with a larger number of students and teachers being involved in the testing of the extended version of the toolkit.

5. Conclusions

Understanding the comprehensiveness of resource acquisition, the real status of renewable resources and alternative energy solutions, their role and importance for human society is one of the basic prerequisites for the effective transformation of the current form of human civilization towards a more sustainable and less environmentally burdensome form. Given that this is a very complex issue overall and crosses a number of scientific disciplines, obtaining relevant information, forming an objective judgement, and taking a sober attitude towards the issues among the majority population are difficult, and it is advisable to focus already on the educational process.

The educational tools presented in the study were designed to overcome the boundaries defining the individual disciplines reflected in the standard teaching subjects, focusing on the secondary school level, and have the added value of attractiveness and appeal of educational form and content. Two tools were developed, one of which was built more on a physical basis and a face-to-face form of delivery, the other on an ICT platform and a more distance or online form.

The experiences presented with the use of both tools show that, overall, they were positively received by the students. In the validation of the second tool, three research questions were defined and the hypotheses based on them were verified by way of a questionnaire survey. The results obtained confirmed the expectations: All three hypotheses, A-H1, B-H1, and C-H1, were accepted by testing answers of students with Non-Parametric Related Samples Wilcoxon Signed Rank Test in SPSS software and indicated that the presented way of obtaining information/knowledge in the field appears to be promising and effective. The knowledge and skills acquired in this way are of a more permanent nature, strengthen the student's self-confidence, and will be a real competitive advantage for the graduate in the labor market. The results of this paper are in line with the studies mentioned in the introduction. The general results are that the innovations in the educational process have a positive impact on students' learning and their acquisition of knowledge in the field of interdisciplinary understanding of the world.

The presented findings could also contribute to the effort to objectively perceive the position of the overall importance of the issue of raw material–energy resource extraction in the sense of increasing the chances for the implementation of the necessary changes in terms of possible reduction of the negative impacts of human activity on the environment and the overall transformation of the current global economy towards a more sustainable form. Last but not least, objectively oriented, educated young people will have a better chance of creating a rational form of society in which both they and future generations will live.

6. Patents

The material platform in this paper, Multilevel hierarchically configured box, as indicated in Section 2.2, is protected under Slovakia Law (PUV 50017-2020, Utility model number: 8953).

The material platform in this paper, Educational set, as indicated in Sections 2.2 and 2.3, is protected under Slovakia Law (PUV 50017-2020, Utility model number: 8933).

Author Contributions: Conceptualization, R.R. and Ľ.G.; methodology, Ľ.G.; software, R.R.; validation, R.R., M.B. and Ľ.G.; formal analysis, Ľ.G.; investigation, L.B.; resources, Ľ.G.; data curation, J.R. and M.B.; writing—original draft preparation, R.R.; writing—review and editing, Ľ.G.; visualization, J.R.; supervision, R.R.; project administration, L.B.; funding acquisition, L.B. All authors have read and agreed to the published version of the manuscript.

Funding: The article was created with the financial support of the project RM@Schools-ESEE. RawMaterials@Schools-ESEE: 19069. This paper was created in connection with the project KEGA—048TUKE-4/2021: Universal educational-competitive platform and project VEGA 1/0290/21 Study of the behavior of heterogeneous structures based on PCM and metal foams as heat accumulators with application potential in technologies for obtaining and processing of the earth resources.

Informed Consent Statement: Informed consent was obtained from all subjects involved in the study.

Data Availability Statement: Not applicable.

Conflicts of Interest: The authors declare no conflict of interest.

References

1. Cook, J.; Oreskes, N.; Doran, P.T.; Anderegg, W.R.L.; Verheggen, B.; Maibach, E.W.; Carlton, J.S.; Lewandowsky, S.; Skuce, A.G.; Green, S.A.; et al. Consensus on consensus: A synthesis of consensus estimates on human-caused global warming. *Environ. Res. Lett.* **2016**, *11*, 048002. [CrossRef]
2. Cook, J.; Nuccitelli, D.; Green, S.A.; Richardson, M.; Winkler, B.; Painting, R.; Way, R.; Jacobs, P.; Skuce, A. Quantifying the consensus on anthropogenic global warming in the scientific literature. *Environ. Res. Lett.* **2013**, *8*, 024024. [CrossRef]
3. Anderegg, W.R.L. Expert credibility in climate change. *Proc. Natl. Acad. Sci. USA* **2010**, *107*, 12107–12109. [CrossRef]
4. Doran, P.T.; Zimmerman, M.K. Examining the scientific consensus on climate change. *EOS Trans. Am. Geophys. Union* **2009**, *90*, 22–23. [CrossRef]
5. List of Worldwide Scientific Organizations Accepted that Climate Change Has Been Caused by Human Action. Available online: http://www.opr.ca.gov/facts/list-of-scientific-organizations.html (accessed on 7 May 2021).
6. Beer, M.; Rybár, R. Development process of energy mix towards neutral carbon future of the Slovak Republic: A review. *Processes* **2021**, *9*, 1263. [CrossRef]
7. Climate Adaptation Summit. Available online: https://www.cas2021.com/ (accessed on 15 November 2021).
8. Valavanidis, A. UN Framework Convention on Climate Change Conference, Glascow, November 2021. Global Challenges for the Agreed Implementation of Reductions of Greenhouse Gases. 2021. Available online: https://www.researchgate.net/ (accessed on 15 November 2021).
9. 2nd International Conference on Environmental Science and Applications (ICESA'21). Available online: https://esaconference.com/ (accessed on 15 November 2021).
10. Bp Statistical Review of World Energy 2021. Available online: https://www.bp.com/content/dam/bp/business-sites/en/global/corporate/pdfs/energy-economics/statistical-review/bp-stats-review-2021-renewable-energy.pdf (accessed on 15 November 2021).

11. IEA. World Energy Investment 2021. Available online: https://iea.blob.core.windows.net/assets/5e6b3821-bb8f-4df4-a88b-e891cd8251e3/WorldEnergyInvestment2021.pdf (accessed on 15 November 2021).
12. Gorman, J. Disobedient Youth: Lessons from the Youth Climate Strike Movement. 2021. Available online: https://pjp-eu.coe.int/documents/42128013/47261800/Gorman-J.-%282021%29-Disobedient-Youth-Lessons-from-the-Climate-Strikes.pdf/b1ec729d-ee2f-1e5d-9de3-a22b68e61bb8 (accessed on 17 November 2021).
13. Renewables 2020 Global Status Report. Available online: https://www.ren21.net/wp-content/uploads/2019/05/gsr_2020_full_report_en.pdf (accessed on 15 November 2021).
14. Worldometer. Available online: https://www.worldometers.info/ (accessed on 15 November 2021).
15. Wittenberger, G.; Cambal, J.; Skvarekova, E.; Senova, A.; Kanuchova, I. Understanding Slovakian gas well performance and capability through ArcGIS system mapping. *Processes* **2021**, *9*, 1850. [CrossRef]
16. Sustainable Consumption and Production. A Handbook for Policymakers. Global Edition. United Nations Environment Programme, 2015. Available online: https://sustainabledevelopment.un.org/content/documents/1951Sustainable%20Consumption.pdf (accessed on 15 November 2021).
17. Carrara, S.; Alves Dias, P.; Plazzotta, B.; Pavel, C. *Raw Materials Demand for Wind and Solar PV Technologies in the Transition Towards a Decarbonized Energy System*; JRC119941; Publications Office of the European Union: Luxembourg, 2020. Available online: https://eitrawmaterials.eu/wp-content/uploads/2020/04/rms_for_wind_and_solar_published_v2.pdf (accessed on 15 November 2021). [CrossRef]
18. Bednárová, L.; Džuková, J.; Grosoš, R.; Gomory, M.; Petráš, M. Legislative instruments and their use in the management of raw materials in the Slovak Republic. *Acta Montan. Slovaca* **2020**, *25*, 105–115.
19. Pacana, A.; Czerwińska, K.; Bednarova, L. Comprehensive improvement of the surface quality of the diesel engine piston. *Metalurgija* **2019**, *58*, 329–332.
20. Williams, S.; Mcewen, L.; Quinn, N. As the climate changes: Intergenerational action-based learning in relation to flood education. *J. Environ. Education.* **2016**, *48*, 154–171. Available online: https://www.researchgate.net/publication/312419951_As_the_climate_changes_Intergenerational_action-based_learning_in_relation_to_flood_education (accessed on 15 November 2021). [CrossRef]
21. Vergara, D.; Antón-Sancho, Á.; Extremera, J.; Fernández-Arias, P. Assessment of virtual reality as a didactic resource in higher education. *Sustainability* **2021**, *13*, 12730. [CrossRef]
22. Hsieh, M.-C. Development and application of an augmented reality oyster learning system for primary marine education. *Electronics* **2021**, *10*, 2818. [CrossRef]
23. Wiyono, B.B.; Wedi, A.; Ulfa, S.; Putra, A.P. The use of information and communication technology (ICT) in the implementation of instructional supervision and its effect on teachers' instructional process quality. *Information* **2021**, *12*, 475. [CrossRef]
24. Bion, W.R. *Learning from Experience*; A Jason Aronson Book: London, UK; Karnac Books: Lanham, MD, USA, 1994.
25. Kosová, B. Selected Chapters from the Theory of Personnel and Social Education for Primary School Teachers. 1st ed. Banská Bystrica: Univerzita Mateja Bela, 1998. 134p. 2nd Edition Published in 2005 (ISBN 80-8083-043-6). Available online: https://publikacie.umb.sk/pedagogicke-vedy/vybrane-kapitoly-z-teorie-personalnej-a-socialnej-vychovy-pre-ucitelov-1-stupna-zs.html (accessed on 15 November 2021).
26. Kovalikova, S.; Olsen, K. *Integrated Thematic Teaching—Model*, 1st ed.; Faber: Bratislava, Slovakia, 1996; 350p, ISBN 8096749269.
27. Larose, S.; Roy, R. Test of reactions and adaptation in college (TRAC): A new measure of learning propensity for college students. *J. Educ. Psychol.* **1995**, *87*, 293–306. [CrossRef]
28. Fisher, R. *Teaching Children to Think and Learn*; Portal: Prague, Czech Republic, 1997; ISBN 8071781207.
29. Petty, G. *Modern Teaching*; Portál: Prague, Czech Republic, 1996; ISBN 8071780707.
30. Kvam, P.H. The effect of active learning methods on student retention in engineering statistics. *Am. Stat.* **2020**, *54*, 136–140. Available online: https://www.researchgate.net/publication/254331243_The_Effect_of_Active_Learning_Methods_on_Student_Retention_in_Engineering_Statistics (accessed on 15 November 2021).
31. Zakaria, E.; Chin, L.C.; Daud, M.Y. The effects of cooperative learning on students' mathematics achievement and attitude towards mathematics. *J. Soc. Sci.* **2010**, *6*, 272–275. Available online: https://www.researchgate.net/publication/49619868_The_Effects_of_Cooperative_Learning_on_Students\T1\textquoteright_Mathematics_Achievement_and_Attitude_towards_Mathematics (accessed on 15 November 2021). [CrossRef]
32. Risnita, R.; Bashori, B. The effects of essay tests and learning methods on students' chemistry learning outcomes. *J. Turk. Sci. Educ.* **2020**, *17*, 332–341. Available online: https://www.tused.org/index.php/tused/article/view/1093/627 (accessed on 15 November 2021).
33. Ponomarenko, T.; Nevskaya, M.A.; Marinina, O.A. Innovative learning methods in technical universities: The possibility of forming interdisciplinary competencies. *Rev. Espac.* **2019**, *40*, 16. Available online: https://www.revistaespacios.com/a19v40n41/a19v40n41p16.pdf (accessed on 15 November 2021).
34. Setiawan, R. The impact of teaching innovative strategy on academic performance in high schools. *Product. Manag.* **2021**, *25*, 1296–1312. Available online: http://repository.petra.ac.id/19048/2/Publikasi4_04045_7050.pdf (accessed on 15 November 2021).
35. Sadeghi, R.; Sedaghat, M.M.; Ahmadi, F.S. Comparison of the effect of lecture and blended teaching methods on students' learning and satisfaction. *J. Adv. Med. Educ. Prof.* **2014**, *2*, 146–150. Available online: https://www.ncbi.nlm.nih.gov/pmc/articles/PMC4235559/ (accessed on 15 November 2021).
36. Rybár, R.; Horodníková, J. Modelling apparatus of the Earth resources laboratory. *Energie* **2012**, *5*, 11–13.

37. Cechovic, M.; Rybar, R. Analysis of the Energy Production of a Wind Turbine Using the Cube Modelling Apparatus. 2017. Available online: https://opac.crzp.sk/?fn=detailBiblioForm&sid=1DDC86077D8354F3A2FE9EA47A7C&seo=CRZP-detail-kniha (accessed on 15 November 2021).
38. Short-Term Energy Outlook, EIA. Available online: https://www.eia.gov/outlooks/steo/report/global_oil.php (accessed on 15 November 2021).
39. Buriánek, D.; Kropáč, K.; Dolníček, Z. Ultrabasic rocks of the teschinite association in the western part of the silesian unit. *Geol. Res. Morav. Sil.* **2013**, *20*, 79–84. [CrossRef]
40. Hampel, C.A. (Ed.) *The Encyclopedia of the Chemical Elements*; Reinhold Book Corporation: New York, NY, USA, 1968; pp. 732–738.
41. Critical Mineral Resources of the United States—Titanium—Professional Paper 1802–T. Available online: https://pubs.usgs.gov/pp/1802/t/pp1802t.pdf (accessed on 15 November 2021).
42. Britannica. Available online: https://www.britannica.com/science/lead-chemical-element (accessed on 15 November 2021).
43. Toshinsky, G.; Petrochenko, V. Modular lead-bismuth fast reactors in nuclear power. *Sustainability* **2012**, *4*, 2293–2316. [CrossRef]
44. Szlugaj, J.; Galos, K. Limestone sorbents market for flue gas desulphurisation in coal-fired power plants in the context of the transformation of the power industry—A case of Poland. *Energies* **2021**, *14*, 4275. [CrossRef]
45. Babel, M.; Schreiber, B.C. Geochemistry of Evaporites and Evolution of Seawater. 2014. Available online: https://www.researchgate.net/publication/237838492_Geochemistry_of_Evaporites_and_Evolution_of_Seawater (accessed on 15 November 2021).
46. Reed, M.L.; Fedder, G.K. 2—photolithographic microfabrication. In *Handbook of Sensors and Actuators*; Fukuda, T., Menz, W., Eds.; Elsevier Science B.V.: Amsterdam, The Netherlands, 1998; Volume 6, pp. 13–61, ISBN 9780444823632. Available online: https://www.sciencedirect.com/science/article/abs/pii/S1386276698800030 (accessed on 15 November 2021).
47. Burkowicz, A.; Galos, K.; Guzik, K. The resource base of silica glass sand versus glass industry development: The case of Poland. *Resources* **2020**, *9*, 134. [CrossRef]
48. Kim, J.; Tae, S.; Kim, R. Theoretical study on the production of environment-friendly recycled cement using inorganic construction wastes as secondary materials in South Korea. *Sustainability* **2018**, *10*, 4449. [CrossRef]
49. Geological Survey Professional Paper, Geographic Names Information Management, U.S. Government Printing Office, 1942. Available online: https://books.google.sk/books?redir_esc=y&hl=sk&id=R7UPAAAAIAAJ&q=washington#v=oneage&q&f=false (accessed on 15 November 2021).
50. Shanmugam, S.P.; Nurni, V.N.; Manjini, S.; Chandra, S.; Holappa, L.E.K. Challenges and outlines of steelmaking toward the year 2030 and beyond—Indian perspective. *Metals* **2021**, *11*, 1654. [CrossRef]
51. La Noce, M.; Lo Faro, A.; Sciuto, G. Clay-based products sustainable development: Some applications. *Sustainability* **2021**, *13*, 1364. [CrossRef]
52. Sofranko, M.; Khouri, S.; Vegsoova, O.; Kacmary, P.; Mudarri, T.; Koncek, M.; Tyulenev, M.; Simkova, Z. Possibilities of uranium deposit Kuriskova mining and its influence on the energy potential of Slovakia from own resources. *Energies* **2020**, *13*, 4209. Available online: https://www.mdpi.com/1996-1073/13/16/4209 (accessed on 15 November 2021). [CrossRef]
53. RMSchools. Virtual Centre. Courses. Available online: https://rmschools.isof.cnr.it/moodle/course/index.php (accessed on 15 November 2021).

 MDPI

Article

The Use of Geothermal Energy for Heating Buildings as an Option for Sustainable Urban Development in Slovakia

Andrea Senova *, Erika Skvarekova, Gabriel Wittenberger and Jana Rybarova

Faculty of Mining, Ecology, Process Control and Geotechnologies, Institute of Earth's Resources, Technical University of Kosice, Park Komenskeho 19, 040 01 Kosice, Slovakia; erika.skvarekova@tuke.sk (E.S.); gabriel.wittenberger@tuke.sk (G.W.); jana.rybarova@tuke.sk (J.R.)

* Correspondence: andrea.senova@tuke.sk; Tel.: +421-55-6022985

Abstract: The use of geothermal energy (GE) and the green economy in the environment of Slovak municipalities and towns is significant, due to the reduction in the negative influences and impacts of human society's constant consumer lifestyle. The authors highlight the use of modern scientific knowledge, practical experience, and ever-improving technologies in the field of renewable energy sources RES. The aim of this contribution is to draw attention to the under-utilization of GE's potential in Slovakia. Given the country's commitment to meeting emission limits under EU carbon neutrality agreements by 2050, the use of this resource is very pertinent. Slovakia has significant geothermal resources that are not currently sufficiently utilized. The article suggests using GE to heat housing units of the housing estate near the geothermal source. Three scenarios (60 °C (pessimistic), 65 °C (conservative), and 70 °C (optimistic)) were considered in our energy balance and economic advantage calculations. The green economy offers a sustainable way of using the earth's resources. The financial calculations regarding the amount of investment, the expected financial return and the possible values of the saved emissions confirm the possibility of the further use of GE technology. The information under consideration can be used in other significant territories, which may be a theme for further research in this field.

Keywords: geothermal energy; renewable technology; thermal energy; investment

Citation: Senova, A.; Skvarekova, E.; Wittenberger, G.; Rybarova, J. The Use of Geothermal Energy for Heating Buildings as an Option for Sustainable Urban Development in Slovakia. *Processes* **2022**, *10*, 289. https://doi.org/10.3390/pr10020289

Academic Editors: Sergey Zhironkin, Radim Rybar and Kody Powell

Received: 16 December 2021
Accepted: 28 January 2022
Published: 31 January 2022

1. Introduction

The necessity of achieving net zero global CO_2 emissions by 2050, to achieve the target set in the Paris Agreement, has stimulated interest in the use of low-carbon energy technologies, including geothermal energy. The constant exploitation of natural resources, excessive waste generation, the emission of harmful substances into the atmosphere (COx, SOx, NOx, etc.), seas, oceans, and soils lead to negative and often permanent changes. Climate change is manifested by an increase in average temperatures, changes in the nature of the climate, more frequent occurrences of extreme weather (melting of glaciers and permafrost, landslides, extreme drought and heatwaves, torrential rains, sudden flooding, and others). There are changes in biodiversity, the migration of animal species, extinction, excessive deforestation, changing soil composition, and the contamination of water resources (increasing the acidity of the oceans). Population migration is another important manifestation of these changes [1]. However, despite various climate agreements, CO_2 emissions reached an all-time high of 35 billion tons per year in 2019. The unprecedented nationwide COVID-19 crisis that began in China at the end of 2019 has rapidly frozen emissions growth. All these reasons, among others, have led to increased interest in the research and development of sustainable and renewable energy (RES) technologies.

Man, through his anthropogenic activity, significantly influences conditions on the planet. At present, the greatest emissions are generated by the burning of coal; for instance, more than 14 billion tons were burned in 2018. Emissions from this source totaled more than 12 billion tons in 2018. This ranking is followed by natural gas and its increasing consumption level of almost 8 billion tons, as well as cement production. The EU is the world's largest importer of energy, relying on imports for 50% of its energy needs. With an energy demand forecast to grow 1–2% a year, this figure will rise, within the coming 20–30 years, up to 70%. Europe's energy needs are growing relatively fast compared to other parts of the world. Climate change has encouraged the inclusion of modern RES technologies in the program, to mitigate the negative impacts of fossil fuel energy production. Europe is being forced to invest in new technologies [2,3]. The overall EU average of RES production in 2019 was 20%, meeting its target for 2020. Iceland has the largest share of RES energy at almost 80%, thus meeting the target above the set limit. This was mainly helped by the geographical relief and nature of the landscape and the wide use of energy from geothermal and water sources—geothermal power plants and hydroelectric power plants. Norway meets about 73% of its energy needs from RES especially due to the hydroelectric power plants made possible by the mountainous profile of the landscape, offering a bountiful supply of rivers with a high falling gradient. Of the other EU countries, Sweden's share is almost 60%, Finland's, about 43%, and Latvia's, 40%, followed by Denmark, Austria, etc. Included in the leading countries that have not (yet) met their commitments is Slovenia, which produced more than 20% of its energy from RES in 2019, but the target for 2020 is 25%. Other such countries are Ireland, Belgium, the Netherlands, Luxembourg, the United Kingdom, and Poland. Poland relies mainly on coal production and, therefore, has long had a problem with air pollution. In 2019, Slovakia produced almost 20% of its total energy from RES, thus meeting its 2020 target [4].

Geothermal energy, as a renewable energy source, can be an important resource for numerous regions of Europe. The development of geothermal energy facilities gives people the potential to gain better control of their own local energy resources and take advantage of a secure, environmentally friendly and domestic source of energy. This energy from within the Earth can be used for different purposes to improve environmental quality and protect public health and safety. The technological and sustainable development of this type of energy will help to solve the world's energy needs and the requisite challenges [5]. For geothermal electricity production, the highest concentration of resources on the European continent is located in Italy, Iceland and Turkey; the present exploited value is only 0.3% of the whole renewable market. The possibilities for geothermal energy to expand its penetration in Europe are mainly from using the enhanced geothermal system (EGS). Some areas have been critically investigated regarding geothermal resource base assessment, recoverable EGS estimates, in-depth research on EGS technologies and the current performance, the designing of suburb-facing systems, drilling technology economics, the conversion of energy using enhanced geothermal systems, the effect of this technology on the environment, and the analysis of enhanced geothermal systems and their sustainability [6,7].

Slovakia should take advantage of the potential geothermal resources it has, which are currently used to a minimal extent. It is important to use energy resources efficiently because the price of energy, in general, is rising [8].

Is geothermal energy renewable? Geothermal energy has often been described as a renewable energy resource. However, on the time scale normally used in human society, geothermal resources are not, strictly speaking, renewable. They are renewable only if the heat extraction rate does not exceed the reservoir replenishment rate. Exploitation through wells, sometimes using down-hole pumps in the case of non-electrical uses, leads to the extraction of very large quantities of fluid, and consequently to a reduction or depletion of the geothermal resources that are in place [9]. Geothermal localities can be subdivided into two categories: springs and deep boreholes. Springs are locations where geothermal water naturally flowed out or is still flowing out from the aquifer onto the earth's surface. The category of deep boreholes accounts for all the localities where there are not, nor have there ever been any natural springs, but where geothermal waters have been found during hydrocarbon exploration and exploitation [10,11].

The drilling works enable geologists to establish the rock composition of an investigated area, along with other data that are connected to the tasks needed in the eventuality of a geothermal energy plant [12]. Geothermal fields, as opposed to hydrocarbon fields, are generally systems with a continuous circulation of heat and fluid, where fluid enters the reservoir from the recharge zones and leaves through discharge areas (hot springs, wells) [7]. Heat production from GE is widespread in many countries of the world. The largest geothermal central heating systems are in the USA and China; in Europe, it is mainly found in France, Germany, Iceland or Turkey [13]. Slovakia has the greatest potential for GE in Central and Eastern Europe [3].

The aim of this article is to demonstrate how to carry out an analysis of an area examined in terms of the availability of a site suitable for GE domestic heating systems, the availability of distribution networks, and energy needs. We provide a case study of how a GE system can be used from the perspectives of investment efficiency, environmental impact, and meeting user needs. Practical information and advice on procedures can help other cities in terms of future perspectives.

2. Materials and Methods

This article addresses the issue of obtaining and using a renewable energy source (RES) for heating purposes, district heating (DH), and domestic hot water (DHW). With this energy-clean technology, it is possible to minimize the production of emission substances and replace the combustion of natural gas during heating.

Data that is publicly available and at the same time processed by us in the case study have been used for the processing of this contribution. The procedures and methods are summarized as follows:

- Based on the available information, we analyzed the possibilities of using geothermal energy in the environment of towns and villages in the area of the High Tatras (Podtatranská basin). To design a model of geothermal energy use for heating, we considered a mining well to serve a housing estate of 5500 households.
- We created a methodology for determining the profitability of a geothermal well using a flow chart. We designed a technology setup for the use of geothermal well energy. The energy potential and energy balance were calculated. The complexity and return on investment were assessed. The volume of emissions of harmful substances that can be saved by using this clean energy was calculated. Based on the findings, we assessed the usability of a geothermal well for the purpose of heating apartments.

2.1. *Hydrogeological and Hydrogeothermal Conditions in Slovakia*

Due to its natural conditions, the Slovak Republic has significant potential geothermal energy. Based on research and surveys to date, the energy potential of GE is important in Slovakia, and its value is at 5538 MWt. Geothermal energy sources are mainly represented by geothermal waters, which are tied to hydrogeological collectors located at depths of 200–5000 m. In Slovakia, the average temperature increase is 3–3.8 °C for every 100 m of the borehole; at a depth of 3 km, the temperature is about 100 °C. Geothermal sources are divided according to temperature (°C), into:

- Low temperature—from 20 °C to 100 °C: these are geothermal sources with a moderate temperature, suitable only for heating and recreational purposes;
- Medium temperature—from 100 °C to 150 °C: these are suitable for heating and using binary cycles, and for electricity generation;
- High temperature—above 150 °C: these are geothermal sources suitable for electricity generation (using water vapor).

In terms of well yield, geothermal sources are distinguished as follows:

- With minimum yield—up to 1.0 $L \cdot s^{-1}$;
- With very little yield– from 1.0 to 5.0 $L \cdot s^{-1}$;
- With a small yield—from 5.0 to 10.0 $L \cdot s^{-1}$;

- With medium yield– from 10.0 to 25.0 L·s^{-1};
- With great yield—from 25.0 to 50.0 L·s^{-1};
- With very high yield—above 50.0 L·s^{-1} [14].

The wells that have been carried out so far (at depths of 92–3616 m) have verified temperatures at the wellhead of 18–129 °C. The yield of free-range wells ranged from tenths of a liter to 100 L·s^{-1}, with a predominantly Na-HCO_3-Cl, Ca-Mg-HCO_3 and Na-Cl water type, with mineralization of 0.4–90 g·L^{-1} [15].

Slovakia, considering its small surface area (49,000 km^2), is very rich in mineral and thermal waters. There are 1200 springs registered on its territory, the equivalent of one spring for every 40 km^2. The rich physicochemical diversity of waters and their even spread throughout the territory is conditioned by the favorable geological–tectonic construction of the territory and its geothermal activity.

From north to south, mineral and thermal waters are linked to sediments of the Flysch zone and the Klippen belt of the Paleogenic and Mesozoic eras, the crystalline rocks of the Paleozoic and Mesozoic, the dolomites and limes of the Mesozoic, and Flysch sediments of Inner Carpathian Paleogene and Neogene origin. The temperature of the waters of the springs is between 15 and 70 °C. Mining works carried out in some areas obtained water with temperatures in the range of 40–130 °C. The yield of natural springs varies between 1 and 40 L·s^{-1}. The well yield is from 5 to 90 L·s^{-1} [16,17]. From the point of view of geothermal water sources, we can only consider as a possibility the Paleogene subsoil in the area of interest, which consists mainly of Triassic carbonates of Krížna Nappe and Choč Nappe, with Karst-fissure permeability. Overall, the geothermal activity of the area of interest can be assessed in terms of the density of the earth's heat flow as average geothermal output (65–70 mW·m^{-2}). Temperatures at the depth of storage of the collectors (1400–3000 m below the surface) are about 45–95 °C. Mineral waters reflect the hydrogeological, hydrological, geological, and structural conditions in the monitored area. The nature of the waters indicates that Triassic structures correlate well with the appearance of Triassic carbonate rocks on the surface around the basin. The nature of the waters in the Paleogene subsoil may have been obviously marine in origin, i.e., waters that have been preserved after the transgression of the Paleogene Sea but are now more or less infiltrated surface waters from the peripheral parts of the basin. The total mineralization of the waters in the Mesozoic subsoil, based on the results of the pumping tests in the surrounding wells, can be expected in the range of 3–5 g·L^{-1}. Information on the hydrogeological conditions of the territory comes mainly from the quaternary sediments of Poprad, where several exploration wells were created. Several deeper wells were created in the Poprad Basin, the results of which provided valuable data on groundwater properties in individual geological units. From this point of view, valuable information can be drawn from the wells in the areas of Stará Lesná (FGP-1), Poprad (PP-1) and Vrbov (VR-1, 2, 2A, 3) and the latest from the area of Veľká Lomnica. Since the PP-1 borehole (Poprad) is relatively close to the site under examination, its results may be analogous to those of the projected well. At the PP-1 well in Poprad, the geothermal waters, of a significantly calcium-magnesium-sulfate-sulphate-hydrogen carbonate type, were verified as having an average mineralization value of 2.88 g·L^{-1}, with a pH value of 6.21. The water is over-gassed with CO_2. The increased sulfuret content is due to the dissolution of plaster stone and anhydrite. The incidence of sodium is low because it is released due to the low pH [18].

Groundwater resources and reserves in Slovakia vary not only depending on the location and time but also in terms of their quality. Although they are regularly renewed, they are not unlimited and only proper use can ensure their relative inexhaustibility. Groundwater, which is a source of quality drinking water, is the most important natural wealth in Slovakia. Therefore, the most fundamental task is getting to know the laws of groundwater creation and flow as well as its protection. This requires very close monitoring, documenting and registering their basic characteristics and parameters. The total usable amounts of groundwater in Slovakia, as documented in 2018 in all categories, represent 77,175.07 L·s^{-1}.

Those usable quantities also include usable amounts of thermal water as an integral part of groundwater and, for the sake of completeness, part of the mineral waters, in particular the usable quantities of mineral waters approved by the Hydrogeological Commission [19]. Figure 1 shows the prospective areas of geothermal waters in the Territory of the Slovak Republic.

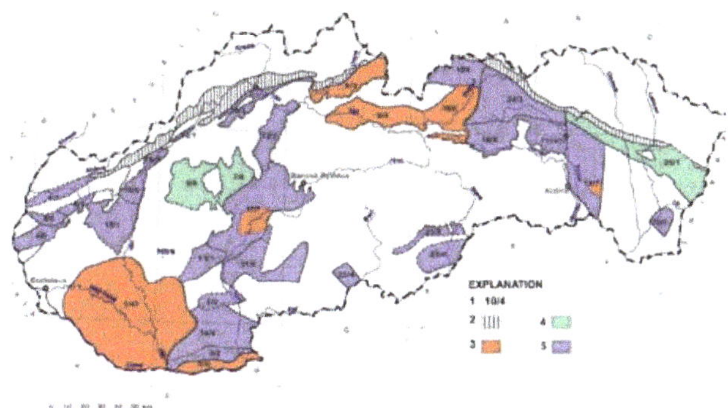

Figure 1. Prospective areas of geothermal waters in Slovakia [19]. (1) 10/4; (2) Klippen belt; (3) prospective areas where hydrogeothermal evaluation has been carried out; (4) prospective areas where hydrogeothermal evaluation is being carried out; (5) prospective areas where hydrogeothermal evaluation has not yet been carried out.

In order to create the best possible conditions for the use of geothermal energy, regional hydrogeothermal evaluations are carried out by determining the quantity of geothermal waters and geothermal energy in the defined 27 hydrothermal areas or structures of Slovakia [20]. Current geothermal conditions in Slovakia are mapped out and reviewed in detail. There are currently 27 prospective geothermal areas that have been defined (Figures 1 and 2).

Figure 2. Partial map of geothermal wells in Slovakia.

A large proportion of geothermal reservoirs provide water with a temperature of up to 135 °C, which is optimal for use for heating buildings or for recreational purposes.

Geothermal energy (GE) is not primarily used for efficient electricity generation. Modern technologies also make it possible to generate electricity using a binary cycle [13].

2.2. Technological Description of Geothermal Well Utilization

The proposed geothermal well is GTK-1 Kežmarok, with an estimated depth of 2800 m; the relevant technological and piping equipment for the purpose of using geothermal energy for DH production and the preparation of DHW in existing block boiler rooms on the housing estate. Currently, housing estate boiler rooms are used for the supply of DH and DHW to apartment buildings, civic facilities, and other buildings in the city. Thermally treated geothermal water will be discharged into a nearby surface flow system. The physicochemical properties of geothermal water and its impact on the material of the pipes are also fully taken into account. Geothermal energy is expected to be used in an open system, with the newly proposed GTK-1 geothermal well serving as a drilling (pumping) well.

The exploitation of geothermal water by free flow has been considered, with water from the well estuary (mouth) entering the separation and storage vessel, where it will be relieved of free gases. Consequently, it will be pumped (if there is not enough pressure at the mouth of the well) and overdrawn in the ground to the isolated pipelines, then carried to an heat exchanger station situated close to the well. In the heat exchanger station, geothermal water transmits its heat to the secondary treated heating system water, by which method geothermal heat will be distributed in a closed and hydraulically separated circuit to 3 boiler rooms, in a ground-stored isolated pipe. Heat exchangers will be added in existing boiler rooms to heat the returning heating water and the preheat/heating of the DHW. After thermal use, geothermal water from the heat exchanger station will be transported under the ground by stored pipes without thermal insulation to a nearby surface stream (see Figure 3).

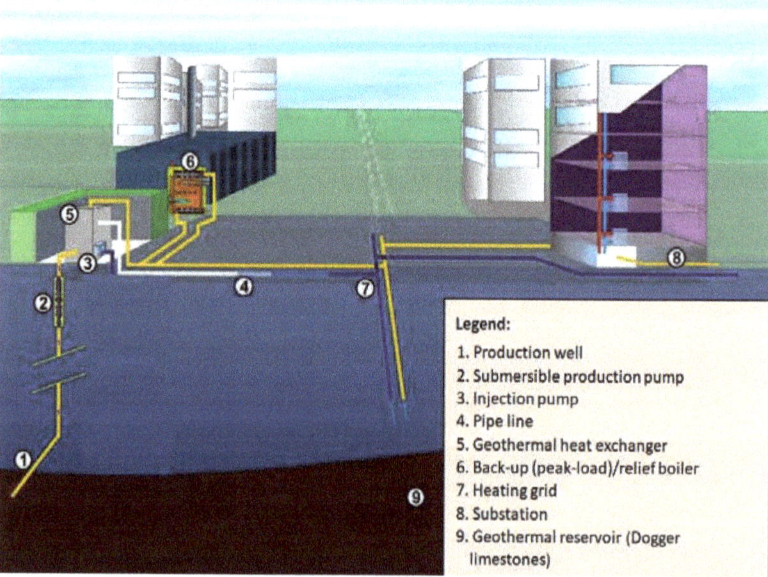

Figure 3. Scheme of district heating using geothermal energy.

2.3. *Methodology for Determining the Suitability of Geothermal Well Usage in the Monitored Area (Podtatranska Basin)*

Hydrogeologically, the projected well is situated in explored territory in the Poprad River Basin. It has a left-hand tributary from the High Tatras and the right-hand tributary is mainly from the Levoča Hills. Water management is important in the area since the surface waters of the tributary of Poprad and the groundwater from its alluvia are often used for drinking water supply. Water quality in the river is influenced by industrial enterprises and local agglomerations. The area of the Sub-Tatran Basin under investigation offers the use of several types of RES in different locations of the territory, in order to reduce negative environmental impacts, especially in terms of reducing emissions or replacing the combustion of fossil fuels. Fossil fuel heating is one of the largest sources of CO_2 emissions. The best solution is the use of thermal energy from geothermal sources.

Geothermal energy is an available local, strong energy source that is characterized by stability of supply, regardless of current climatic conditions. Geothermal energy is a long-term and sustainable energy source. Based on the geological construction of the surrounding area and the conditions of geothermal wells that have already been realized, it is possible to expect a well yield in the range of 20–30 $L{\cdot}s^{-1}$, with a total mineralization of about 3–5 $g{\cdot}L^{-1}$ and a water temperature at the surface of 60–70 °C. One definite point of uncertainty, according to the study, may be the depth of the collectors; therefore, the study recommends counting on the final mining depth being 2800 m [21].

Based on analyses of the available data, we assume that there is potential for the practical use of GE in the monitored Podtatranska Basin. GE may replace the combustion of natural gas in the supply of housing units under the current model. This creates ideal conditions for:

- Limiting the use of energy derived from fossil and conventional fuels;
- Reducing CO_2 emissions, (NOx, CO, SO_2, TZL)
- Stabilizing heat prices,
- Obtaining a stable, green, and renewable energy source.

The aim of the methodological procedure (Figure 4) is to choose the appropriate technology to cover the energy needs of the chosen location—a housing estate—based on sustainability, local availability, and affordability, and with a positive impact on the environment, as an exemplary model of energy independence for towns and villages.

The methodology begins with the search for theoretical knowledge in the field of RES energy. This draws attention to the call for a transition to renewable and sustainable energy sources within the European Union, which aims to achieve carbon neutrality by 2050.

Following a subsequent evaluation and the selection of appropriate information, the individual available RES technologies were evaluated for:

- The area of energy coverage of household needs;
- The types and principle of operation; and
- A large-scale and stable supply for the population throughout the year.

From the information found, geothermal energy appears to be the most suitable form of energy. With the subsequent selection of the site and the examination of existing technologies covering the energy needs of the housing estate, it is possible to proceed with an evaluation of the most appropriate RES technology. Our research has shown that the technology used has the best potential and the appropriateness of the subsequent investment is confirmed by further calculation.

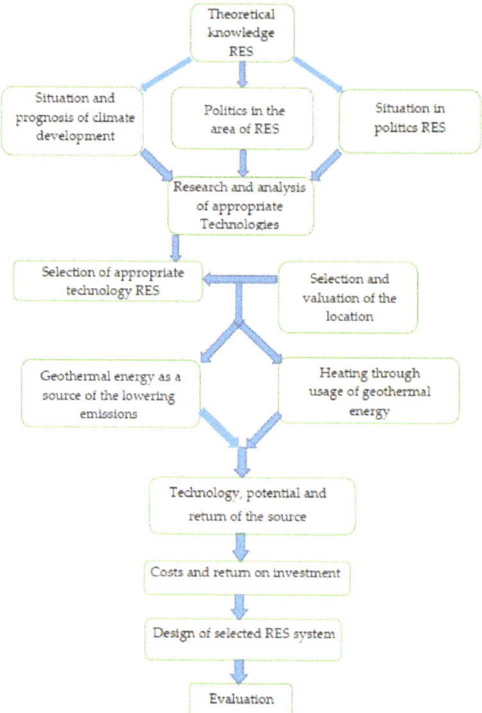

Figure 4. Flowchart methodology for the work of detecting a suitable GE source.

The processing of the geothermal contribution was based on data from the technical study of the geothermal well GTK-1, which is publicly available [21]. A specific site (GTK-1, Figure 5) has been designated in which to carry out the geothermal well under investigation. Residential houses (in a housing estate with a population of 5500) were selected that met the requirements for the use of geothermal energy from the borehole. To determine the yield of a geothermal source for heating the estimated capacity needed for the source was compared with the heating volume of 3 boiler rooms in previous years (average of years 2015–2017).

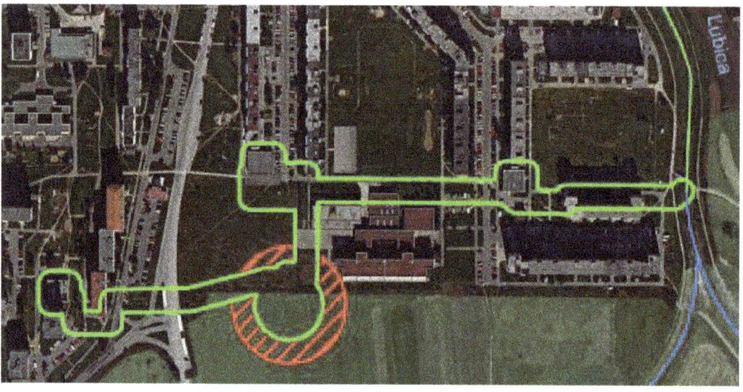

Figure 5. Location of the proposed GTK-1 well area under investigation [21].

The calculation of the energy potential of the geothermal source was based on the following parameters: yield, mineralization, and water temperature at the surface. Given that it is not possible to determine the exact temperature of geothermal water on the basis of current knowledge, three variants of geothermal water temperature (60 °C, 65 °C and 70 °C) have been considered when calculating the energy potential.

2.4. Procedure for Calculating the Chosen Technology's Potential Using Geothermal Energy

A methodological procedure was used to evaluate the use of the suggested possibilities regarding suitable RES technology with the best potential and lowest subsequent investment, with a recalculation, determination, and proposal as to whether the given system (investment) is suitable. The procedure for calculating the investment technology chosen, using geothermal energy, is as follows:

- Calculation of the Return on Investment (ROI)

ROI is the most frequently used parameter for the assessment of the economic efficiency of investments. It is the total income that results from concrete investments, divided by the amount of investment funds. This indicator is completely time-independent [22].

$$\text{Return on Investment ROI} = \frac{\text{cumulative incomes}}{\text{total investments}} \quad (1)$$

- Calculation of Cash Flow (CF)

Calculation of the annual cash flow is carried out on the basis of the following relationship [22]:

$$\text{CF} = \text{incomes investments} - \text{production costs} \quad (2)$$

CF, or annual cash flow (annual income), is the product of the unit price of natural gas (EUR 0.0219/kWh, for 2021) and the amount of heat produced (DH and DHW) in kWh (MWh) with a difference in estimated annual operating costs (EUR).

The cash flow of a project is the sum of positive and negative items, incomes and costs, connected with a certain activity. The sum of all financial flows, which results from the investment into a project, is called the cash flow produced by capitalized investments.

- Calculation of Net Present Value (NPV)

$$\text{NPV} = \sum_{i=1}^{n} \left\{ (-\text{I} + \text{CF}_i) / (1 + a)^i \right\} \quad (3)$$

where:

NPV—net present value;
I—investments;
CF—cash-flow;
a—update rate;
i—current year;
n—project duration.

If the NPV of the first project is higher than the NPV of the second project, and vice versa—the ROI of the first project is smaller than the ROI of the second project (NPV1 > NPV2; ROI1 < ROI2). In this case, there is no precise mathematical formula defining which of the two projects is better. The volume of investments and the risk level of the project will probably play the most important role. The intuition and experience of the project evaluator, as well as other arguments, can influence investment decisions.

- Calculation of Payback Period (pBp)

A payback period is the project's duration, from its beginning until the point when the cumulative cash flow becomes positive. Although in the case of some projects, the assessment results based on the payback period may seem interesting, this indicator does not say anything about the project's future course from the viewpoint of its cash-flow development. This could be either positive or negative.

- Calculation of Emission Values for Natural Gas Heating

The CO_2 emission factors needed to calculate CO_2 emissions from the operation of buildings (heating, hot water preparation and the operation of other appliances) are country-specific or operational (and also different for each IPCC1 category) and are derived from specific fuel characteristics. Average CO_2 emission factors are used for natural gas, hard coal, lignite by region of origin (Slovak, Ukrainian and Czech), and coke. Due to these reasons, emission factors should be revised each year [23,24].

The values of the weighted arithmetic mean of the qualitative parameters of natural gas, distributed in the territory of the Slovak Republic by SPP—distribúcia, a.s, were used according to the method followed in [25]. Density, calorific value, combustion heat and Wobbe number are given for the business unit, i.e., m^3 at 15 °C, pressure 0.101325 MPa, and relative humidity $\varphi = 0$. The formula used for the conversion of units: 1 kWh = 3.6 MJ.

Annual consumption of MNG NG at a calorific value of 34,848 MJ/m^3 [26]:

$$M_{NG} = \frac{\text{annual heat consumption}}{\text{calorific value} \cdot \text{boiler efficiency}} \quad (4)$$

- Calculation of Energy Potential

The energy potential is an elementary indicator of the possibility of using geothermal water. The energy potential of water, which is heated by the action of the earth's core, is its heat output [27].

$$P_t = m \cdot c_v \cdot \Delta t \quad (5)$$

where:

Pt—thermal power (kW)
m—weight ($kg \cdot s^{-1}$)
c_v—specific water heat ($m^{-3} \cdot K^{-1}$)
Δt—temperature difference (K)

- Determination of the Energy Balance

Three variants of geothermal water temperature (60 °C, 65 °C and 70 °C) were considered when determining the energy balance. Natural gas savings were assessed by comparing the average natural gas consumption for 2015–2017 and the energy balance of geothermal energy as a percentage; thus, it was also possible to assume an approximate reduction in CO_2 production.

When processing the energy balance of the GTK-1 geothermal source, we will take into account the following assumptions:

- Geothermal energy will be used for heating the housing estate of the DH and for the preparation of DHW.
- Heat loss or a temperature drop of 4 K due to the use of heat exchangers is considered.
- The daily operating time of boiler rooms is assumed to be 16 h.
- A maximum heating water temperature of 70 °C is assumed.
- A reduction factor in the use of geothermal energy for each whole 1 °C rise in the external temperature, which takes into account differences in outdoor temperatures during the day and resulting fluctuations in the temperature of the heating water. The reduction factor varies for different geothermal water temperatures.
- Thermally used geothermal water will be discharged into a nearby stream.

The energy balance will be processed for individual scenarios: the pessimistic scenario (geothermal water temperature of 60 °C), the conservative scenario (geothermal water temperature of 65 °C) and the optimistic scenario (geothermal water temperature of 70 °C).

To process the energy balance, it is necessary to know the following values:

- Number of days of temperature duration from 13 °C to −16 °C;
- Temperature of the supply and return heating water;
- Average heat output per DH;
- Amount of heat produced per DH;
- Amount of heat produced per DHW;
- Usable thermal output of the geothermal source for DH;
- Reduction factor for the use of geothermal energy for DH.

3. Results and Discussion

In the context of rising energy (heat/electricity) prices and societal pressure on environmental impacts in terms of the green economy, the interest of business owners/population/society in ecological sources of energy production (heat/electricity) is increasing. The design of a GE system for heating towns and villages is problematic, due to several of the input parameters of the calculation. The timing mismatch, the suitability of site selection in terms of the time optimization of production, the capacity of the distribution network, daily consumption, and the possibility of using GE as an energy source for building heating systems are data that has to be mutually optimized when dimensioning a balanced energy system.

3.1. Determination of the Parameters of the Proposed Geothermal Well

The predicted physicochemical properties of geothermal water from the proposed GTK-1 borehole are similar to geothermal boreholes already created in the vicinity of the Podtatranska Basin (wells Vr-1 and Vr-2 Vrbov, VL-1 Veľká Lomnica).

The mineralization type of Ca-Na-HCO_3-SO_4 is assumed, i.e., a composition similar to Vrbov; the dominant cationic components will be calcium (up to 600 mg·L^{-1}), sodium (up to 300 mg·L^{-1}) and magnesium (about 150 mg·L^{-1}), with anionic components of bicarbonate concentrations (about 2000 mg·L^{-1}) and sulfites (up to 700 mg·L^{-1}). Total mineralization is estimated to be up to 3 g·L^{-1}. The concentration of strontium (approx. 10 mg·L^{-1}) and potassium (80 to 100 mg·L^{-1}) will also be increased. Radioactivity will also be slightly increased compared to the natural background. Trace amounts of sulfide can also be expected. The expected phase ratio (dissolved gas content) will be up to 1 $m^3.m^{-3}$ of geothermal water, the main component being carbon dioxide (95 to 98% by volume), the remainder is mainly nitrogen and methane, with a small admixture of ethane, propane, isobutene, etc. [22].

Table 1 shows the predicted parameters of the geothermal borehole in the explored site of the Podtatranska Basin, based on an analogy of geothermal wells located nearby.

Table 1. Predicted parameters of the proposed geothermal well.

Scenario	Pessimistic	Conservative	Optimistic
Geothermal water temperature at the mouth of the well (°C)	60	65	70
Yield geothermal well (L·s^{-1})	25	25	25
Mineralization (g·L^{-1})		3–5 g·L^{-1}	
Usable amount of geothermal water per year (L·s^{-1})	788,400	788,400	788,400
Theoretically usable energy potential (kW)	4710.4	5233.8	5757.1
Theoretically usable annual amount of heat (MWh)	41,262.9	45,847.7	50,432.4

Table 2 shows the values of total annual heat production and NG consumption for the years 2015 to 2017.

Table 2. Total heat production and consumption of natural gas in the selected location, (source: elaborated by authors).

Year	2015		2016		2017	
Type of (MWh)	DH 17,964	DHW 9565	DH 19,084	DHW 9565	DH 18,975	DHW 9290
Total DH + DHW (MWh)	27,529		28,649		28,264	
Total consumption (NG) Total DH + DHW (kWh)	15,021,319		15,558,809		15,474,532	
Total consumption (NG) DH + DHW (m^3)	1,398,633		1,448,678		1,440,831	

- Calculation of saved emission values using geothermal energy

Table 3 shows the specific values of individual emission substances that pollute the environment by burning natural gas. The recalculation was carried out according to the literature [23]; these pollutants would be eliminated by using the thermal energy of the geothermal well. The replacement of NG with geothermal energy minimizes the greenhouse effect.

Table 3. Calculation of emissions from natural gas heating.

Annual Emissions from Natural Gas Heating (kg)	
TZL	270.8
SO_2	32.5
NOx	5958
CO	1997
TOC	253.9
Total emissions (kg)	8512.2

These pollutants would be eliminated annually for a housing estate with about 5500 residents. The calculation of emissions in NG heating was recalculated according to the literature [23] (see Table 3).

- Energy balance of geothermal well

Geothermal wells offer a number of positive options in terms of reducing negative environmental impacts in the long term. The presumed energy potential of the well could supply the housing estate with hot water intended for heating and hot service water preparation in full coverage, but a certain amount of natural gas usage is still envisaged by heating, to cover possible failures of the energy system.

We considered the following values:

- The specific weight of salt water of 1025 kg·m$^{-3;}$
- Specific water heat c_v = 4.18 MJ m^{-3} K^{-1};
- Temperature difference (for pessimistic, conservative, and optimistic scenarios) Δt = 42, 45 and 55 K.

The energy balance of a geothermal well at 3 different geothermal water temperatures was determined as shown in Table 4.

Table 4. Energy balance of geothermal well GTK–1.

Geothermal Water Temperature	60 °C	65 °C	70 °C
Annual amount of heat produced for DH (MWh)	8538.8	8538.8	8538.8
Annual amount of heat produced for DHW (MWh)	4676.5	4676.5	4676.5
Annual production of heat TOTAL (MWh)	13,215.3	13,215.3	13,215.3
- geothermal energy (MWh)	9763.1	11,070.3	12,091.4
- geothermal energy (%)	74	84	91
- natural gas (MWh)	3452.2	2,45	1123.9
- natural gas (%)	26	16	9
Annual average efficiency of geothermal energy (%)	24	24	24

The energy potential of geothermal water from the proposed well has been calculated. Since the energy potential of water heated by the earth's core is its heat output, the heat output formula has been used (2). The energy potential has been calculated for three temperature scenarios. Table 4 shows that as the geothermal water temperature rises, the energy potential and annual amount of thermal energy increase. Due to the large range of processed data, the evaluation of the energy potential by the authors in a conservative scenario of the implementation of a geothermal well with a water temperature of 65 °C is given in Table 5.

Table 5. Evaluation of the energy potential of a geothermal well in a conservative scenario with a water temperature of 65 °C.

Geothermal Water Temperature (°C)	65
Amount of heat produced annually from geothermal energy for DH (MWh)	7939.75
Amount of heat produced annually from geothermal energy for DHW (MWh)	4275.25
Total amount of annual heat produced from geothermal energy (MWh)	12,215
Natural gas savings (m^3)	1,136,279
Natural gas savings (kWh)	12,215,000
Price of saved natural gas (EUR/year)	354,235

The results of Tables 4 and 5 clearly show how the energy potential and the annual amount of thermal energy increase with increasing geothermal water temperature. Using data from the technical study of the geothermal well, the energy balance of the geothermal source GTK-1 was calculated for three temperature scenarios (pessimistic, conservative, and optimistic) of geothermal water.

3.2. Assessment of the Economic Feasibility of Using GE and the Payback Period of the Implemented Project

The assessment of economic profitability was recalculated by calculating the net present value, and the return-on-investment method was used. We considered two variants—a pessimistic and an optimistic scenario.

In the calculations, we considered the following input data:

1. Quantity delivered to GE per year;
2. Credit financing up to 90% of the realized investment;
3. Interest rate—0.75%;
4. Loan maturity—12 years;
5. Annual increase in operating costs—by 1% per year;
6. Discount rate—7.5%.

In Table 6, the recalculation of the evaluation of the economic efficiency of the implemented geothermal project for the pessimistic scenario is shown.

Based on the economic assessment, and assuming Tables 6 and 7 have the considered inputs, we came to the following partial conclusions:

- The project would achieve a return of 16.4 years in a pessimistic scenario.
- The project would achieve a negative cash flow in the years 2022 to 2025 until, in 2026, the cash flow would reach a positive value, which would increase in subsequent years.
- The project would achieve a return of 7.4 years in an optimistic scenario.
- In an optimistic scenario, the project would achieve a high positive cash flow from the beginning, which would increase in subsequent years.

Table 6. Calculation of the evaluation of the economic efficiency of the implemented geothermal project for the pessimistic scenario.

	Pessimistic Scenario										
1.	Total Investment Costs (EUR)					2,459,687					
2.	Credit Share of Total Investment					90%					
3.	Credit Amount (EUR)					2,213,718					
4.	Interest Rate					0.75%					
5.	Credit Payment Period (Years)					12					
6.	Annual Operating Costs (EUR)					29,129					
7.	Annual Savings on NG (EUR)					228,843					
8.	Discount Rate					7.5%					
9.	Year	2022	2023	2024	2025	2026	2027	2028	2029		2031
10.	Number of Year	1	2	3	4	5	6	7	8	9	10
11.	Year Annual Sales/Operating CF (EUR)	228,843	228,843	228,843	228,843	228,843	228,843	228,843	228,843	228,843	228,843
12.	Annual Increase in Operating Costs	-	1%	1%	1%	1%	1%	1%	1%	1%	1%
13.	Operating Costs (EUR)	29,129	29,421	29,715	30,012	30,312	30,615	30,921	31,231	31,543	31,858
14.	Annual Depreciation (EUR)	187,252	187,252	187,252	187,252	187,252	187,252	128,745	128,745	90,307	90,307
15.	Credit Repayment (EUR)	184,477	184,477	184,477	184,477	184,477	184,477	184,477	184,477	184,477	184,477
16.	Credit Balance (EUR)	2,213,718	2,029,242	1,844,765	1,660,289	1,475,812	1,291,336	1,106,859	922,383	737,906	553,430
17.	Annual Interest on the Credit (EUR)	16,603	15,219	13,836	12,452	11,069	9685	8301	6918	5534	4151
18.	CF (EUR)	−4141	−3049	−1960	−873	210	1291	60,876	61,950	101,459	102,527
19.	NOPAT (CF–Income Tax 19%) (EUR)	−4141	−3049	−1960	−873	170	1046	49,310	50,180	82,182	83,047
20.	Discounted FCF (EUR)	−1270	−236	657	1425	2052	2476	−3871	−3113	−6253	−5397
21.	NPV (EUR)	−247,239	−247,475	−246,818	−245,393	−243,341	−240,865	−244,736	−247,849	−254,102	−259,498
22.	Payback Period (Years)					16.4					

Table 7. Calculation of the evaluation of the economic efficiency of the implemented geothermal project for the optimistic scenario.

	Optimistic Scenario										
1.	Total Investment Costs (EUR)	2,459,687									
2.	Credit Share of Total Investment	90%									
3.	Credit Amount (EUR)	2,213,718									
4.	Interest Rate	0.75%									
5.	Credit Payment Period (Years)	12									
6.	Annual Operating Costs (EUR)	29,129									
7.	Annual Savings on NG (EUR)	283,420									
8.	Discount Rate	7.5%									
9.	Year	2022	2023	2024	2025	2026	2027	2028	2029	2030	2031
10.	Number of Year	1	2	3	4	5	6	7	8	9	10
11.	Year Annual Sales/Operating CF (EUR)	283,420	283,420	283,420	283,420	283,420	283,420	283,420	283,420	283,420	283,420
12.	Annual Increase Operating Costs	-	1%	1%	1%	1%	1%	1%	1%	1%	1%
13.	Operating Costs (EUR)	29,129	29,421	29,715	30,012	30,312	30,615	30,921	31,231	31,543	31,858
14.	Annual Depreciation (EUR)	187,252	187,252	187,252	187,252	187,252	187,252	128,745	128,745	90,307	90,307
15.	Credit Repayment (EUR)	184,477	184,477	184,477	184,477	184,477	184,477	184,477	184,477	184,477	184,477
16.	Credit Balance (EUR)	2,213,718	2,029,242	1,844,765	1,660,289	1,475,812	1,291,336	1,106,859	922,383	737,906	553,430
17.	Annual Interest on the Credit (EUR)	16,603	15,219	13,836	12,452	11,069	9685	8301	6918	5534	4151
18.	CF (EUR)	50,436	51,528	52,617	53,704	54,787	55,868	115,453	116,527	156,036	157,104
19.	NOPAT (CF-Income Tax 19%) (EUR)	40,853	41,738	42,620	43,500	44,378	45,253	93,517	94,387	126,389	127,254
20.	Discounted FCF (EUR)	40,585	38,519	36,542	34,651	32,845	31,121	22,775	21,674	16,805	16,053
21.	NPV (EUR)	−205,384	−166,865	−130,323	−95,672	−62,827	−31,706	−8931	12,743	29,548	45,601
22.	Payback Period (Years)	7.4									

The payback period of the project in the optimistic scenario is very attractive and is advantageous for the specified parameters. A higher share of credit financing has a significant effect on the payback period.

The presented data create a synergy for the final conclusion that the implementation of this project will be an effective investment in the future.

4. Conclusions

Geothermal wells offer a number of positive options for reducing negative environmental impacts in the long term. In theory, the projected energy potential of the well would be able to supply the modeled housing estate with warm water intended for heating and the preparation of hot service water. One of the most burdensome aspects of today's society is energy. In particular, energy in households, especially heating and cooling, contributes significantly to negative impacts on the country as a whole. In the case of Slovakia, there is currently a system of supported RES and limited use of fossil resources. In the case of the specific location of the Podtatranska Basin, we propose to use the potential of geothermal energy for heating domestic housing units (housing estates). This will help to limit the combustion of natural gas and, in the long term, save money on the purchase of NG and achieve increased energy security and stability. The use of geothermal energy for housing estate heating will bring a number of benefits to this area. GE can also be used in a combined way (heating and electricity production) for greenhouse heating for growing crops and flowers, for fish farming, etc. Primarily, it will offer an ecological benefit in the form of significant emission reductions in the combustion of processes (NOx, SO_2, CO, TZL). Currently, the city's thermal economy burns natural gas and coal. Geothermal energy is a highly ecological source; immediately after heating use, it will be drained off to a nearby surface stream. Another benefit will be that it will reduce dependence on the imports of primary heat sources (NG and coal) and replace them with a renewable energy source. Finally, the use of geothermal energy will lead to a lower heating bill for the inhabitants of the housing estate.

Currently, the aim is to use RES technology that contributes as much as possible to reducing the negative consequences of energy use on the global climate system. The topic is vital, according to current developments in energy policy in the EU and around the world. At the end of October 2021, the UN World Climate Summit was held in Glasgow; it highlighted the need to address the issue of reducing emissions in energy production. Our contribution, in terms of the use of geothermal energy, highlights a clean renewable resource and is therefore highly relevant for future generations.

Author Contributions: Conceptualization: A.S. and E.S.; methodology: A.S. and E.S.; software: G.W. and J.R.; validation: A.S. and E.S.; formal analysis: A.S. and E.S.; investigation and resources: A.S. and G.W.; data curation: A.S. and E.S.; writing—original draft preparation: J.R. and G.W.; writing—review and editing: A.S. and E.S.; visualization: A.S. and E.S.; supervision: A.S. and E.S.; project administration and funding acquisition: G.W. and J.R. All authors have read and agreed to the published version of the manuscript.

Funding: This research received no external funding.

Institutional Review Board Statement: Not applicable.

Informed Consent Statement: Not applicable.

Data Availability Statement: Data sharing is not applicable to this article.

Acknowledgments: This work is supported by the Scientific Grant Agency of the Ministry of Education, Science, Research, and Sport of the Slovak Republic, project KEGA 048TUKE−4/2021, Universal educational—competitive platform.

Conflicts of Interest: The authors declare no conflict of interest.

References

1. IPCC. *2018: Global Warming of 1.5 °C. An IPCC Special Report on the Impacts of Global Warming of 1.5 °C above Pre-Industrial Levels and Related Global Greenhouse Gas Emission Pathways, in the Context of Strengthening the Global Response to the Threat of Climate Change, Sustainable Development, and Efforts to Eradicate Poverty*. Masson-Delmotte, V., Zhai, P., Pörtner, H.-O., Roberts, D., Skea, J., Shukla, P.R., Pirani, A., Moufouma-Okia, W., Péan, C., Pidcock, R., et al., Eds.; Available online: https://www.ipcc.ch/site/assets/uploads/sites/2/2019/06/SR15_Full_Report_High_Res.pdf (accessed on 15 September 2021).
2. Braunmiller, G.; Horbaj, P.; Jasminská, N. Geothermal energy and power generation in Germany. *Communications* **2009**, *11*, 64–66.
3. Rybár, R.; Kudelas, D.; Beer, M. Selected problems of classification of energy sources—What are renewable energy sources? *Acta Montan. Slovaca* **2015**, *20*, 172–180.
4. Eurostat. Renewable Energy Statistics. 2020. Available online: https://ec.europa.eu/eurostat/statistics-explained/index.php/Renewable_energy_statistics (accessed on 10 September 2021).
5. Sowizdzal, A. Geothermal Energy Resources in Poland–Overview of the Current State of Knowledge. *Renew. Sustain. Energy Rev.* **2018**, *82*, 4020–4027. [CrossRef]
6. Wilberforcea, T.; Baroutajib, A.; ElHassana, Z.; Thompsona, J.; Soudane, B.; Olab, A.G. Prospects and challenges of concentrated solar photovoltaics andenhanced geothermal energy technologies. *Sci. Total Environ.* **2019**, *659*, 851–861. [CrossRef] [PubMed]
7. Sliwa, T.; Sapinska-Sliwa, A.; Gonet, A.; Kowalski, T.; Sojczynska, A. Geothermal Boreholes in Poland—Overview of the Current State of Knowledge. *Energies* **2021**, *14*, 3251. [CrossRef]
8. Koščo, J.; Tauš, P.; Taušová, M.; Jeňo, M. Geothermal energy—One of the resources of tourism expansion in Slovakia. *Acta Montan. Slovaca* **2016**, *21*, 171–179.
9. Barbier, E. Geothermal energy technology and current status: An overview. *Renew. Sustain. Energy Rev.* **2002**, *6*, 3–65. [CrossRef]
10. Hudeček, V.; Zapletal, P.; Stoniš, M.; Sojka, R. Results from dealing with rock and gas outburst prevention in the Czech Republic. *Arch. Min. Sci.* **2013**, *58*, 779–787.
11. Borovič, S.; Markovič, I. Utilization and tourism valorisation of geothermal waters in Croatia. *Renew. Sustain. Energy Rev.* **2015**, *44*, 52–63. [CrossRef]
12. Cehlár, M.; Jurkasová, Z.; Kudelas, D.; Tutko, R.; Mendel, J. Geothermal Power Plant in Conditions of Geological and Hydrological Characteristics. *Adv. Mater. Res.* **2014**, *1001*, 63–74. [CrossRef]
13. Ferenc, Š. *Geotermálna Energia a jej Využitie*, 1st ed.; Belanium Univerzity Mateja Bela v Banskej Bystrici: Banská Bystrica, Slovakia, 2015; p. 154.
14. STATISTA. Historical Carbon Dioxide Emissions from Global Fossil Fuel Combustion and Industrial Processes from 1758 to 2020. 2020. Available online: https://www.statista.com/statistics/264699/worldwide-co2-emissions/ (accessed on 25 September 2021).
15. Rybár, P. Zdroje geotermálnej energie a možnosti ich využívania. *Acta Montan. Slovaca* **2010**, *12*, 31–41. Available online: https://core.ac.uk/download/pdf/25941789.pdf (accessed on 25 September 2021).
16. Tometz, L.; Dugáček, D. Potenciál podzemných vôd Slovenska ako obnoviteľných zdrojov energie. *Acta Montan. Slovaca* **2010**, *15*, 116–125. Available online: https://core.ac.uk/download/pdf/25955231.pdf (accessed on 15 September 2021).
17. Barbacki, A.; Miecznik, M.; Tomaszewska, B.; Skrzypczak, R. Assessment of the Lower Carboniferous-Devonian Aquifer as a Source of Geothermal Energy in the Silesian–Kraków Region (Poland). *Energies* **2020**, *13*, 6694. [CrossRef]
18. Fendek, M.; Hanzel, V.; Bodiš, D.; Nemčok, J. *Hydrotermálne Pomery Popradskej Kotliny. Manuskript.Archív Tatra Thermálu, a.s.*; Archív Tatra Thermál a.s. Poprad: Poprad, Slovakia, 1992; p. 99.
19. Franko, O.; Remšík, A.; Fendek, M. a kol. Atlas Geotermálnej Energie Slovenska, Bratislava: Štátny Geologický Ústav Dionýza Štúra. 2010. Available online: http://apl.geology.sk/atlasge (accessed on 3 January 2022).
20. Slovenská Agentúra Životného Prostredia. Kjótsky Protokol k Rámcovému Dohovoru OSN o Zmene Klímy. Available online: https://www.enviroportal.sk/dokumenty/medzinarodne-dohovory/dohovor/2 (accessed on 25 September 2021).
21. Enviroportál MŽP SR. Available online: file:///C:/Users/Erika/AppData/Local/Temp/soh_geotermalny_vrt_kezmarok_optimised_asice_soh_geotermalny_vrt_kezmarok_optimised_redigovane-.pdf (accessed on 3 January 2022).
22. Rybár, P.; Drebenstedt, C.; Cehlár, M.; Domaracká, L.; Khouri, S.; Dietze, T. *Mining Investment*; Vydavatelství a nakladatelství Aleš Čeněk: Plzeň, Czech Republic, 2019; 250p.
23. Kvantifikácia Emisií: Metodický Postup Pre Tvorbu Regionálnych Nízkouhlíkových Stratégií, Priatelia Zeme-CEPA. 2020. Available online: https://cepa.priateliazeme.sk/images/publikacie/EVS_vystupy/M10_web.pdf (accessed on 28 September 2021).
24. Zloženie Zemného Plynu, Spaľovacie Teplo a Výhrevnosť, Emisný Factor. Available online: https://www.spp.sk/sk/velki-zakaznici/zemny-plyn/o-zemnom-plyne/emisie/ (accessed on 30 September 2021).
25. zlRESnie-zemneho-plynu-a-emisny-faktor. Available online: https://www.spp-distribucia.sk/dodavatelia/informacie/zlRESnie-zemneho-plynu-a-emisny-faktor/ (accessed on 30 September 2021).
26. Všeobecné Emisné Závislosti a Všeobecné Emisné Factory. Available online: https://www.minzp.sk/ovzdusie/ochrana-ovzdusia/zdroje-znecistovania-ovzdusia/novy-adresar/vseobecne-emisne-zavislosti-vseobecne-emisne-faktory.html (accessed on 30 September 2021).
27. Specific Heat Formula. Available online: https://www.toppr.com/guides/physics-formulas/specific-heat-formula/ (accessed on 3 January 2022).

 processes

MDPI

Article

Numerical Study for Determining the Strength Limits of a Powered Longwall Support

Dawid Szurgacz [1,2], Konrad Trzop [3], Jan Gil [4], Sergey Zhironkin [5,6,7,*], Jiří Pokorný [8] and Horst Gondek [9]

1. Center of Hydraulics DOH Ltd., 41-906 Bytom, Poland; dawidszurgacz@vp.pl
2. Polska Grupa Górnicza S.A., ul. Powstańców 30, 40-039 Katowice, Poland
3. KWK Ruda Ruch Bielszowice, ul. Halembska 160, 41-717 Ruda Śląska, Poland; konrad.trzop.kt@gmail.com
4. PGG Oddział Zaklad Remontowo-Produkcyjny, ul. Granitowa 132, 43-155 Bieruń, Poland; j.gil@pgg.pl
5. Department of Trade and Marketing, Siberian Federal University, 79 Svobodny Av., 660041 Krasnoyarsk, Russia
6. Department of Open Pit Mining, T.F. Gorbachev Kuzbass State Technical University, 28 Vesennya St., 650000 Kemerovo, Russia
7. School of Core Engineering Education, National Research Tomsk Polytechnic University, 30 Lenina St., 634050 Tomsk, Russia
8. VSB—Faculty of Safety Engineering, Technical University of Ostrava, Lumírova 13/630, 700 30 Ostrava, Czech Republic; jiri.pokorny@vsb.cz
9. VSB—Department of Machine and Industrial Design, Technical University of Ostrava, 17 Listopadu 2172/15, 708 00 Ostrava, Czech Republic; horst.gondek@gmail.com

* Correspondence: zhironkinsa@kuzstu.ru

Abstract: The process of designing a longwall powered support is extremely complex and requires many operations related to the creation of a complete machine. The powered support section is one of the basic elements of the longwall system. It acts as protection for the working space and takes part in the process of excavating and transporting the spoil. The implementation of the support that meets the guidelines of the manufacturer and user requires an endurance analysis at the design stage conducted according to the regulations in force. The main objective of this research, pursued by the authors, was to perform the analysis of the ultimate strength of selected elements of the designed powered support section. The research was carried out with the use of special software that uses the finite element method. This article presents the analysis of the strength limits conducted with the help of the finite element method, determining the strength of selected elements of the longwall support section. The solutions proposed by the authors include changes in the structure and properties of the material in the design process. The aim of the proposed solution was to obtain a model with strength value that meets safety standards. The research results are a valuable source of knowledge for designers. Solutions of this type set examples for spatial models of the longwall support section being designed currently. The analysis presented in the article, together with the results of the research and the conclusions resulting from them, may improve the safety and effectiveness of powered supports.

Keywords: finite element method (FEM); mining; numerical modeling; power roof supports; safety

Citation: Szurgacz, D.; Trzop, K.; Gil, J.; Zhironkin, S.; Pokorný, J.; Gondek, H. Numerical Study for Determining the Strength Limits of a Powered Longwall Support. *Processes* **2022**, *10*, 527. https://doi.org/10.3390/pr10030527

Academic Editor: Jean-Pierre Corriou

Received: 31 January 2022
Accepted: 4 March 2022
Published: 7 March 2022

1. Introduction

Currently, the dynamically developing area of automation and robotization of machines in accordance with the idea of modern technologies [1–3] brings about changes in the processes of their design and production. These changes force producers to use modern methods and techniques [4–6], influencing the improvement of safety and reduction of costs and time [7–9]. At the stage of design, the fulfillment of these assumptions allows using model research [10,11] on virtual objects with the use of numerical methods [12–14]. Numerical modeling techniques based on differential equations [15] enable the creation of a virtual three-dimensional model. The model, thanks to dedicated software, can be subjected to processes determining changes in the behavior of physical parameters, including determination of yield strengths. Determining the yield strengths is an important issue from the point of view of designing machine structure elements [16–19], as it entails the determination of the state of stress and deformations resulting from the impact of external forces [14,16]. At this stage, it is also worth taking into account the working environment of the designed machines [20–22].

Numerical tests of powered support sections conducted by using the finite element method in the design process shorten the time of creating a prototype [23–25]. They enable the use of various design variants [26,27] in order to choose the most optimal solution based on the analysis [28,29]. Thanks to this, the final product, in this case a powered support, will meet the quality and safety requirements [30,31]. Due to the significant role of powered support in the production process, it is important for it to exhibit a high safety index [32–34]. The support sections are responsible for securing the longwall excavation against the fall of roof rocks and for moving the mining machine along the face conveyor towards the face of the wall [35–37]. It is one of the three main machines included in the longwall complex, which forms a set of mutually dependent machines [38–40].

Elements of the support section are constructed according to specific safety standards that the support must meet [40–43]. The powered support section consists of basic elements (Figure 1), which include all the parts carrying load caused by the pressure of roof rocks [44]. The additional elements are the parts that do not transfer the load resulting from the pressure of the roof rocks, but are necessary for the functioning of the support [45]. The working environment of the longwall support in underground mining plants is extremely difficult and dangerous. Complicated natural (environmental) conditions make the process of coal mining inextricably associated with both natural and mechanical hazards—resulting from improper use of machines and devices. The accumulation of many threats, such as methane, rock bursts, the risk of gas and rock outbursts, inclination of underlay, depth of the selected deposit, water leaks, tectonic disturbances, and coal and rock cracks force machine designers to take into account these extremely difficult mining and geological conditions [46].

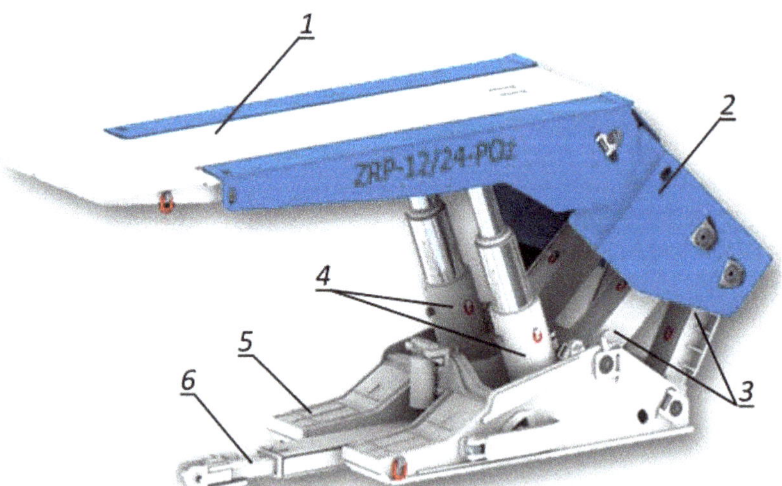

Figure 1. Section of the powered roof support: 1—canopy, 2—shield support, 3—lemniscate tie rods, 4—leg, 5—floor base, 6—shifting system.

The powered support section opposes the occurrence of dynamic phenomena in the rock mass, the effect of which is the formation of tremors. The tremors occurring in the rock mass can be divided into natural and mining, i.e., caused by improper roof control [47].

The conditions of cooperation between the support and the rock mass are analyzed on the basis of the processes related to the subsidence of the roof layers. This method has been widely used to determine the factor characterizing the maintenance of the roof. The support is influenced by some dynamic phenomena occurring in the rock mass, such as vibrations, rapid clamping of the adjacent rocks, impact of crushed and ejected rocks and an increase in load force resulting from a fractured rock mass. All this collaboration is extremely complex. It depends on many mining and geological factors influencing the value of the load on the powered support by the rock mass. These parameters, apart from the previously mentioned depth of the works, are structure and strength of the surrounding rocks, exploitation history, height, length, span of the wall and water accumulation [47].

At the stage of designing, planning and conducting research on a powered support section, each of the elements should be thoroughly analyzed prior to the interaction between the support and the rock mass under real conditions [48–50]. The design stage is based on the selection of adequate technical solutions and on determining mutual dependency between them in order to obtain the final effect [51,52].

The first stage involves the preparation of a flat model, which is subject to analysis in order to determine its geometric and kinetostatic features [53]. The performed analysis must meet the design guidelines. The data are analyzed in the form of dimensioned elements of the powered support section. The second stage covers the scope of strength analysis and determination of boundary conditions. An adequately prepared model is generated in dedicated software [54], where it is subjected to stimulation. The third stage includes the preparation of the prototype and testing in a notified body [47,53].

After obtaining a positive opinion from the research laboratory, the production of powered support begins. The importance of the powered support section in the entire process [55,56] triggers the need to conduct a series of tests using modern methods [57–59] to optimize its strength parameters with the aim of improving safety in accordance with the PN-EN standards, as well as improving quality, efficiency and reduction of production cost.

This publication presents model strength tests of powered support sections with the use of numerical calculations. This work contains the examples of design solutions for the formation of the support in accordance with the standards and safety regulations. The analysis also took into account the working environment [60] of the machine and the conditions of interaction with other machines.

The purpose of this article is to determine the strength limits of the elements of powered support sections by means of numerical calculations. The tests will make it possible to get rid of defects and will allow design features to be given to the machine as early as at the design stage. Such activities will take into account the reduction of the costs of creating a prototype in accordance with the manufacturer's and the user's guidelines.

2. Materials and Methods

The load acting on the body may cause deformation of the structure. In case of reaching a load value exceeding the permissible values of individual elements of the structure, it may lead to destruction or irreversible deformation. The impact of the occurring deformations and stress depends on multiple factors regarding the type of material, the force affecting the structures and the condition of the machine [61–63].

Operations using calculation methods simulate the effect of load force on individual elements of the machine structure. These elements are subjected to load carrying capacity assessment, that is, strength assessment and deformation assessment, in accordance with the standards, regulations and manufacturer's recommendations [64,65]. All these operations are meant to guarantee safety [66] and rigidity as well as limitation of economic losses of the machine. The computational methods use the principles of solid mechanics, where they use mathematical calculations [66,67] to describe the influence of the load by determining the state of deformations and stress of the machine and mutual interactions, taking into account the material properties of the structure [68,69]. The determination of the properties of continuity, homogeneity, isotropy and elasticity of the material allows specifying mathematical dependencies [70]. When it takes into account the mechanical properties of the material and the analysis of computational methods, the machine's operation is able to limit the formation of permanent deformations.

According to the guidelines concerning the strength of the materials, the body subjected to a load force is deformed. The deformations can be divided into elastic and plastic. The elastic deformations do not change the shape of the structure elements after removing the load force affecting the mechanical properties. In case of plastic deformation, the permanent deformation which affects the structure from the very beginning occurs. Only after reaching the permissible values, called the limit of elasticity, it damages the structure [71]. When focusing on deformations in steel structures, it is taken into consideration that due to the deformation mechanisms, the material is a part of the elastic–plastic group.

The deformation intensity, which is a measure of deformation, determines the changes in the dimensions of the length and is given by the formula:

$$\varepsilon = \lim_{l \to 0} \frac{\Delta l}{l} \quad (1)$$

where:

ε—deformation (%);

l—distance between the assumed points before material deformation (mm);

Δl—the sum of the distance between the points created after the deformation of the material (mm), and changes in the dimensions of the angle:

$$\gamma = \lim_{CE \to 0.DE \to 0} CDE - C'D'E' \quad (2)$$

where:

γ—value of the deformation angle (°);

CDE—points marked on the material before deformation;

$C'D'E'$—the difference in the position of the points marked on the material after deformation.

The change in length is the result of a loosening of the body structure, and the change in form is the result of a slip, i.e., shifting the layer of atoms one after another. The state of deformation in the vicinity of the point, e.g., O, will be described by the quantities ε and γ in all directions, with point O as the reference point [72–74]. In addition, the deformations can be purely of the volumetric type while meeting the following conditions:

$$\gamma_{xy} = \gamma_{yz} = \gamma_{xx} = 0 \quad (3)$$

where:

$\gamma_{xy,yz,xx}$—measures of component angles (°).

$$\varepsilon_x = \varepsilon_y = \varepsilon_z = \varepsilon \quad (4)$$

where:

$\varepsilon_{x,y,z}$—components of the deformations (%);

and purely figural under the following conditions:

$$\cos\gamma_{xy} = \cos\gamma_{yz} = \cos\gamma_{xx} \to 1 \text{ czyli } \Delta V = 0 \quad (5)$$

where:

$\cos\gamma_{xy,yz,xx}$—cos angles of deformation components;

ΔV—the sum of the displacement (mm).

$$\varepsilon_x = \varepsilon_y = \varepsilon_z = \varepsilon \quad (6)$$

where:

$\varepsilon_{x,y,z}$—components of the deformations (%).

The powered support section is constructed in such a way that the structural elements balance the load force, affecting its operational elements. The rock mass, which in addition is an anisotropic body, and the activities related to the functioning of the machine, have an influence on the operation of the support. The force load is divided into permanent, variable and resulting from the environment. It is obvious that during the analysis from the set of possible load combinations, one should take into account the sets of load force causing the most unfavorable systems of force and moments, both for the components and the entire structure [72–74].

2.1. Calculation Model for the Limit of Elasticity

The calculation model for the limits of elasticity takes into account the stress of the force impacting the powered support and shows the creation of elastic deformations in the model. The position of elementary body particles exposed to external force has an influence on the stress value. In addition, deformations are conditioned by the characteristics of the material from which the structure is made. This leads to following arbitrary formulas:

$$\sigma = f(\varepsilon) \tag{7}$$

where:

σ—stress values (MPa);

$f(\varepsilon)$—values of the deformation function (%).

Data on the functions f of the interdependencies of σ on ε are obtained by conducting adequate strength tests, in accordance with the standards and regulations for the safe use of machines and devices [74].

Elastic–plastic materials, in this case elements of the powered support section, are subjected to the following tests:

- Static load tests:
    - Under bending load capacity test;
    - Under compressive and tensile load capacity test;
    - Elements for mounting props and cylinders under compressive and pulling load capacity test;
    - Working load capacity;
    - In the case of a mine crib support with a horizontal load;
    - In the case of a shielding support with a horizontal load;
    - With an asymmetric load.
- Cyclic endurance tests:
    - Bending;
    - Twisting [45].

From the tests of the dependence $\sigma = f(\varepsilon)$, certain material properties, called the material strength properties, can be observed if they relate to the entire test or the material elasticity properties and if they refer to the area of elastic deformations.

2.2. Calculation Model for the Yield Point

The implementation of the computational model including the yield limits allows one to obtain the values describing the average properties of the material at the design stage and describes with a certain accuracy the phenomena that occur in the material in reality. Exceeding the yield point leads to permanent deformations as a result of which the machine is damaged and prevents its further functioning in the longwall excavation. The yield point is one of the values of a material's endurance [61].

The diagram of the limits of elasticity (Figure 2) presents the way in which a given material behaves when impacted by load force. The values of the behavior of the tested steel were superimposed on the basic values of the elasticity limits of a given material by presenting a preview of the moment of exceeding the permissible values of elasticity and plasticity of structural elements of the powered support section. The operation was carried out for the purpose of comparison.

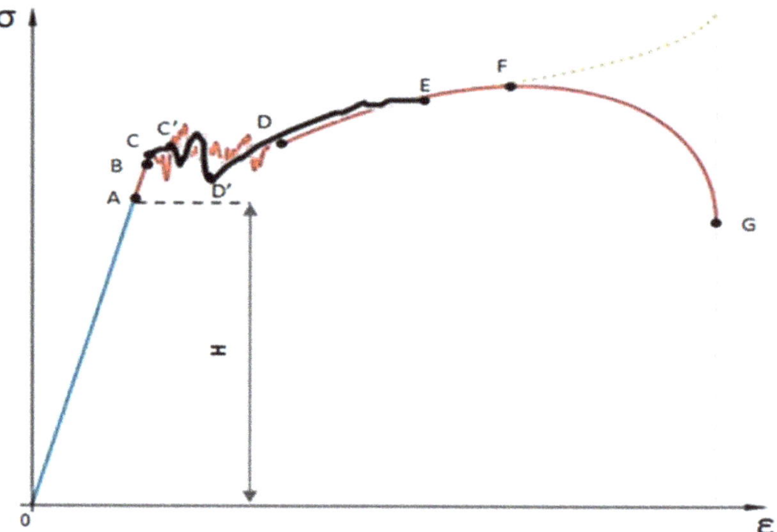

Figure 2. Graph of the strength limits of the tested material, where A—proportionality limit, B—elasticity limit, C—yield point, C′—yield point for steel, D—material hardening, D′—steel hardening, E—temporary steel endurance, F—ad hoc endurance, G—breaking the material, H—Hook's Law, Blue Line—scope of Hook's Law, Red Line—range of nonlinear unsteady and plastic deformations, Yellow Line—Stress with consideration of the constriction, Black Line—The range of nonlinear intractable and ductile deformations for steel.

The limit of proportionality determines the moment (A) in which there was no change in the structure of material and that maintains the linearity between deformation and stress. In contrast, the elasticity limit (B) shows the lack of linearity between stress and deformation. The sample deforms but returns to its original shape. This means that the elastic energy is completely returned. For the yield point (C) and for steel (C′), at the moment of applying stress, the plastic deformation becomes visible, which after removing the stress does not return to its original state. The area between points CD, and for the value of steel C′-D′, is the material flow moment. Strengthening of the material (D) and steel (D′) takes place at the moment of a certain inhibition of the creation of slips. During this time, from point D and D′, it is necessary to increase the stress in order to increase the deformation or destruction of the material. The phenomenon of temporary strength for steel (E) and for reference material (F) is the point of state at which the stress is no longer homogeneous.

Point (G) marks the point of rupture, the loss of physical properties of the material. The range we are interested in for material strength is the one highlighted in blue (Blue Line). This is the range of elastic deformation of a material, i.e., where the material will return to its form after the external force disappears. The range (Red/Black Line) defines the elastic limits. In this area, if the stresses exceed the permissible value, permanent deformation will occur in the material. The range (Yellow Line) defines the absence of a momentary increase in stress; we have the material transitions to a plastic state. When the stress is further increased, it causes a nonlinear increase in strain until a noticeable local constriction called a neck occurs.

In the tested samples, there is a concentration of slips in one place, visualized in the form of a local narrowing [57,64]. Avoiding the phenomenon in point E, F requires the designer to take measures to prevent exceeding the permissible values in order to meet the safety requirements of the support structure.

Taking into account the above calculation models for the limits of elasticity and plasticity of the support, the support was subjected to numerical tests. The model determines geometric and kinetostatic features. The model of the 3D section is made in dedicated software [54]. It defines the design guidelines and material properties for the production of individual elements of the support at the design stage. The simulation included applying load force on individual elements, and it indicated strategic places where there was the greatest probability of exceeding the permissible values. Withdrawal from the design procedures may lead to permanent damage, the result of which is the inability to use the support section for further operation.

3. Results

Based on the data regarding the powered support, which are presented in Table 1, model tests of strength limits of its elements were performed with the use of dedicated software [54]. The tested powered support is a two-rack chock shield support with a lemniscate mechanism for guiding the canopy. The basic parameters of the support are shown below.

Table 1. Parameters of the tested section of powered support.

Name	Value	Unit
Height range	1.0 ÷ 3.50	(m)
The support's operation range—for non-bursting coal seams	1.70 ÷ 3.40	(m)
The support's operation range—for bursting coal seams	1.80 ÷ 3.40	(m)
Set division	1.50	(m)
The support's movement	to 0.8	(m)
Longwall inclination	to 35	(°)
Lateral longwall inclination	±20	(°)
Initial load capacity of the props for 25 MPa (30 MPa)	2 × 1 767 (2 × 2 120)	(kN)
Nominal load capacity of the props for 43 MPa	2 × 3039	(kN)
Unit pressure on the floor	1.850 ÷ 2.083	(MPa)
Unit pressure on the roof	0.782 ÷ 1.060	(MPa)
Set displacement force for 30 MPa	603	(kN)
Conveyor travel force for 30 MPa	291	(kN)
Supply pressure	25 ÷ 30	(MPa)
Mass	~20°500	(kg)

While conducting the project research, a modern prototyping system was used [53]. The assumed result of the design research is achieving the highest possible strength values of the structure elements of the powered support tested section in the 3D model, which is shown in Figure 3. For this purpose, a model of the support section was developed and due to existing experience and knowledge, was enriched with elements and features allowing achieving the appropriate level of safety according to the standards [67] of the prototype.

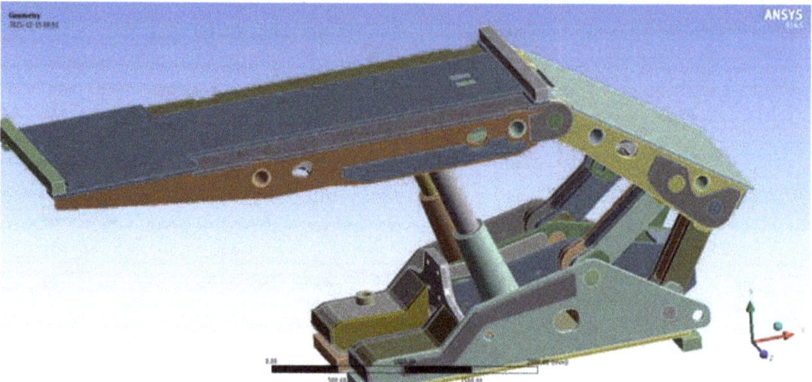

Figure 3. The 3D model of the support section.

3.1. Development of a Model for Analysis

The first stage of the numerical analysis of the powered support section was to prepare a flat model in order to determine its geometric and kinetostatic features. After determining the dimensions of the structure, taking into account the operation range, the support's model was validated, and the geometry of the kinematic chain was calculated [56]. The physical parameters of the support were determined: load carrying capacity, internal forces affecting the support, the angle of the canopy, the angle of friction and the force affecting the shield support, respectively. Taking into account the calculations of the flat model [53], the 3D model was made in special software, shown in Figure 4, enabling a computer-based solution of the problem [54].

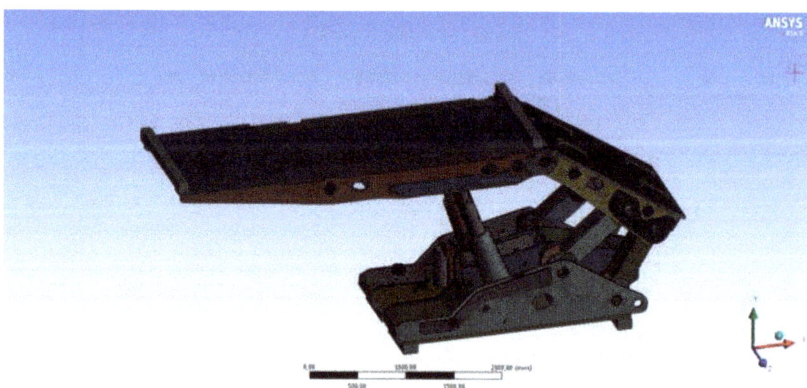

Figure 4. The 3D model of the support section with a numerical grid.

This software made it possible to verify visually and kinematically the operation of the support in its full height range [58,75]. It also helped to verify a possible collision of the support's components with other machines of the longwall complex. The analysis had to meet the design guidelines. The data were subjected to numerical calculations in the form of dimensioned elements of the powered support section. The geometry of the calculation area was set, then a numerical grid covering the calculation area was generated (Figure 4) and the type of boundary conditions was determined [54].

The 3D model prepared in this way made it possible to calculate the strength value of the support section at any point in the calculation area.

3.2. Research Analysis of the Research Model

The conducted analysis of the model of a powered support section assumed the application of forces acting on its elements with the minimum to maximum value, in accordance with the guidelines included in the standards [76]. The methods of applying load force and the number of cycles, which must be performed by the support before being allowed to operate, have been defined [47,53] according to the standard [76]. The requirements set by the standard guidelines reflect the load requirements in real conditions. The support was subjected to simulation (Figure 5) in dedicated software, which allowed for the calculation of reduced stress at a selected point according to the Huber–Mises–Hencky hypothesis. This hypothesis assumes that the measure of the stress of material subjected to load force is the energy of the shear deformation associated with the change in the shape of material [54,75].

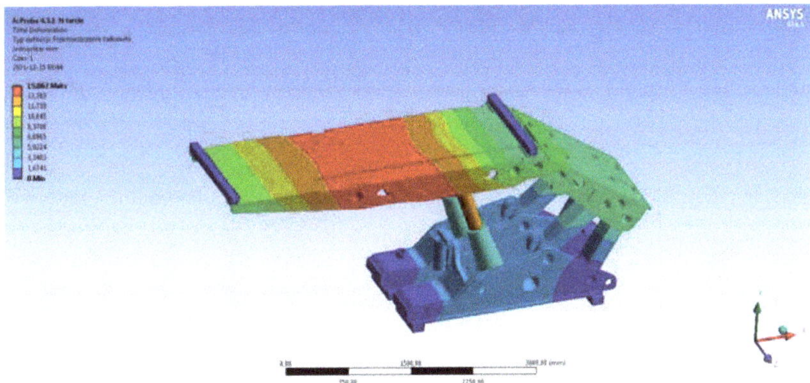

Figure 5. Load force simulation of the support's canopy.

The results obtained in the simulation of the canopy element did not meet the safety requirements defined by the standards [76] and the manufacturer. In order to improve the strength of the canopies, geometric adjustments were made, which allowed for more effective use of the material and increased its strength. The effect of these operations is presented in Figure 6.

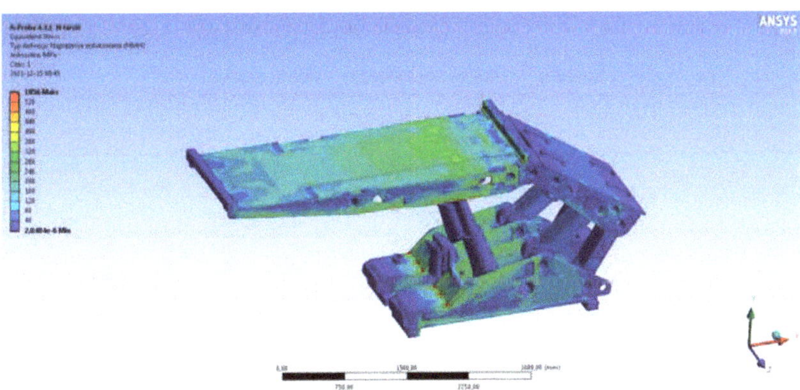

Figure 6. Load force simulation of the support's canopy—correction.

Properly selected elements were fastened in place with beam nodes and a grid was created (Figure 7). The force values were successively assigned at the appropriate angle. The result of the strength analysis was the creation of a map drawn on a finite element grid. These maps illustrated the number of cycles after which a crack will occur in a given area (Figure 8).

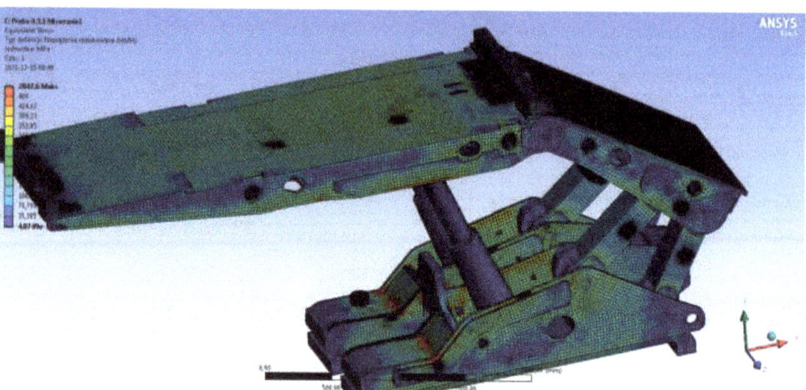

Figure 7. Stress analysis on a finite element grid.

Figure 8. Stress analysis on a finite element grid—support's canopy.

The results of the analysis showed places exposed to plasticization or material breaking in the part of the canopy and floor [61,71]. These points are located in the places where elements connect, such as welds and notches of individual elements of the support section (Figures 9 and 10).

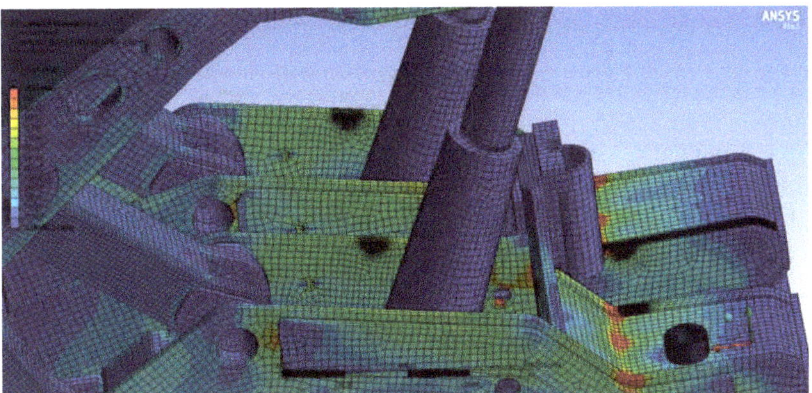

Figure 9. Stress analysis on a finite element grid—floor.

Figure 10. Stress analysis on a finite element grid—floor.

The strength analysis is able to determine a possible change in terms of structure or material [47]. The procedures related to the change of material or geometric and construction elements required repeating computational analysis to obtain a result that meets the requirements. The purpose of the analysis is to define guidelines ensuring failure-free operation of the powered support section, taking into account at the same time the costs of production and operation. The performed tests are supposed to reflect the underground operating conditions of the powered support section.

The result of the performed design activities is the creation of a prototype of the housing section, which is subjected to laboratory tests. These tests are carried out in accordance with the requirements of the Machinery Directive and other standards [45]. A positive test result would be the basis for the implementation of the prototype and the commencement of the bench tests.

4. Discussion

Initially, the research was based on giving the prepared flat model geometric and kinetostatic features in order to create a spatial model of the powered support section. The 3D model was subjected to calculations specifying the limits of elasticity and plasticity. For this purpose, the method of numerical calculations was used, taking into account the basic formulas included in the article. The calculations were made with the use of dedicated software, which allowed us to model and simulate the effect of stress [54]. The elements of the powered support section were subjected to load force that may occur in real conditions.

Before starting the model endurance tests, the working hypothesis was enriched with specification of the type of material used and the location of additional structural elements. The hypothesis resulted from the experience of designers and the knowledge acquired while designing powered supports with similar parameters.

The canopy of the powered support section was subjected to simulation, as shown in Figure 6. Higher strength values of this element were achieved. The type of material the support would be made of has been changed, thus meeting safety standards. The effect of this operation is shown in Figure 7. The section elements are connected in space with nodes and a grid is created (Figure 8). This narrows the area of applying the force affecting particular places in the model machine. The force values applied at an appropriate angle were used to perform an endurance analysis of the given cycles after which possible changes in the elasticity or plasticity limits of the material would occur. The tests were performed until the model met the permissible number of cycles, which were required by the safety standards. During the analysis, the places marked in red turned out to be (Figures 9 and 10) most exposed to stress in real conditions. The resulting model, having been subjected to numerical analysis while meeting the manufacturer's guidelines and regulations, is the result of making a prototype, which was previously subjected to bench tests. The development of the study model and the analysis of the results during the conducted research were divided into several stages, which are shown in Figure 11:

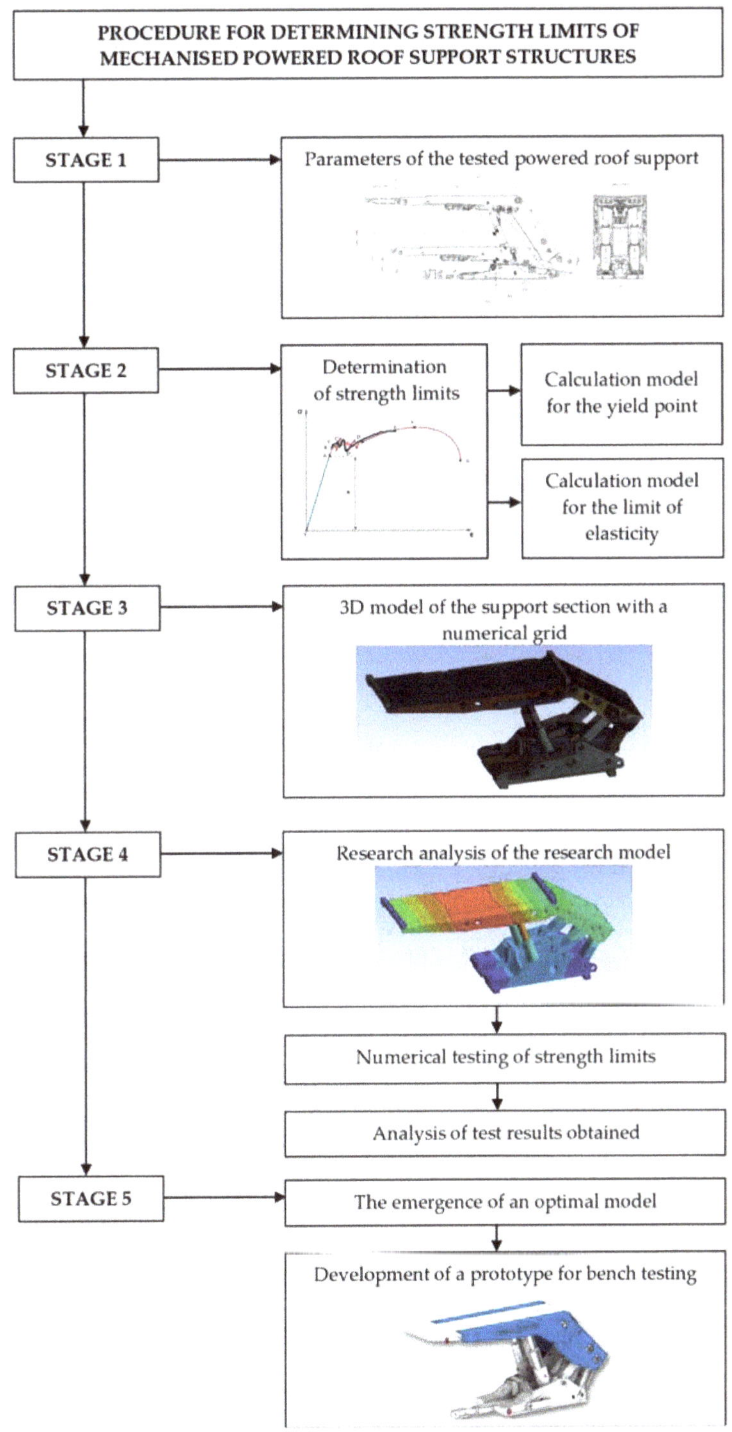

Figure 11. Steps in the analysis conducted.

5. Conclusions

The analyzed spatial model of the powered support presupposed additional structural elements and given material properties in order to increase the strength. These factors were specified by numerical calculations. The places most exposed to possible damage resulting from exceeding the allowable stress values were designated. The data from the performed support's load force cycles showed that the machine meets the strength requirements.

The main disadvantage of the MES test performed is the need to control the numerical error. This error can be due to grid density, changes in boundary conditions and material properties, time step, and others. This drawback, however, is acceptable for all numerical methods. The limitations were due to the ability to analyze the test model, with only one program to determine the strength limits.

The keynote of the model endurance tests was to achieve safety and quality requirements while reducing production costs. The structural elements of the machine were designed in such a way that they would be able to meet the requirements of safety standards. The reduction of production costs included the implementation of one prototype subjected to bench tests in a research laboratory. It is possible only when the model is properly designed.

The performance of model endurance tests already at the design stage showed us the advantages and disadvantages of the structure. Eliminating defects and providing design features made it possible to obtain the optimal model. On the basis of the numerical model, a prototype was created, which was successfully subjected to bench tests.

Author Contributions: Conceptualization, D.S. and K.T.; methodology, D.S. and S.Z.; software, K.T.; validation, J.P., H.G. and J.G.; formal analysis, D.S.; investigation, D.S.; resources, K.T. and J.G.; data curation, S.Z. and D.S.; writing—original draft preparation, D.S., K.T. and S.Z.; writing—review and editing, D.S. and S.Z.; visualization, K.T.; supervision, D.S.; project administration, D.S. and S.Z.; funding acquisition, D.S. and S.Z. All authors have read and agreed to the published version of the manuscript.

Funding: This research received no external funding.

Institutional Review Board Statement: The study was conducted according to the guidelines of the Declaration.

Informed Consent Statement: Not applicable.

Data Availability Statement: Not applicable.

Conflicts of Interest: The authors declare no conflict of interest.

References

1. Bortnowski, P.; Gładysiewicz, L.; Król, R.; Ozdoba, M. Energy Efficiency Analysis of Copper Ore Ball Mill Drive Systems. *Energies* **2021**, *14*, 1786. [CrossRef]
2. Wajs, J.; Trybała, P.; Górniak-Zimroz, J.; Krupa-Kurzynowska, J.; Kasza, D. Modern Solution for Fast and Accurate Inventorization of Open-Pit Mines by the Active Remote Sensing Technique—Case Study of Mikoszów Granite Mine (Lower Silesia, SW Poland). *Energies* **2021**, *14*, 6853. [CrossRef]
3. Wodecki, J.; Góralczyk, M.; Krot, P.; Ziętek, B.; Szrek, J.; Worsa-Kozak, M.; Zimroz, R.; Śliwiński, P.; Czajkowski, A. Process Monitoring in Heavy Duty Drilling Rigs—Data Acquisition System and Cycle Identification Algorithms. *Energies* **2020**, *13*, 6748. [CrossRef]
4. Borkowski, P.J. Comminution of Copper Ores with the Use of a High-Pressure Water Jet. *Energies* **2020**, *13*, 6274. [CrossRef]
5. Góralczyk, M.; Krot, P.; Zimroz, R.; Ogonowski, S. Increasing Energy Efficiency and Productivity of the Comminution Process in Tumbling Mills by Indirect Measurements of Internal Dynamics—An Overview. *Energies* **2020**, *13*, 6735. [CrossRef]
6. Zimroz, P.; Trybała, P.; Wróblewski, A.; Góralczyk, M.; Szrek, J.; Wójcik, A.; Zimroz, R. Application of UAV in Search and Rescue Actions in Underground Mine—A Specific Sound Detection in Noisy Acoustic Signal. *Energies* **2021**, *14*, 3725. [CrossRef]
7. Ziętek, B.; Banasiewicz, A.; Zimroz, R.; Szrek, J.; Gola, S. A Portable Environmental Data-Monitoring System for Air Hazard Evaluation in Deep Underground Mines. *Energies* **2020**, *13*, 6331. [CrossRef]
8. Krauze, K.; Mucha, K.; Wydro, T.; Pieczora, E. Functional and Operational Requirements to Be Fulfilled by Conical Picks Regarding Their Wear Rate and Investment Costs. *Energies* **2021**, *14*, 3696. [CrossRef]

9. Kotwica, K.; Stopka, G.; Kalita, M.; Bałaga, D.; Siegmund, M. Impact of Geometry of Toothed Segments of the Innovative KOMTRACK Longwall Shearer Haulage System on Load and Slip during the Travel of a Track Wheel. *Energies* **2021**, *14*, 2720. [CrossRef]
10. Kumar, R.; Singh, A.K.; Mishra, A.K.; Singh, R. Underground mining of thick coal seams. *Int. J. Min. Sci. Technol.* **2015**, *25*, 885–896. [CrossRef]
11. Jixiong, Z.; Spearing, A.J.S.; Xiexing, M.; Shuai, G.; Qiang, S. Green coal mining technique integrating mining-dressing-gas draining-backfilling-mining. *Int. J. Min. Sci. Technol.* **2017**, *27*, 17–27.
12. Bazaluk, O.; Slabyi, O.; Vekeryk, V.; Velychkovych, A.; Ropyak, L.; Lozynskyi, V. A Technology of Hydrocarbon Fluid Production Intensification by Productive Stratum Drainage Zone Reaming. *Energies* **2021**, *14*, 3514. [CrossRef]
13. Adach-Pawelus, K.; Pawelus, D. Influence of Driving Direction on the Stability of a Group of Headings Located in a Field of High Horizontal Stresses in the Polish Underground Copper Mines. *Energies* **2021**, *14*, 5955. [CrossRef]
14. Zabiciak, A.; Michalczyk, R. Wyznaczenie Stanów Naprężeń i Odkształceń w Mechanistycznych Projektowaniu Nawierzchni Poddatnych (Computation of Stress and Strain States for Mechanistic Design of Flexible Pavements). 2011. Available online: https://www.researchgate.net/publication/263414437 (accessed on 30 January 2022).
15. Klishin, V.I.; Klishin, S.V. Coal Extraction from Thick Flat and Steep Beds. *J. Min. Sci.* **2010**, *46*, 149–159. [CrossRef]
16. Dlouhá, D.; Dubovský, V. The improvement of the lake Most evaporation estimates. *Inż. Miner.* **2019**, *21*, 159–164.
17. Bazaluk, O.; Velychkovych, A.; Ropyak, L.; Pashechko, M.; Pryhorovska, T.; Lozynskyi, V. Influence of Heavy Weight Drill Pipe Material and Drill Bit Manufacturing Errors on Stress State of Steel Blades. *Energies* **2021**, *14*, 4198. [CrossRef]
18. Woźniak, D.; Hardygóra, M. Method for laboratory testing rubber penetration of steel cords in conveyor belts. *Min. Sci.* **2020**, *27*, 105–117. [CrossRef]
19. Bajda, M.; Błażej, R.; Hardygóra, M. Optimizing splice geometry in multiply conveyor belts with respect to stress in adhesive bonds. *Min. Sci.* **2018**, *25*, 195–206. [CrossRef]
20. Grzesiek, A.; Zimroz, R.; Śliwiński, P.; Gomolla, N.; Wyłomańska, A. A Method for Structure Breaking Point Detection in Engine Oil Pressure Data. *Energies* **2021**, *14*, 5496. [CrossRef]
21. Patyk, M.; Bodziony, P.; Krysa, Z. A Multiple Criteria Decision-Making Method to Weight the Sustainability Criteria of Equipment Selection for Surface Mining. *Energies* **2021**, *14*, 3066. [CrossRef]
22. Janus, J.; Krawczyk, J. Measurement and Simulation of Flow in a Section of a Mine Gallery. *Energies* **2021**, *14*, 4894. [CrossRef]
23. Huang, P.; Spearing, S.; Ju, F.; Jessu, K.V.; Wang, Z.; Ning, P. Control Effects of Five Common Solid Waste Backfilling Materials on In Situ Strata of Gob. *Energies* **2019**, *12*, 154. [CrossRef]
24. Prostański, D. Empirical Models of Zones Protecting Against Coal Dust Explosion. *Arch. Min. Sci.* **2017**, *62*, 611–619. [CrossRef]
25. Mo, S.; Tutuk, K.; Saydam, S. Management of floor heave at Bulga Underground Operations—A case study. *Int. J. Min. Sci. Technol.* **2019**, *29*, 73–78. [CrossRef]
26. Juganda, A.; Strebinger, C.; Brune, J.F.; Bogin, G.E. Discrete modeling of a longwall coal mine gob for CFD simulation. *Int. J. Min. Sci. Technol.* **2020**, *30*, 463–469. [CrossRef]
27. Dlouhá, D.; Dubovský, V.; Pospíšil, L. Optimal calibration of evaporation models against Penman-Monteith Equation. *Water* **2021**, *13*, 1484. [CrossRef]
28. Dubovský, V.; Dlouhá, D.; Pospíšil, L. The calibration of evaporation models against the Penman-Monteith equation on lake Most. *Sustainability* **2021**, *13*, 313. [CrossRef]
29. Hu, S.; Ma, L.; Guo, J.; Yang, P. Support-surrounding rock relationship and top-coal movement laws in large dip angle fully-mechanized caving face. *Int. J. Min. Sci. Technol.* **2018**, *28*, 533–539.
30. Szurgacz, D.; Zhironkin, S.; Čehlár, M.; Vöth, S.; Spearing, S.; Liqiang, M. A Step-by-Step Procedure for Tests and Assessment of the Automatic Operation of a Powered Roof Support. *Energies* **2021**, *14*, 697. [CrossRef]
31. Świątek, J.; Janoszek, T.; Cichy, T.; Stoiński, K. Computational Fluid Dynamics Simulations for Investigation of the Damage Causes in Safety Elements of Powered Roof Supports—A Case Study. *Energies* **2021**, *14*, 1027. [CrossRef]
32. Rajwa, S.; Janoszek, T.; Prusek, S. Influence of canopy ratio of powered roof support on longwall working stability—A case study. *Int. J. Min. Sci. Technol.* **2019**, *29*, 591–598. [CrossRef]
33. Xiaozhen, W.; Jialin, X.; Weibing, Z.; Yingchun, L. Roof pre-blasting to prevent support crushing and water inrush accidents. *Int. J. Min. Sci. Technol.* **2012**, *22*, 379–384.
34. Szurgacz, D. Dynamic Analysis for the Hydraulic Leg Power of a Powered Roof Support. *Energies* **2021**, *14*, 5715. [CrossRef]
35. Buyalich, G.; Buyalich, K.; Byakov, M. Factors Determining the Size of Sealing Clearance in Hydraulic Legs of Powered Supports. *E3S Web Conf. Second. Int. Innov. Min. Symp.* **2017**, *21*, 3018. [CrossRef]
36. Buyalich, G.; Byakov, M.; Buyalich, K. Factors Determining Operation of Lip Seal in the Sealed Gap of the Hydraulic Props of Powered Supports. *E3S Web Conf. IIIrd Int. Innov. Min. Symp.* **2018**, *41*, 1045. [CrossRef]
37. Buyalich, G.; Byakov, M.; Buyalich, K.; Shtenin, E. Development of Powered Support Hydraulic Legs with Improved Performance. *E3S Web Conf. VIrd Int. Innov. Min. Symp.* **2019**, *105*, 3025. [CrossRef]
38. Wang, J.; Wang, Z. Systematic principles of surrounding rock control in longwall mining within thick coal seams. *Int. J. Min. Sci. Technol.* **2019**, *29*, 591–598. [CrossRef]
39. Ji, Y.; Ren, T.; Wynne, P.; Wan, Z.; Zhaoyang, M.; Wang, Z. A comparative study of dust control practices in Chinese and Australian longwall coal mines. *Int. J. Min. Sci. Technol.* **2016**, *25*, 687–706. [CrossRef]

40. Peng, S.S.; Feng, D.; Cheng, J.; Yang, L. Automation in U.S. longwall coal mining: A state-of-the-art review. *Int. J. Min. Sci. Technol.* **2019**, *29*, 151–159. [CrossRef]
41. Ralston, J.C.; Reid, D.C.; Dunn, M.T.; Hainsworth, D.W. Longwall automation: Delivering enabling technology to achieve safer and more productive underground mining. *Int. J. Min. Sci. Technol.* **2015**, *25*, 865–876. [CrossRef]
42. Ralston, J.C.; Hargrave, C.O.; Dunn, M.T. Longwall automation: Trends, challenges and opportunities. *Int. J. Min. Sci. Technol.* **2017**, *27*, 733–739. [CrossRef]
43. Stoiński, K.; Mika, M. Dynamics of Hydraulic Leg of Powered Longwall Support. *J. Min. Sci.* **2003**, *39*, 72–77. [CrossRef]
44. Frith, R.C. A holistic examination of the load rating design of longwall shields after more than half a century of mechanised longwall mining. *Int. J. Min. Sci. Technol.* **2015**, *26*, 199–208. [CrossRef]
45. *CSN-EN 1804-1*; Machines for Underground Mining. Safety Requirements for Powered Roof Support—Part 1: Support Units and General Requirements. Czech Office for Standards, Metrology and Testing: Praha, Czech Republic, 2020.
46. Doroszuk, B.; Król, R. Conveyor belt wear caused by material acceleration in transfer stations. *Min. Sci.* **2019**, *26*, 189–201. [CrossRef]
47. Stoiński, K. *Mining Roof Support in Hazardous Conditions of Mining Tremors*; Collective Work for Editing Kazimierza Stoińskiego; The Central Mining Institute: Katowice, Poland, 2018.
48. Zhao, X.; Li, F.; Li, Y.; Fan, Y. Fatigue Behavior of a Box-Type Welded Structure of Hydraulic Support Used in Coal Mine. *Materials* **2015**, *8*, 6609–6622. [CrossRef]
49. Zhou, R.; Meng, L.; Yuan, X.; Qiao, Z. Research and Experimental Analysis of Hydraulic Cylinder Position Control Mechanism Based on Pressure Detection. *Machines* **2022**, *10*, 1. [CrossRef]
50. Juárez-Ferreras, R.; González-Nicieza, C.; Menéndez-Díaz, A.; Álvarez-Vigil, A.E.; Álvarez-Fernández, M.I. Measurement and analysis of the roof pressure on hydraulic props in longwall. *Int. J. Coal Geol.* **2008**, *75*, 49–62. [CrossRef]
51. Toraño, J.; Diego, I.; Menéndez, M.; Gent, M. A finite element method (FEM)—Uzzy logic (Soft Computing)—Virtual reality model approach in a coalface longwall mining simulation. *Autom. Constr.* **2008**, *17*, 413–424. [CrossRef]
52. Ren, H.; Zhang, D.; Gong, S.; Zhou, K.; Xi, C.; He, M.; Li, T. Dynamic impact experiment and response characteristics analysis for 1:2 reduced-scale model of hydraulic support. *Int. J. Min. Sci. Technol.* **2021**, *31*, 347–356. [CrossRef]
53. Qiao, S.; Zhang, Z.; Zhu, Z.; Zhang, K. Influence of cutting angle on mechanical properties of rock cutting by conical pick based on finite element analysis. *J. Min. Sci.* **2021**, *28*, 161–173.
54. ANSYS. Approval for Use in Accordance with the License. Released to the User of the Company. 2016. Available online: https://www.ansys.com/2021.01.28 (accessed on 30 January 2022).
55. Król, R.; Kisielewski, W. Research of loading carrying idlers used in belt conveyor-practical applications. *Diagnostyka* **2014**, *15*, 67–74.
56. Szurgacz, D.; Zhironkin, S.; Vöth, S.; Pokorný, J.; Spearing, A.J.S.; Cehlár, M.; Stempniak, M.; Sobik, L. Thermal Imaging Study to Determine the Operational Condition of a Conveyor Belt Drive System Structure. *Energies* **2021**, *14*, 3258. [CrossRef]
57. Bajda, M.; Hardygóra, M. Analysis of Reasons for Reduced Strength of Multiply Conveyor Belt Splices. *Energies* **2021**, *14*, 1512. [CrossRef]
58. Gładysiewicz, L.; Król, R.; Kisielewski, W.; Kaszuba, D. Experimental determination of belt conveyors artificial friction coefficient. *Acta Montan. Slov.* **2017**, *22*, 206–214.
59. Kawalec, W.; Suchorab, N.; Konieczna-Fuławka, M.; Król, R. Specific energy consumption of a belt conveyor system in a continuous surface mine. *Energies* **2020**, *13*, 5214. [CrossRef]
60. Pokorny, J.; Dlouhá, D.; Kucera, P. Study of the necessity of use virtual origin in assessment of selected fire plume characteristics. *MM Sci. J.* **2016**, *5*, 1424–1428. [CrossRef]
61. Brzoska, Z. *Strength of Materials*; PWN: Warsaw, Poland, 1972; pp. 33–49.
62. Leyko, J. *General Mechanics*; PWN: Warsaw, Poland, 2021; pp. 53–59.
63. Uth, F.; Polnik, B.; Kurpiel, W.; Baltes, R.; Kriegsch, P.; Clause, E. An innovate person detection system based on thermal imaging cameras dedicate for underground belt conveyors. *Min. Sci.* **2019**, *26*, 263–276. [CrossRef]
64. Dlouhá, D.; Pokorný, J.; Dlouhá, K. Necessity of knowledge about math in safety engineering. In Proceedings of the 14th Conference E-Learning: Unlocking the Gate to Education around the Globe, Prague, Czech Republic, 20–21 June 2019; pp. 380–386.
65. Dlouhá, D.; Kozlová, K. Knowledge assessment of student's high school mathematics. In Proceedings of the 17th Conference on Applied Mathematics (APLIMAT 2019), Bratislava, Slovak Republic, 5–7 February 2019; Volume 1, pp. 243–252.
66. Hamříková, R.; Dlouhá, D. Video tutorials for students of the master's program. In Proceedings of the 12th Conference Open Education as a Way to a Knowledge Society, Prague, Czech Republic, 26–27 June 2017; pp. 446–451.
67. Ji, Y.; Zhang, Y.; Huang, Z.; Shao, Z.; Gao, Y. Theoretical analysis of support stability in large dip angle coal seam mined with fully-mechanized top coal caving. *Min. Sci.* **2020**, *27*, 73–87. [CrossRef]
68. Kawalec, W.; Błażej, R.; Konieczna, M.; Król, R. Laboratory Tests on e-pellets effectiveness for ore tracking. *Min. Sci.* **2018**, *25*, 7–18. [CrossRef]
69. Baiul, K.; Khudyakov, A.; Vashchenko, S.; Krot, P.V.; Solodka, N. The experimental study of compaction parameters and elastic after-effect of fine fraction raw materials. *Min. Sci.* **2020**, *27*, 7–18. [CrossRef]

70. Dlouhá, D.; Hamříková, R. Interactive distance materials of mathematics for VŠB-TU Ostrava. In Proceedings of the 13th Conference Overcoming the Challenges and the Barriers in Open Education, Prague, Czech Republic, 25–26 June 2018; pp. 67–72. Available online: https://www.fast.vsb.cz/230/cs/Veda-a-vyzkum/Publikace/2018/ (accessed on 30 January 2022).
71. Andrzejewski, S. *Mechanical Engineer's Guidebook*; Scientific and Technical Publishing House: Warsaw, Poland, 1968; pp. 73–89.
72. Dlouhá, D.; Hamříková, R. Our experience with the involvement of students in the creation of study materials. In Proceedings of the 17th Conference on Applied Mathematics (APLIMAT 2019), Bratislava, Slovak Republic, 5–7 February 2019; Volume 1, pp. 301–308.
73. Pokorny, J.; Mozer, V.; Malerova, L.; Dlouhá, D.; Wilkinson, P. A simplified method for establishing safe available evacuation time based on a descending smoke layer. *Commun. Sci. Lett. Univ. Zilina* **2018**, *20*, 28–34. [CrossRef]
74. Misiak, J. *Technical Mechanics*; Scientific and Technical Publishing House: Warsaw, Poland, 1999; pp. 22–41.
75. Siemieniec, A.; Wolny, S. *Strength of Materials*; Academy of Mining and Metallurgy Publishing House: Krakow, Poland, 1996; pp. 87–96.
76. *CSN-EN 1804-2*; Machines for Underground Mining—Safety Requirements for Hydraulic Powered Roof Support—Part 2: Power Set Legs and Rams. Czech Office for Standards, Metrology and Testing: Praha, Czech Republic, 2020.

MDPI

Article

Strengthening the Mitigation of Climate Change Impacts in Slovakia through the Disaggregation of Cultural Landscapes

Jana Rybárová, Ľubomíra Gabániová *, Lucia Bednárová , Radim Rybár and Martin Beer

Institute of Earth Resources, Faculty of Mining, Ecology, Processing Control and Geotechnology, Technical University of Kosice, 04001 Kosice, Slovakia; jana.rybarova@tuke.sk (J.R.); lucia.bednarova@tuke.sk (L.B.); radim.rybar@tuke.sk (R.R.); martin.beer@tuke.sk (M.B.)

* Correspondence: lubomira.gabaniova@tuke.sk

Abstract: This article presents the results of research on the possibilities of fragmentation of cultural, and especially agricultural, landscapes in a selected locality in eastern Slovakia, which is currently characterized by a high proportion of large-scale soil units used for growing cereals and crops subsequently used as energy sources (maize and oilseed rape, among others). Slovakia, as the country with the largest average field area in the European Union (EU), is facing a process of fragmentation of these units to counter climate change and increase the resilience of the landscape to erosion, soil fertility, and biodiversity loss. This paper presents a fragmentation method based on the restoration of former dividing lines, mainly formed by dirt roads, based on the historical mapping. The results show that in this way it is possible to achieve denser landscape fragmentation, to create dividing green belts, to increase the resilience of the environment to water and wind erosion, and to create an environment for pollinator resources and a background for plants and animals in the landscape, while respecting the ergometric routing of the dividing lines and the ownership relations of the land. Last but not least, benefits have been quantified in the form of carbon capture, as well as in the construction of a network of recreational or hiking trails.

Keywords: field fragmentation; land use; biodiversity; climate change; renewable energy sources (RES)

Citation: Rybárová, J.; Gabániová, Ľ.; Bednárová, L.; Rybár, R.; Beer, M. Strengthening the Mitigation of Climate Change Impacts in Slovakia through the Disaggregation of Cultural Landscapes. *Processes* **2022**, *10*, 658. https://doi.org/10.3390/pr10040658

Academic Editor: Roberto Alonso González Lezcano

Received: 26 February 2022
Accepted: 25 March 2022
Published: 28 March 2022

1. Introduction

Anthropogenic climate changes have recently become a phenomenon that influences current events and changes in society, human behavior, economics, politics, local and global perceptions of the environment, and the position of humans and human society within it [1], as very comprehensively elaborated in [2] and in earlier works such as [3,4]. The possible impacts of climate change with a focus on the territory of Slovakia have been elaborated in the study [5]. The area with the manifestation of these impacts is also the rural landscape, which in the conditions of the EU, as well as Slovakia, is mostly in the form of a cultural landscape with agricultural or forestry use. The importance of solving these problems in the conditions of Slovakia is reinforced by the fact that Slovakia is the country with the largest average area of monoculture fields in the whole EU [6]. This is a parameter that worsens the resistance of the landscape formed in this way, on the one hand, against water and wind erosion, and on the other hand, against the depletion of the soil of organic matter and living organisms. According to [7], this also leads to the overall loss of biodiversity of the landscape in general. As stated by [6], large fields not covered by strips of vegetation, among other things, exacerbate the effects of drought and contribute to overheating of the landscape. The need to reduce large-scale agricultural areas in Slovakia (similarly in the Czech Republic) has recently been addressed by state bodies and agencies such as the Ministry of Agriculture, the Ministry of Environment via the Institute of Environmental Policy, e.g., in [6]; and the Slovak Environmental Agency, e.g., in environmental organizations and associations, research, and academic bodies.

For illustration, according to [6], the average field size in the Slovak Republic is 12 ha; the EU average is 3.9 ha (in neighboring countries it is: Czech Republic (CR): 9.5 ha; Hungary(HU): 8.5 ha; Poland (PL): 3.7 ha; and Austria (AT): 2.8 ha). Last but not least, as an EU member, the Slovak Republic must meet certain environmental criteria to which the EU subscribes by adopting the Fit For 55 agenda (a package of directives). As stated in [8], the Fit for 55 agenda talks about increasing the targets by 2030, with the main objectives being to increase the share of RES to 40% (previously the target was 32%), and to reduce CO_2 emissions (eq) compared to 1990 by 55% (with a view to carbon neutrality by 2050). Each member country must respond to these targets by adopting a new or updated Integrated National Energy and Climate Plan 2021–2030 [9]. The plan that is currently in force foresees a target of 19.2% RES (calculated for the original EU target of 32%), while it already foresees measures such as new RES capacities (at a level of about 1200 MW). Additionally, measures related to the LULUCF (Land Use, Land Use Change and Forestry) sector are foreseeable, which also includes the topic of agricultural land (building block of the cultural rural landscape). This plan is currently being reworked under Fit For 55, and the final values of the individual quotas have not yet been determined. According to MoE sources [9], 24% is being considered for RES, while the European Commission (EC) would like to introduce a quota of at least 30% for each country. The change regarding the reduction in CO_2 emissions (eq.) is not yet known. The current figure is 46% compared to 1990 [10,11], but here it should be said that the most work has been contributed in the run up to 2009 when the easiest available measures were implemented, but any further percentage will be a more difficult task. The new plan for Slovakia should be submitted to the EC for approval by June 2023.

The impacts of climate change on rural landscapes, mostly in terms of their agricultural form and function, and consequently the impact on biomass production, have already been addressed in earlier works from the EU perspective, e.g., in [12], in Slovakia in [13–15]. According to [16], the cultural landscape can be understood as a hybrid open natural–anthropogenic system, which is the result of the action of humans and human society in space and time. As stated by [17], land use = land cover + land utilization, which means that land use results from the knowledge of the landscape cover, represented by natural as well as man-made and man-modified objects.

The consequences of such management and the landscape configured in this way are the increasing need for pesticides and artificial fertilizers [18], decreasing soil permeability, decreasing humus content in the soil, decreasing number of soil organisms, deteriorating conditions for insects, pollinators, and higher animals [19] and the associated reduction in biodiversity in the landscape [20]; water and wind erosion [21]; and loss of water resources [22]. According to [23], 47.7% of agricultural land in Slovakia is potentially threatened by water erosion. As reported by [24], in Slovakia and the Czech Republic, 10 to 100 m^3 of sediment are added to reservoirs annually from 1 km^2 of the catchment area. The result is a landscape with an overall simpler/poorer structure and a lower level of diversity, to which other factors also contribute, such as the disappearance of features that increase the diversity of the landscape such as, according to [25], dirt roads, tree avenues, solitary trees, copses, and thickets, or historical objects such as technical, architectural, and cultural works/monuments, etc. Such landscapes ultimately lose the protective mechanisms that would enable them to resist the pressure caused by climate change and may ultimately be further degraded to the point of losing their original ecosystem and economic production functions, which, according to [26,27], will ultimately be reflected in economic indicators as well.

Within the Slovak Republic, the issue under study is regulated by the Slovak Environmental Agency (SEA) Green Infrastructure agenda within the Territorial Ecological Stability System (TES) package, according to [28], to address the overall provision of ecological stability of the landscape in Slovakia, the interconnection of natural areas, and the protection of habitats and representative species in their natural environment. The issue and status of the SUE are regulated by several laws, as mentioned in [28].

Currently, the issue is to be addressed in the Slovak Republic in the framework of the preparation of the Strategic Plan of the Common Agricultural Policy 2021–2027, based on which the Slovak Republic will set its priorities according to its conditions. According to [29], the European Commission in its legislative proposal for the future Common Agricultural Policy has listed nine common strategic objectives, three economic, three environmental, and three social objectives, namely:

- Contribute to climate change mitigation and adaptation, as well as to the promotion of sustainable energy;
- Promote sustainable development and efficient management of natural resources such as water, soil, and air;
- Contribute to the protection of biodiversity, improve ecosystem services, and conserve habitats and landscapes.

The object of the research—the proposal of the method of fragmentation of the cultural landscape formed mainly by large-scale fields of monoculture crops—is directly related to these environmental objectives, due to the fact that it returns to the landscape natural elements diversifying the landscape and creates conditions supporting the development of positive phenomena such as water retention in the landscape or the creation of conditions for the increase in biodiversity in the territory.

In the context of the above, there is a growing need to find mechanisms to meet the commitments and targets in question, one of the important tasks being the objective to divide large-scale fields into smaller ones, whereby each field should not exceed 30 ha [6], which is so far the consensually established limit for the area of a field with one monoculture [6], while fields with an area greater than 30 ha represent more than half of the total area of agricultural land in Slovakia.

The paper presents a proposal for a method of dividing fields, which is based on the fact that, in the past, fields were fragmented, separated by landscape features—roads drawn depicted according to [30] in the old military and orthographic maps. For the solution, a site in eastern Slovakia was selected, which represents a sample of a normal rural landscape with undulating relief, without any major peculiarities. Map documents were used which record the appearance of the area in the 18th century (first military mapping) and the middle of the 20th century (orthophoto maps). Linear features (roads and property boundaries) were designed and identified and transferred to the map base reflecting the current form of the area. The fragmentation of the fields took into account aspects such as soil protection from water and wind erosion, the implementation of trees and green belts in the area, and the overall increase in carbon sequestration potential. Aspects of potential tourist and recreational use of the fragmented landscape were also taken into account. Fragmentation was proposed at several levels. Parameters such as the overall reduction in the average area of fields, the length of new road-dividing elements, the area of green belts, the length of newly created tree avenues, the number of trees, the amount of biomass produced, etc., were quantified. The aim was to find an effective way of fragmenting the cultural landscape while achieving the desired reduction in the area of individual fields, implementation of landscape elements (trees, thickets/shrubs, avenues, and green belts), increase in the amount of biomass in the landscape, increase in the amount of grazing for bees, and implementation of carbon-capturing elements, while respecting the cultural and historical aspects and the possibility of using the landscape for recreation and tourism.

2. Territory of Interest

The area of interest was selected; the criterion for selection was the coverage of the landscape representing the cultural landscape of the central part of eastern Slovakia, where the East Slovak lowland meets as a horizontally low-lying landscape with predominantly agricultural land with the Vihorlat Mountains—a vertically indented relief with forestation. The landscape in the selected area has an undulating morphology and contains human settlements in the form of five villages. The morphology of the area is also a prerequisite for the existence of phenomena related to possible wind and water erosion. There are large-scale agricultural fields in the area of interest and the area has a history of pre-collectivization agricultural use and subsequent processing activities (milling, etc.) and the presence of old mining and metallurgical activities/production (old tunnels and smelters).

The area of interest is located in the north-eastern part of the Košice Region (The Nomenclature of Territorial Units for Statistics—NUTS 3), on the northern edge of the East Slovak Lowland, in the area bounded by the villages of Oreské, Staré, Zbudza, Trnava pri Laborci, and Vínne, and the Vihorlat Mountains (Figure 1). The area is made up of a cultural landscape with predominantly agricultural use, in which human settlements are concentrated in villages and, according to the One Soil portal [31], is one of the regions with an average area of monoculture fields of 10–15 ha, and this territory includes 32 fields of more than 30 ha, with a total area of 1865 ha. The area is partly covered by continuous forest cover and contains elements that can be the subject of leisure use or have a tourist and recreational function (medieval castles, cottages, or lakes). According to [32], the cadastral area of the municipalities is 8°394.81 ha.

Figure 1. Map of the Košice Region (48.346260–49.012527 north latitude, 20.190700–22.380257 east longitude) showing the location of the area of interest (marked in red) (pink colour represents towns, blue—water areas).

In terms of the historical development of the landscape, the current form of the agriculturally used rural landscape in most of Slovakia is a remnant of the consolidation of individual peasants' land within the process of collectivization and association, as shown, for example, by [33]. Figure 2 shows an example of a selected section (8.58 ha) of the current field (37.78 ha) in the area of interest, which consisted of 15 plots in the pre-aggregation (1950 state), the largest of which measured 232 m × 54 m (1.02 ha).

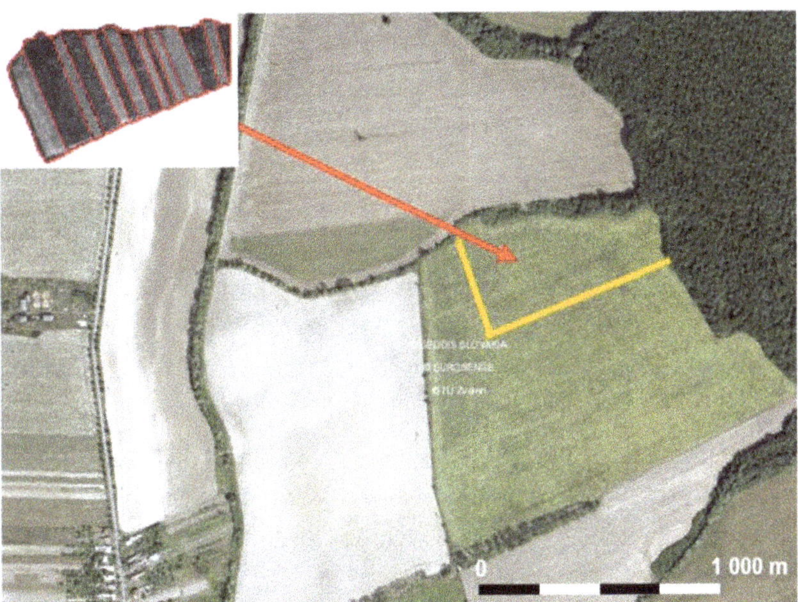

Figure 2. Example of a selected section of the current field in the area of interest, which consisted of 15 plots in the pre-aggregation (1950 state). (Own elaboration according to [34]).

According to [35], the first steps towards the creation of joint management within associations/societies in Slovakia occurred in the middle of the 19th century. After the onset of the totalitarian regime in 1948, a massive acclimatization took place, with the process itself being based on Act No. 69/1949 Coll. on UPC (Unified Peasant Cooperatives) [36]. As stated in [37], the establishment of the UPC led to the liquidation of private ownership. After the collapse of the totalitarian regime in 1989, the agricultural cooperatives were transformed [38], but the large-scale nature of the fields remained intrinsically unchanged.

It is clear from previous research that the current form of the cultural, human-used rural landscape in Slovakia is related to the predominance of intensive farming, predominantly on an industrial scale, on large-scale fields, where monocultures of cereals, energy, and technical crops are grown. The orthophotos in Figure 3 show examples of border areas, on the left SR (SK) and Austria (AT), area near the border crossing Bratislava-Jarovce/Kittsee (48.070883, 17.072833), on the right SR (East Slovakian Lowland) and Hungary (HU), near the village Malý Horeš (48.395207, 21.950210), where the difference in the structure or mosaic of the cultural landscape is obvious. While the picture on the left in Slovakia is dominated by a triangular field 1 km long with an area of 106 ha, in Austria it is a mosaic consisting of narrow fields up to 50 metres wide, which at the same length have an area of up to 5 ha. Similarly, the picture on the right shows fields on the Slovak side with an area of about 15 ha to 103 ha, just over the border in Hungary there are fields of about 1–10 ha, many bordered by small draws.

From the climatic point of view, eastern Slovakia is situated at the transition between the maritime and inland climate of Europe, with the continental influence prevailing. The intensity of solar radiation of the locality falling within the geological formation of the East Slovak Lowland exceeds the national average, as does the number of clear days per year or the highest annual temperatures. The vegetation in the monitored area is more thermophilic, the forest cover is oak and period-hornbeam, gradually changing into hornbeam-hornbeam in the foothills of the Vihorlat Mountains. Cultivated agricultural crops include wheat, oilseed rape, and maize.

(a) (b)

Figure 3. Examples of border areas. On the left (**a**) Slovakia (SK) and Austria (AT), on the right (**b**) Slovakia (East Slovakian Lowland) and Hungary (HU). The border between the countries is highlighted in red.

3. Materials and Methods

Fragmentation proceeded in three stages/levels of division. Map bases were used:

1. Historical map 1764—First military mapping [38].
2. A 1950 orthophoto map from a mapping application [34].
3. A 2017–2019 orthophoto mosaic [34].
4. Tourist maps of the region from mapy.cz portal [31].
5. Google Earth map platform.
6. Map application One Soil [39].
7. GSAA (Geospatial support request) [40].
8. ZB-GIS—a spatial, object-oriented database that is the reference basis of the national spatial information infrastructure [41].

In order to clarify the issues presented and the methods used in the context of broader interdisciplinary contexts, it is necessary to cite some related relevant sources that cover each aspect in more detail. The creation of maps based on modelling the spatial distribution of linear landscape features across Europe was elaborated in [42]. Comprehensively worked out problems of image material analysis and overall photosciences are presented in [43–46]. From the point of view of biological corridors, this problem has been addressed in [47]. The effects of the presence of selected linear features in the landscape on the presence of insects and their possibilities of the free movement were addressed in [48]. Aspects of tourism in relation to landscape features were addressed in [49] and in the context of insolation and heat load in [50].

The issue of using military maps for the analysis of landscape features in the neighbouring Czech Republic with conditions analogous to the Slovak Republic has been described in [51]. Possibilities of identification of these elements on the territory of the Slovak Republic were described in [52], while in [53] attention was focused on agricultural cultural landscapes. From the point of view of cultural landscape diversity and its varied composition, the issue was presented in [54,55]. The issue of water erosion and anti-erosion tillage in Slovakia was treated in [56]. Areas of the condition of some selected localities of eastern Slovakia (Medzibodrožie) were dealt with in [57].

As reported in [58], information on the size of agricultural parcels in Slovakia is available from the LPIS (Land-parcel identification system), GSAA, and IACS (Integrated administration and control system) information systems, which keep records at the level of land parcels broken down by crops (LPIS) and also at the level of parcels (user hectares) with a specific crop grown, which have been declared by applicants for payments (GSAA, IACS).

As reported by [59], according to GSAA and IACS, when considering the number of parcels below and above 30 ha in 2019, single-cropped parcels below 30 ha dominated (91% of the total number of parcels registered as arable land). However, the total area of single-crop parcels under 30 ha accounted for only 47% of the total area of declared arable land. Thus, up to 53% (706,600 ha) of the area of declared arable land in Slovakia is cultivated in units larger than 30 ha. Approximately 516,700 ha of arable land is cultivated in units of 50 ha or more.

The situation in the Czech Republic is mapped in sources [60,61]. According to the data published in [62], from 2020 the condition of limiting monoculture cultivation to a maximum of 30 ha of continuous area (set by the Good Agricultural and Environmental Condition standard GAEC ((Good Agricultural and Environmental Condition), 7d) has been introduced. As reported by [6,63], the GAEC 7d standard requires that no more than 30 ha of arable land should be under continuous single cropping on an arable land block. A continuous area of a single crop within a section of a land block is defined as areas sown or planted with that crop that are not visibly separated from each other by a buffer zone sown with forage or crops approved for the buffer zone under Government Regulation No. As stated in [6], in the German province of Rhineland-Palatinate, the local government also financially supports 'non-productive' green elements on arable land. Payments are linked to biodiversity indicators such as species diversity on the land. Fragmentation of agricultural land is also recommended by the FAO (Food and Agriculture Organization of the United Nations).

The area of interest was documented based on current satellite maps, topographic maps, and physical field reconnaissance. Parameters such as the size of individual monoculture fields (here based on GSAA and One Soil data), number of fields, crops grown, slope gradient (based on topographic maps), exposure to prevailing airflow (based on wind directional rosette data valid for the site and the presence of orographic effects), the presence of roads, watercourses, linear energy structures, and the presence of landscape features (trees, avenues of trees, trees, draws, uncultivated green belts, and permanent grassland).

The methods of retrospective analysis according to [64] and the method of visual interpretation according to [65] were used in the study of the historical map documents.

When delineating individual fields in the GSAA application, each delineated unit is associated with LPIS information, consisting of a field code, field area in hectares, site name, and classification by use. These data were used in the recording and processing of individual fields as well as the determination of their area.

The first step was based on map base 1 (Historical map from 1764), the origin and background of which are described in [66]. Roads were identified—roads of different differentiated levels (carriage roads (historical name), roads in the floodplain, roads in the embankment, bridges (wooden/stone), and other landscape features/objects).

The polygon areas of the subdivided portions of the original fields, as well as the lengths of the subdivided sections, were surveyed using ZB-GIS [67].

All identified roads were entered into the resulting map built on the map base from the google earth environment (satellite imagery). The part of the roads that led outside the currently existing roads (through the fields) represents the proposed dividing lines of the fields.

In the second step, the map base was used as a basis for map 2 (Orthophotomap from 1950). The roads—roads of different levels—were identified. Some of the identified roads were identical to the roads identified in the first step (intersection of the set of roads of Step I and Step II)), while some represented the set of new/additional dividing lines. The identified paths/dividing lines were entered into the map from Step I.

The maps that formed the basis for Steps I and II come from different historical periods. Some roads are found in both maps at the same time, others only in one or the other, or some roads may have partially changed their position, so two separate steps were followed.

The third step was based on the same map base 2, identifying the field boundaries from the pre-collectivization period, which also represented the access roads to each group of land/fields (dedicated or access roads). At this stage, only fields with an area of more than 30 ha were taken into account in the selection of the dividing lines (in some cases, re-division was necessary). In the selection of specific field boundaries, aspects related to soil erosion protection (especially slope gradient) were taken into account according to the mapping documents presenting the areas affected and threatened by water erosion [68,69]. Meanwhile, the area of interest generally falls within the area of medium risk of torrential rainfall and according to the exposure to wind flow. The average wind speed at the site at 10 m above the ground surface, according to [65] is 3.8 ms^{-1}. The area of interest is located in the area of medium risk of torrential rainfall with a predominant southeast and north flow direction. According to [70], the area of interest falls within an area of medium wind erosion hazard. When selecting possible dividing lines, those were chosen which led to the shortening of declining and open sections of fields in the sense of eliminating the LS factor (the influence of relief) according to the methodology for determining the real threat to the soil from water erosion processes, taking into account the current vegetation cover and the method of farming presented in [71]. The selected identified roads were entered into the next map layer from step II.

The steps described result in three layers of maps capturing the cumulative fragmentation within each step.

Subsequently, parameters were evaluated for each level:

- the size of each field after fragmentation, frequency of fields, average field size, number of fragmented/number of newly created fields, and length of fragmentation lines (proposed).
- the degree of correspondence of roads from different historical periods, respectively.

For the proposed dividing lines, the alternatives of tree avenues (different tree species) and green belts were proposed. The design was based on the methodology of technical conditions of vegetation treatments for roads, according to [72].

According to the different design proposals, the outputs were determined:

- Biomass production (hay);
- The amount of fixed carbon;
- Shortening of declining water-erosion-prone sections of fields;
- Shortening of open, exposed, wind-eroded sections of fields;
- The number of bee colonies that can be fed;
- The length of potentially usable routes for soft tourism, as defined in [68,69].

4. Results and Discussion

Based on the analysis of the map documents [70] and the inspection, the detailed delimitation of the area of interest was proceeded by its delimitation by natural boundaries (the Laborec River, the Šírava Canal, and the bank of the Zemplínska Šírava water reservoir) and boundary lines leading along with dirt or forest roads so that the boundaries of the area of interest do not intersect the whole fields (Figure 4). The area of interest thus defined has a total area of 4011 hectares, of which the area of agricultural land is 2769.41 hectares, with a total of 137 fields. In the figure, the boundaries of individual fields (single crop plots cultivated as a single unit) are plotted in red.

There are five villages in the area of interest, in whose intravilane there are 40 fields with an area ranging from 0.17 to 5.94 ha. These fields have not been subject to fragmentation and have not been taken into account in the subsequent solution. The fields in question do not constitute agricultural land used for industrial purposes.

If we do not take into account the fields located in the intramural area (allotments), the size of the area under analysis is 2731.71 hectares. There are 97 fields in this area, of which 32 have an area greater than 30 ha. The individual parameters relating to the part of the area of interest subject to the analysis are given in Table 1.

Figure 4. Boundaries of the area of interest (yellow line) with boundary lines formed by natural boundaries and leading along with dirt and forest roads, marking the intravital of the villages and highlighting the boundaries of individual fields (48.789516–48.868967 north latitude, 21.856570–21.986668 east latitude).

Table 1. Parameters of the fields in the area of interest were analysed.

Indicator	Value	Unit
Number of fields	97	-
Area of the area of interest	4011	ha
The perimeter of the territory	39,887	km
Largest field	122.37	ha
Smallest field	0.52	ha
Average field	28.16	ha
Number of fields over 30 ha	34	-

The recording of individual fields was based on the LPIS designation. In this way, it is possible to identify each field individually on the map as well as assign additional attributes as required.

In a first step, 15 roads with a total length of 35,354 km were identified in the map documents from the first military mapping of 1764–1787.

Figure 5 shows the routes/dividing lines of all three levels of subdivision. The markings are as follows: first level—red, second level—blue, and third level—yellow; the sections marked XY=AB are the intersection of the first and second levels of division, i.e., these road sections/dividing lines can be found in both map documents. In the first level of division, 17 fields have been divided.

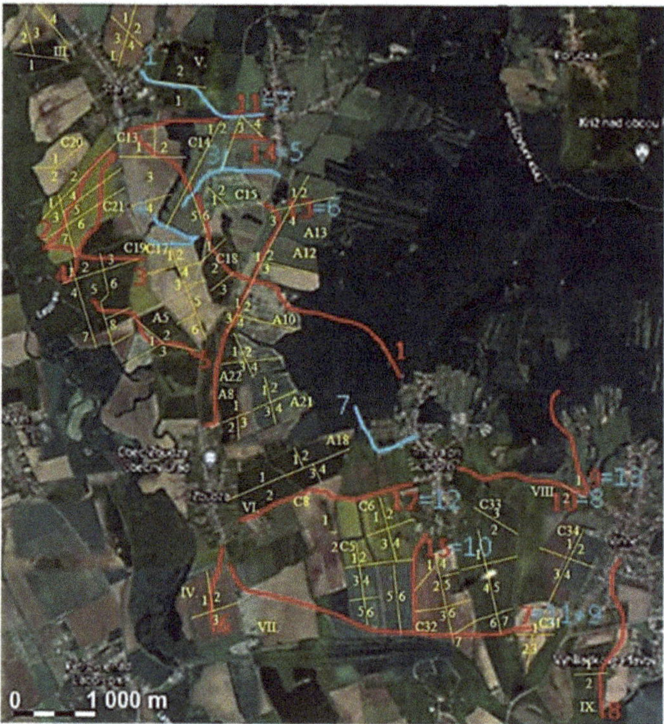

Figure 5. Routes/dividing lines of all three levels of division (first level—red, second level—blue, and third level—yellow; sections marked XY = AB are the intersection of the first and second levels of division).

In the second step, 13 roads with a total length of 5904 km were identified in the 1950 orthophoto map.

In a third step, 28 field access roads/field block boundaries with a total length of 15,079 km were identified in the 1950 orthophoto maps.

Figure 6 shows a graphical representation of the average field size at the beginning and after each fragmentation step.

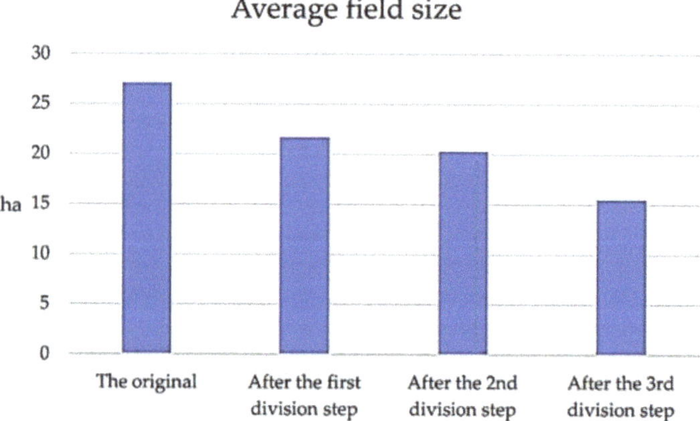

Figure 6. Average field size at the beginning and after each fragmentation step.

In the first step, 17 fields with a total area of 1083.56 ha were divided, in the second step were 2 fields with a total area of 77.39 ha, and the third step had 13 fields with a total area of 631.34 ha. The number of newly created fields was 44, 4, and 49, respectively. The number of fields above 30 ha was 15 after the first step, 13 after the second step, and 0 after the third step (Figure 7).

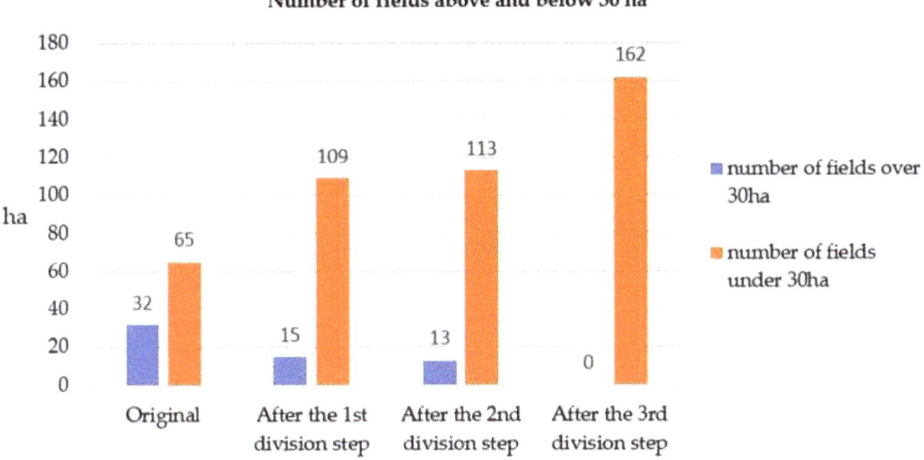

Figure 7. Representation of fields below and above 30 ha in the individual stages of the solution.

The share of the divided fields for the individual division levels and the share of the resulting fields for the individual division levels expressed as a percentage are shown in the graphs in Figure 8.

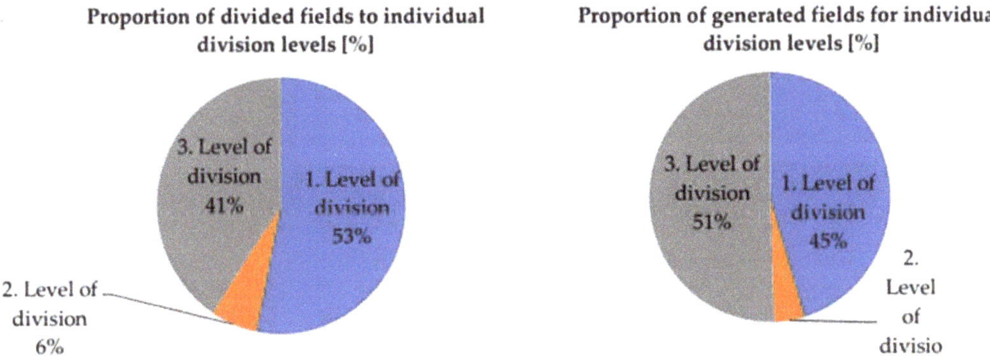

Figure 8. The proportion of divided fields to individual division levels and the share of generated fields to individual division levels are expressed as a percentage.

As some sections of the roads identified within each subdivision step were identical to the field boundaries, they did not represent new subdivision lines in those locations; therefore, the total lengths of the subdivision lines are less than the lengths of the individual roads carried over from the original mapping documents.

The total length of the dividing lines was 16,109 km after the first subdivision step, 1868 km after the second subdivision step, and 15,079 km after the third subdivision step. The number of dividing lines per division step is shown graphically in Figure 9.

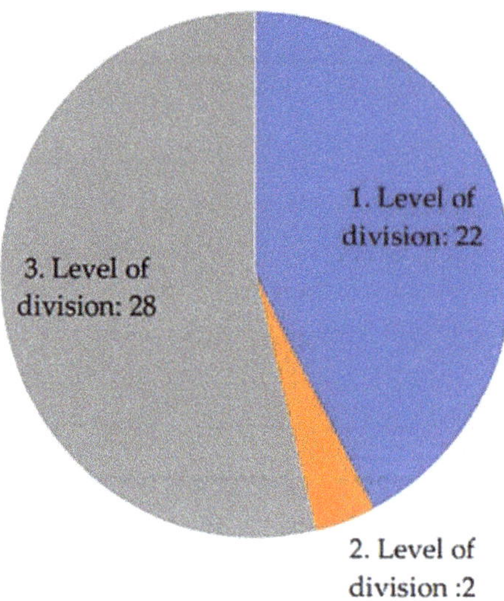

Figure 9. The number of dividing lines per division step.

The proportion of roads that match in the maps from both periods, i.e., from the first and second steps, represents 45.9% of the total length of the identified roads. In the third step, only the division lines based on field boundaries and off-road access roads from steps one and two were considered.

Based on the results in the form of the length of the dividing lines, it was considered to use them in the form of green belts with a width of 22 m according to [71]. The areas of green belts that can be sown with meadow vegetation according to [72] in the individual steps are as follows: for step I, 23.3 ha; for step II, 3.2 ha; and for step III, 33 ha. If meadow vegetation is considered, the biomass obtained can be used energetically as an input to a biogas plant or non-energetically as livestock feed. In terms of botanical composition, regional grass–herb mixtures are suitable. As reported by [68], their high species richness and higher representation of herbs at the expense of grasses are advantageous for overall biodiversity enhancement. The biological value of sown grassland supports the diversity of animals that are tied to the presence of herbs, especially insects. According to [68], it is possible to obtain 3.5 to 6 tons of hay per hectare, depending on the crop (with or without fertilization). For a total area of 59.67 ha, it is then possible to obtain from 179.01 to 358.02 tons of hay (268 t on average). The share attributable to the individual stages of division is shown graphically in Figure 10.

If the planting of woody plants, i.e., shrubs and trees, is considered, taking into account the climatic conditions and species composition in the locality, it is possible to consider shrubs mainly with hawthorn, and trees mainly with deciduous trees—summer oak, apple tree (wild species), and horse chestnut [65]. Single or multi-row strips of trees and shrubs could be used as windbreaks, with multiple layers of shrubs and trees from the ground to the tops of the trees. When using vegetation as windbreaks, care should be taken to situate them perpendicular to the direction of the prevailing winds.

The total number of shrubs up to 3 m in height at a stub of 3 m is 54,333. For shrubs up to 1.5 m tall at 1.5 m height and 1.5 m clamp, the figure is 108,669, and 163,002 at 1 m clamp. For a single line planting of trees in the 8–6 m buffer, the number is 6792 to 9056. For a double line—double-sided avenue around the road—this amounts to between 13,584 and 18,112 pcs at the same spacing. For a triple line, which can be placed in a 22 m wide strip, the number of trees will be from 20,376 to 27,168. According to [65], such planting in a 22 m wide green belt (a planting of shrubs and trees) can act as a protection for agricultural land against wind and soil erosion, depending on the conditions.

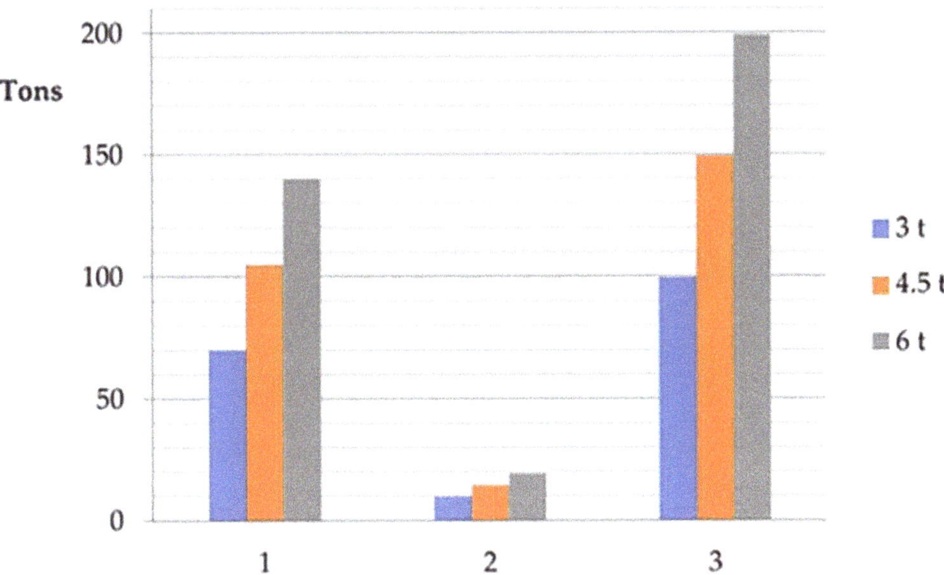

Figure 10. The proportion of biomass in the form of hay belonging to individual stages of division, for individual yields per hectare.

From the point of view of obtaining grazing for pollinators, especially bees, one colony per 10 ha can be considered according to local conditions [69], which represents a total of six bee colonies.

From the point of view of carbon sequestration in the wood mass on the area obtained in the defragmentation process, according to [70], it is possible to consider carbon sequestration of 0.41 t of carbon per tree, which represents a total of 2784.7 to 11,138.8 tons of carbon sequestered gradually and in the long term in the wood mass. In terms of carbon sequestration, the further design will also need to take into account the benefits in terms of carbon sequestered in the soil, not only in the green belts but also in the adjacent field margins, which will be affected by factors increasing the proportion of humus in the soil. This corresponds well with the data published in [71], which states that there are about 75 billion tons of soil organic carbon stored in the EU, with EU CO_2 production in 2017 of 4.5 billion tons.

When planting or implementing greenery in green belts of dividing lines, care must be taken to eliminate invasive species, such as *Solidago* sp. (goldenrod) and *Reynoutria* sp. (syn. *Fallopia* sp.—wingwort), among others, which could lead to silting up of surrounding fields and make it more difficult to replant desirable native/natural species.

When protecting against water erosion, priority was given to running the dividing lines perpendicular to the recession line to shorten long recession sections of fields, especially in the north-eastern part of the area of interest, on the foothills of the Vihorlat Mountains. A significant part of the protection against topsoil runoff or protection against leaching of the humus component from the soil is due to the fact that, according to [72], the old (historical) roads were deliberately routed in the landscape usually along straight, often elevated routes that drain well, avoid muddy sections, etc. At the same time, these roads were routed to minimize the effort and exertion involved in traversing the elevation (both upwards and downwards), from the perspective of man and beasts of burden.

The dividing lines in the western, open part of the area of interest were designed to eliminate mainly north and south-easterly wind flow in this part of the area by dividing the sections in the NE–SW direction.

The implementation and further design of the specific form of the dividing strips will have to take into account the demands of the existing farming equipment (the use of the so-called wide-harvesting equipment is questionable), the working and transport requirements, and the connectivity and transport accessibility of all the fields. From the farmer's or operator's point of view, such a fragmented landscape may incur additional fuel costs as well as an increase in the time required for the work (number of passes and downtime). These aspects will also have to be taken into account and adequately analysed and reflected in the forthcoming legislation.

The design of the dividing lines in relation to the correct addition of woody vegetation, such as trees and shrubs, protects these features, which are also areas of ecological interest, from being ploughed back by agricultural machinery.

From a contemporary point of view, with the decline of large-scale tourism and the desire to move towards local soft forms, an important part of the proposal is the creation of a network of hiking trails—routes that can serve both to improve accessibility to landmarks in the countryside (e.g., cairns, noble graves, and Lake Vinné), but also as a rich network of variously routed interconnected paths for recreation/recreation in the countryside itself. Table 2 shows the lengths of the different hiking trails that were created in the first and second steps of the fragmentation process.

Table 2. Proposed hiking trails.

Route	Number	Unit
Number of hiking trails	34	pcs
Number of connected municipalities	5	pcs
Length of hiking trails	117.1	km

The analogically conceived problem of fragmentation of the cultural landscape, i.e., agricultural land in Slopensko, was also presented in the case study "Land Use in the Dropie Protected Area" published by the Institute of Environmental Policy of the Ministry of the Environment of the Slovak Republic as part of a broader study [6]. The average area of the fields in the locality was 40.5 ha.

A method of dividing a selected area in the south of Western Slovakia on the border with Austria was proposed, whereby the creation of sections of grass or woody greenery in unproductive strips or the creation of dirt roads was considered. The primary tool for fragmentation was not the routing of roads and their use in the design of dividing lines, but the division of large units into smaller ones and their subsequent structuring with green elements. The area under consideration was a relatively small unit with no villages and no major through roads (border area), and it was also a bird area, so the objectives pursued were different. Last but not least, the differences between the two approaches were also related to the fact that, in the case of the Dropie Protected Area, it was a specific area—a protected bird area—and the following aspects, according to [6], were prioritized to reduce the value of the acreage of the fields and to increase the diversity of the crops grown, where 70% of the area of agricultural land in 2017 consisted of the two crops maize and wheat.

5. Conclusions

The proposed method of fragmentation of the cultural landscape mainly made up of large-scale fields of agricultural land is based primarily on two steps of restoration of historical roads and subsequent thickening of the network of dividing lines by restoring the boundaries of field blocks/access roads to these fields from the past. This method demonstrated the possibility of using the former infrastructure in the landscape and thus the possibility of respecting the historical and natural factors in solving the current environmental problems associated with the impacts of climate change and decades of intensive industrial farming in the cultural landscape of the Slovak countryside. At the same time, the proposed method of fragmentation leads to a homogeneously distributed network of dividing lines that are interconnected and can create a transport infrastructure useful both for access to agricultural land and for promoting local soft tourism in the place of residence of the inhabitants. In principle, such soft tourism is not oriented towards creating an alternative use of the landscape from agricultural production towards recreational or leisure use, but creates another more fully fledged parallel use of the landscape. The relevance of the above is linked to the trend in recent years to strengthen local tourism, which does not require travelling, thus putting a strain on the transport infrastructure. Such tourism serves the local (national) population or the population of the surrounding area, which is mostly accessible without the use of motorised means of transport, i.e., by bicycle and on foot. Its basic premise is a positive perception of the immediate surroundings and at the same time the creation of the necessary network of communications, without the need to implement additional elements, burdening the environment and creating the need for additional services. Last but not least, the recreational function implemented in the landscape reinforces the relationship between man and the environment and its protection. The productive function of such a fragmented rural landscape can remain paramount and can thus take on a more sustainable and climate-resilient form.

An important element of the proposal is the combination of dividing lines based on native roads and green belts with meadows, small woods, or woody vegetation with cumulative functions in terms of biomass production (e.g., hay), grazing for pollinators, erosion control, and overall improvement of ecosystem diversity. Important potential benefits, taking into account how green areas are created in the sense of [6], are protection against solar radiation; protection against wind; psychological and aesthetic effects; creation of variability in the landscape; and an insulating role (background for animals). Hygienic benefits are noise mitigation; dust reduction; and reduction in exposure to emissions. The results of the solution may provide a basis for the preparation of a future legislative framework, and it will be necessary to extend the research to include aspects falling within the field of landscape biology, taking into account its prospective development. This is particularly the case for respecting the components of nature and landscape protection, national park protection zones, landscape conservation zones, hygiene, aesthetics, and climatic conditions of the landscape. Property and legal issues, which go beyond the scope of the authors' coverage of the presented problem, will also be an integral part of the solution.

In favour of the above is the fact that the method of fragmentation, based on the restoration of historically justified dividing lines, largely respects the distribution of the ownership structure of the land, which has evolved in an analogous way, respecting a similar mechanism to the creation of parcel boundaries. From this point of view, it is possible to expect a coincidence between, on the one hand, the process of fragmentation of fields and, on the other hand, the consolidation of fragmented plots in terms of their ownership, which is a problem that needs to be solved in order to be able to apply the forthcoming legislative changes effectively in practice.

The presented work showed that many substantial parts of the main dividing lines have their basis in ancient history, represented in this case by a historical map from 1764 (First Military Mapping). These remained preserved until the period of collectivization in the mid-20th century, documented by an orthophoto map from 1950. This was reflected in the predominance of the first step of division over the second, where there were only a small number of "new paths". An important factor here is the connection to the municipality and other infrastructure, which has been preserved in the form of historical and technical monuments. The possibility of multiple fragmentations of large-scale fields was brought about by the third step of division, based on the original "pre-collectivization" structure of the country. What is important is the interconnectedness between all the steps of division, which gives them spatial and temporal continuity. The presented results can be, among other things, an important basis for further processing of layers in the GIS environment, as another part of existing services and applications.

Author Contributions: Conceptualization, J.R. and Ľ.G.; methodology, J.R.; validation, J.R.; investigation, J.R.; resources, J.R. and M.B.; writing—original draft preparation, R.R.; writing—review and editing, Ľ.G. and L.B.; visualization, Ľ.G.; supervision, L.B.; project administration, M.B.; and funding acquisition, L.B. All authors have read and agreed to the published version of the manuscript.

Funding: The article was created with the financial support of the project 19069—RM@Schools-ESEE. RawMaterials@Schools-ESEE.

Institutional Review Board Statement: Not applicable.

Informed Consent Statement: Not applicable.

Data Availability Statement: Not applicable.

Acknowledgments: This paper was created in connection with the project KEGA—048TUKE-4/2021: Universal Educational-Competitive Platform.

Conflicts of Interest: The authors declare no conflict of interest.

References

1. Beer, M.; Rybár, R. Development Process of Energy Mix towards Neutral Carbon Future of the Slovak Republic: A Review. *Processes* **2021**, *9*, 1263. [CrossRef]
2. Letcher, T.M. *The Impacts of Climate Change: A Comprehensive Study of Physical, Biophysical, Social and Political Issues*, 1st ed.; Elsevier: Amsterdam, The Netherlands, 2021; ISBN 9780128223734.
3. Trenberth, K.E.; Jones, P.D.; Ambenje, P.; Bojariu, R.; Easterling, D.; Tank, A.K.; Parker, D.; Rahimzadeh, F.; Renwick, J.A.; Rusticucci, M.; et al. Observations: Surface and atmospheric climate change. In *Climate Change 2007: The Physical Science Basis*; Contribution of Working Group I to the Fourth Assessment Report of the Intergovernmental Panel on Climate Change; Cambridge University Press: Cambridge, UK; New York, NY, USA, 2007; Chapter 3; pp. 235–336.
4. Brysse, K.; Oreskes, N.; O'Reilly, J.; Oppenheimer, M. Climate change prediction: Erring on the side of least drama? *Glob. Environ. Change* **2013**, *23*, 327–337. [CrossRef]
5. Sobocká, J.; Šurina, B.; Torma, S.; Dodok, R. *Klimatická Zmena a Jej Možné Dopady na Pôdny Fond Slovenska*; VÚPOP: Bratislava, Slovakia, 2005.
6. Gális, M. Na Poliach Pusto-o Škodlivosti Rozľahlých Monokultúr na Ornej Pôde a Možných Riešeniach. Inštitút Environmentálnej Politiky, Ministerstvo Životného Prostredia SR. Available online: https://www.minzp.sk/files/iep/2020_5_na_poliach_pusto.pdf (accessed on 22 February 2022).
7. Janák, M.; Černecký, J.; Saxa, A. *Monitoring Živočíchov Európskeho Významu v Slovenskej Republike. Výsledky a Hodnotenie za Roky 2013–2015*; The State Nature Conservancy of the Slovakia: Banská Bystrica, Slovakia, 2015.
8. Jenčová, I. Klimatické Reformy: Čo Prináša Legislatívny Balík Fit for 55. EURACTIV. Available online: https://euractiv.sk/section/klima/special_report/klimaticke-reformy-co-prinasa-legislativny-balik-fit-for-55/ (accessed on 22 February 2022).
9. MHSR. Integrovaný Národný Energetický a Klimatický Plán na Roky 2021–2030 Spracovaný Podľa Nariadenia EP a Rady (EÚ) č. 2018/1999 o Riadení Energetickej Únie a Opatrení v Oblasti Klímy. 2019. Available online: https://www.mhsr.sk/uploads/files/zsrwR58V.pdf (accessed on 22 February 2022).
10. Jasminska, N. Evaluation of hydrogen storage capacities on individual adsorbents. In *Measurement*; Elsevier: Amsterdam, The Netherlands, 2014.
11. Brestovic, T. Measurement of boundary conditions for the numerical solution of temperature fields of metal hydride containers. In *Measurement*; Elsevier: Amsterdam, The Netherlands, 2015.
12. Harrison, P.A.; Butterfield, R.E.; Downing, T.E. *Climate Change and Agriculture in Europe: Assessment of Impacts and Adaptation*; Research Report No. 9; Environmental Change Unit, University of Oxford: Oxford, UK, 1995.
13. Špánik, F.; Tomlain, J. *Klimatické Zmeny a Ich Dopad na Poľnohospodárstvo*; SPU Nitra: Nitra, Slovakia, 1997.
14. Šiška, B.; Mindáš, J.; Škvarenina, J.; Takáč, J. Zmeny podnebia, extrémy počasia a pôdohospodárstvo. In Proceedings of the Climate Change, Weather Extremes, Organisms and Ecosystems, International Bioclimatological Workshop, Trebišov, Slovakia, 23–26 August 2004.
15. Šiška, B.; Takáč, J.; Igaz, D. Môžeme očakávať zmeny v rozdelení výšky úrod obilnín v oblasti Podunajskej nížiny v dôsledku klimatickej zmeny? In Proceedings of the Climate Change Weather Extremes, Organisms and Ecosystems; International Bioclimatological Workshop, Trebišov, Slovakia, 23–26 August 2004.
16. Žigrai, F. Dimensions and Attributes of the Cultural Landscape. *Životné Prostr.* **2000**, *34*, 229–233.
17. Feranec, J.; Oťahel, J. *Krajinná Pokrývka Slovenska/Land Cover of Slovakia*; VEDA: Bratislava, Slovakia, 2001; ISBN 978-80-224-0663-5.
18. Eurostat. Agri-Environmental Indicator—Mineral Fertiliser Consumption. 2017. Available online: http://ec.europa.eu/eurostat/statistics-explained/index.php/Agri-environmental_indicator_-_mineral_fertiliser_consumption (accessed on 23 February 2022).
19. Enviroportál. Stav Druhov Rastlín a Živočíchov Európskeho Významu. 2013. Available online: https://www.enviroportal.sk/indicator/detail?id=182 (accessed on 23 February 2022).

20. Enviroportál. Stav Biotopov Európskeho Významu. 2013. Available online: https://www.enviroportal.sk/indicator/detail?id=183 (accessed on 23 February 2022).
21. Šúri, M.; Cebecauer, T.; Hofierka, J.; Fulajtár, J.E. Soil Erosion Assessment of Slovakia at a Regional Scale Using GIS. *Ecology* **2002**, *21*, 404–422.
22. Enviroportal. Pitná Coda. 2016. Available online: www.enviroportal.sk/indicator/detail?id=441 (accessed on 23 February 2022).
23. Baran, T. Fragmentácia Polí Vyžaduje Detailný Prístup. Naše Pole. 2020. Available online: https://nasepole.sk/fragmentacia-poli-vyzaduje-detailny-pristup/ (accessed on 23 February 2022).
24. Ilavská, B.; Jambor, P.; Lazúr, R. *Identifikácia Ohrozenia Kvality Pôdy Vodnou a Veternou Eróziou a Návrhy Opatrení*; Výskumný Ústav Pôdoznalectva a Ochrany Pôdy: Bratislava, Slovakia, 2005.
25. Moravcik, F.; Benova, A. Analýza výskytu vybraných krajinných prvkov v časti kopaničiarskeho regiónu Myjava. Kartografické listy. *Cartogr. Lett.* **2020**, *28*, 15–29.
26. Horvathova, J.; Mokrisova, M.; Suhanyiova, A. Analysis of Cost of Equity Models in Calculating Economic Value Added of Slovak Businesses, Sgem, Political Sciences, Law, Finance, Economics and Tourism, Vol II, Book Series. In *International Multidisciplinary Scientific Conferences on Social Sciences and Arts*; Stef92 Technology Ltd.: Sofia, Bulgaria, 2014; pp. 35–42. ISSN 2367-5659.
27. Bandlerová, A. Podnikanie formou družstva na Slovensku. In *Zborník Vedeckých Prác Medzinárodné Vedecké Dni: Ekonomické A Manažérske Aspekty Trvalo Udržateľného Rozvoja Poľnohospodárstva*; Slovak University of Agriculture in Nitra: Nitra, Slovakia, 2001; pp. 263–268, ISBN 80-7137-867-4.
28. EPI. Nařízení Vlády č. 50/2015 Sb. 2015. Available online: https://www.epi.sk/zzcr/2015-50 (accessed on 23 February 2022).
29. Skaloš, J.; Weber, M.; Lipský, Z.; Trpáková, I.; Šantrůčková, M.; Uhlířová, L.; Kukla, P. Using old military survey maps and orthophotography maps to analyze long-term land cover changes—Case study (Czech Republic). *Appl. Geogr.* **2011**, *31*, 426–438. [CrossRef]
30. Vojenský Kartografický Ústav Harmanec. Mapová Podklady 1:50,000. Available online: https://www.vku-mapy.sk/ (accessed on 15 November 2021).
31. TUZVO. Historická Ortofotomapa Slovenska. Available online: https://mapy.tuzvo.sk/hofm (accessed on 23 February 2022).
32. Hrnčiarová, T. Reprezentatívne historické prvky krajiny a ich manažment. *Folia Geogr.* **2010**, *16*, 79–86.
33. Stankoviansky, M. *Geomorfologická Odozva Environmentálnych Zmien na Území Myjavskej Pahorkatiny*; Univerzita Komenského v Bratislave: Bratislava, Slovakia, 2003.
34. Jambor, P.; Ilavská, B. *Metodika Protieróznеho Obrábania Pôdy*; Výskumný Ústav Pôdnej Úrodnosti: Bratislava, Slovakia, 1998.
35. Kunaková, L.; Terek, J. Akvatické krajinné prvky Medzibodrožia na Slovensku. *Folia Geogr.* **2016**, *58*, 70–80.
36. eAgronom. The Only Sustainable Solution for Your Farm—Build Up Your Soils and Profits. Available online: https://eagronom.com/cs/blog/rozdelovani-poli-nad-30-ha/ (accessed on 23 February 2022).
37. Pamiatkový Úrad SR. Prvé Vojenské Mapovanie. 2011. Available online: http://www.pamiatky.sk/sk/page/mapy-1-2-3-vojenskeho-mapovania (accessed on 23 February 2022).
38. OneSoil. Available online: https://onesoil.ai/en/ (accessed on 23 February 2022).
39. MPRV SR/PPA: Ortofotomozaika 2018–2020. Available online: https://gsaa.mpsr.sk/2021/ (accessed on 23 February 2022).
40. ZBGIS. Available online: https://zbgis.skgeodesy.sk/mkzbgis/sk/kataster?pos=48.694053,21.250734,10 (accessed on 23 February 2022).
41. MPRV SR. Správa o Poľnohospodárstve a Potravinárstve v Slovenskej Republike za Rok 2020. Available online: https://www.mpsr.sk/zelena-sprava-2021/122---17379/ (accessed on 23 February 2022).
42. Van der Zanden, E.H.; Verburg, P.H.; Mücher, C.A. Modelling the spatial distribution of linear landscape elements in Europe. *Ecol. Indic.* **2013**, *27*, 125–136. [CrossRef]
43. Seidlitz, H.; Thiel, S.; Krins, A.; Mayer, H. Comprehensive series in photosciences, Elsevier. In *Modeling Solar Radiation at the Earth's Surface*; Springer: Berlin/Heidelberg, Germany, 2001; ISSN 1568-461X/9780444508393.
44. Kovanič, L.; Kovanič, L.; Blistan, P.; Urban, R.; Štroner, M.; Blišťanová, M.; Bartoš, K.; Pukanská, K. Analysis of the Suitability of High-Resolution DEM Obtained Using ALS and UAS (SfM) for the Identification of Changes and Monitoring the Development of Selected Geohazards in the Alpine Environment—A Case Study in High Tatras, Slovakia. *Remote Sens.* **2020**, *12*, 3901. [CrossRef]
45. Bartoš, K.; Pukanská, K.; Repáň, P.; Kseňak, Ľ.; Sabová, J. Modelling the Surface of Racing Vessel's Hull by Laser Scanning and Digital Photogrammetry. *Remote Sens.* **2019**, *11*, 1526. [CrossRef]
46. Van Geert, A.F.; van Rossum, F.; Triest, L. Do linear landscape elements in farmland act as biological corridors for pollen dispersal? *J. Ecol.* **2010**, *98*, 178–187. [CrossRef]
47. Krewenka, K.M.; Holzschuh, A.; Tscharntke, T.; Dormann, C.F. Landscape elementsas potential barriers and corridors for bees, wasps and parasitoids. *Biol. Conserv.* **2011**, *144*, 1816–1825. [CrossRef]
48. Cheng, C.H.K.; Kuo, H.Y. Bonding to a new place never visited: Exploring the relationship bet-ween landscape elements and place bonding. *Tour. Manag.* **2015**, *46*, 546–560. [CrossRef]
49. Žižlavská, N.; Mikita, T.; Patočka, Z. The Effects of Roadside Woody Vegetation on the Surface Temperature of Cycle Paths. *Land* **2021**, *10*, 483. [CrossRef]
50. Brůna, V.; Buchta, I.; Uhlířová, L. *Identifikace Historické Sítě Prvků Ekologické Stability Krajiny na Mapách Vojenských Mapování*; Univerzita J.E. Purkyně: Ústí nad Labem, Czech Republic, 2002.
51. Petrovič, F.; Bugár, G.; Hreško, J. A list of landscape features mappable in Slovakia. *GEO Inf.* **2009**, *5*, 112–124.

52. Klikušovská, Z.; Sviček, M. *Environmentálne Indexy a Indikátory Analýzy a Hodnotenia Krajiny 2009 (Terénny Prieskum, Modelovanie a Diaľkový Prieskum Zeme ako Alternatívne Zdroje Údajov)*; Zborník Príspevkov z Vedec-Kého Seminára; VÚPOP: Bratislava, Slovakia, 2009.
53. Hrnčiarová, T. Historické krajinné prvky a mozaiky –súčasť diverzity kultúrnej krajiny. In *Smolenická Výzva–IV. Zborník Príspevkov*; Ústav Krajinnej Ekológie SAV: Bratislava, Slovakia, 2008.
54. Suhanyiova, A.; Suhanyi, L. Selected Aspects of Subsistence Minimum in Slovakia. In Proceeding of the CBU International Conference on Innovations in Science and Education, Prague, Czech Republic, 22–24 March 2017.
55. Martuliak, P. *150 Years of Slovak Co-Operatives 1845–1995*; Knižná Publikácia: Nitra, Slovakia, 1995.
56. Madarász, D. Historický vývoj poľnohospodárskych družstiev na Slovensku od 19. storočia až po súčasnosť. In *Podnikanie na Vidieku (Obchodné Právo EÚ I)*; SPU: Nitra, Slovakia, 2011; pp. 174–180.
57. SEA. Územný Systém Ekologickej Stability (ÚSES). Available online: https://www.sazp.sk/zivotne-prostredie/starostlivost-o-krajinu/zelena-infrastruktura/uzemny-system-ekologickej-stability-uses.html (accessed on 23 February 2022).
58. Kaňuk, J.; Gallay, M.; Hofierka, J. Generating time series of virtual 3-D city models using a retrospective approach. *Landsc. Urban Plan.* **2015**, *139*, 40–53. [CrossRef]
59. Vojtko, R.; Reichwalder, P. *Interpretácia Leteckých A Kozmických Snímok v Geológii. Vysoko-Školské Skriptá*; Prírodovedecká Fakulta Univerzity Komenského: Bratislava, Slovakia, 2016.
60. Jakubík, J. Development of military cartography in territory of Slovakia/ Vývoj vojenskej kartografie na území Slovenska. Kartografické listy. *Cartogr. Lett.* **2012**, *20*, 28–38.
61. AKCR. Podrobné Informace k Omezení Pěstování Monokultur na Max. 30 ha Souvislé Plochy Prostřednictvím Standardu tzv. DZES 7d). 2019. Available online: http://www.akcr.cz/txt/podrobne-informace-k-omezeni-pestovani-monokultur-na-max-30-ha-souvisle-plochy-prostrednictvim-standardu-tzv-dzes-7d (accessed on 23 February 2022).
62. Šúri, M.; Cebecauer, T.; Fulajtár, E.J.; Hofierka, J. Actual water erosion (map 1:500,000). In *Landscape Atlas of the Slovak Republic*, 1st ed.; Ministry of the Environment of the Slovak Republic: Bratislava, Slovakia, 2002.
63. Šúri, M.; Cebecauer, T.; Fulajtár, E.J.; Hofierka, J. Potential water erosion (according to W.H. Wischmeier and D.D. Smith; map 1:1 mil.). In *Landscape Atlas of the Slovak Republic*, 1st ed.; Ministry of the Environment of the Slovak Republic: Bratislava, Slovakia, 2002.
64. Global Wind Atlas. Available online: https://globalwindatlas.info/ (accessed on 23 February 2022).
65. Ministerstvo Dopravy, Pôšt a Telekomunikácií SR Sekcia Cestnej Dopravy a Pozemných Komunikácií: Technické Podmienky Vegetačné Úpravy Pri Pozemných Komunikáciách. TP 04/2010. Available online: https://www.ssc.sk/files/documents/technicke-predpisy/tp/tp_035.pdf (accessed on 23 February 2022).
66. Bacsi, Z. Łęcka; Tóth, É. Word Heritage Sites as soft tourism destinations—Their impacts on international arrivals and tourism receipts. *Bull. Geogr. Socio-Econ. Series.* **2019**, *45*, 25–44. [CrossRef]
67. Weaver, D.B.; Lawton, L.J. Twenty years on: The state of contemporary ecotourism research. *Tour. Manag.* **2007**, *28*, 1168–1179. [CrossRef]
68. Martincová, J.; Kizeková, M.; Michalec, M. *Ekologická Obnova Trávnych Porastov Metodická Príručka*; Národné Poľnohospodárske a Potravinárske Centrum: Lužianky, Slovakia, 2017.
69. Moje Včely. Včelia Pastva a Výroba Medu. 2016. Available online: https://mojevcely.sk/vcelia-pastva-a-vyroba-medu/ (accessed on 23 February 2022).
70. Global Climate Change. Examining the Viability of Planting Trees to Help Mitigate Climate Change. 2019. Available online: https://climate.nasa.gov/news/2927/examining-the-viability-of-planting-trees-to-help-mitigate-climate-change/ (accessed on 23 February 2022).
71. European Environment Agency: Land and Soil in Europe. *Why We Need to Use These Vital and Finite Resources Sustainably*; Publications Office of the European Union: Luxembourg, 2019. Available online: https://www.eea.europa.eu/publications/eea-signals-2019-land (accessed on 23 February 2022).
72. Bolina, P.; Klimek, T.; Cílek, V. *Staré Cesty v Krajině Středních Čech*; Academia: Prague, Czech Republic, 2018; ISBN 9788020028310.

www.ingramcontent.com/pod-product-compliance
Lightning Source LLC
LaVergne TN
LVHW071021170726
843515LV00006B/1173
* 9 7 8 3 0 3 6 5 6 2 0 3 2 *